PHYSICS AND CHEMISTRY OF METAL CLUSTER COMPOUNDS

Physics and Chemistry of Materials with Low-Dimensional Structures

VOLUME 18

Editor-in-Chief

F. LÉVY, *Institut de Physique Appliquée, EPFL, Département de Physique, PHB-Ecublens, CH-1015 Lausanne, Switzerland*

Honorary Editor

E. MOOSER, *EPFL, Lausanne, Switzerland*

International Advisory Board

J. V. ACRIVOS, *San José State University, San José, Calif., U.S.A.*

S. BARISIC, *University of Zagreb, Department of Physics, Zagreb, Croatia*

J. G. BEDNORZ, *IBM Forschungslaboratorium, Rüschlikon, Switzerland*

C. F. van BRUGGEN, *University of Groningen, Groningen, The Netherlands*

R. GIRLANDA, *Università di Messina, Messina, Italy*

D. HAARER, *University of Bayreuth, Germany*

A. J. HEEGER, *University of California, Santa Barbara, Calif., U.S.A.*

H. KAMIMURA, *Dept. of Physics, University of Tokyo, Japan*

W. Y. LIANG, *Cavendish Laboratory, Cambridge, U.K.*

P. MONCEAU, *CNRS, Grenoble, France*

J. ROUXEL, *CNRS, Nantes, France*

M. SCHLÜTER, *AT&T, Murray Hill, N.J., U.S.A.*

I. ZSCHOKKE, *Universität Basel, Basel, Switzerland*

The titles published in this series are listed at the end of this volume.

PHYSICS AND CHEMISTRY OF METAL CLUSTER COMPOUNDS

MODEL SYSTEMS FOR SMALL METAL PARTICLES

Edited by

L. J. DE JONGH

Kamerlingh Onnes Laboratory,
Leiden University,
The Netherlands

KLUWER ACADEMIC PUBLISHERS
DORDRECHT / BOSTON / LONDON

Library of Congress Cataloging-in-Publication Data

```
Physics and chemistry of metal cluster compounds : model systems for
  small metal particles / edited by L.J. de Jongh.
      p.   cm. -- (Physics and chemistry of materials with low
  -dimensional structures ; v. 18)
    Includes index.
    ISBN 0-7923-2715-2 (acid free paper)
    1. Metal crystals.  2. Metal-metal bonds.  I. Jongh, L. J. de.
  II. Series.
  QD921.P48  1994
  546.3--dc20                                                93-49928
```

ISBN 0-7923-2715-2

Published by Kluwer Academic Publishers,
P.O. Box 17, 3300 AA Dordrecht, The Netherlands.

Kluwer Academic Publishers incorporates the publishing programmes
of D. Reidel, Martinus Nijhoff, Dr W. Junk and MTP Press.

Sold and distributed in the U.S.A. and Canada
by Kluwer Academic Publishers,
101 Philip Drive, Norwell, MA 02061, U.S.A.

In all other countries, sold and distributed
by Kluwer Academic Publishers Group,
P.O. Box 322, 3300 AH Dordrecht, The Netherlands.

Printed on acid-free paper

All Rights Reserved
© 1994 Kluwer Academic Publishers
No part of the material protected by this copyright notice may be reproduced or
utilized in any form or by any means, electronic or mechanical,
including photocopying, recording or by any information storage and
retrieval system, without written permission from the copyright owner.

Printed in the Netherlands

TABLE OF CONTENTS

PREFACE ix

1. L. J. DE JONGH / Introduction to Metal Cluster Compounds:
From Molecule to Metal! 1
1.1. The impact of cluster-science 1
1.2. Structural characteristics of metal clusters 4
1.3. Electronic energy-level structures 10
1.4. Brief introduction to the physical properties of
metal cluster compounds 19
1.5. Conductivity studies 25
1.6. Application of the Anderson–Hubbard approach 31
References 37

2. A. CERIOTTI, R. DELLA PERGOLA AND L. GARLASCHELLI /
High-Nuclearity Carbonyl Metal Clusters 41
2.1. Introduction 41
2.2. Synthesis and reactivity 49
2.3. Structural aspects 67
2.4. Electron counting for clusters 93
References 101

3. G. SCHMID / Ligand-Stabilized Giant Metal Clusters and Colloids 107
3.1. Strategy for making giant metal clusters 107
3.2. Synthetic and structural examples 109
    3.2.1. Ligand-stabilized colloids 109
    3.2.2. Ligand-stablized clusters 116
3.3. Chemical properties 122
3.4. Catalysis 127
3.5. Outlook 131
References 133

4. D. E. ELLIS / Theory of Electronic Properties of Metal
Clusters and Particles 135
4.1. Why are metal particles interesting? 135
4.2. Model Hamiltonians 136

4.3. Traditional quantum chemical methods	141
4.4. Density functional approaches	144
4.4.1. Alkali metal clusters	145
4.4.2. Cluster compounds	146
4.4.3. Transition metal particles: free and embedded	148
4.5. Summary	153
References	155

5. R. ZANONI / X-Ray Photoelectron Spectroscopy Applied to Pure and Supported Molecular Clusters	159
5.1. Introduction	159
5.2. Generalities of photoemission spectroscopy applied to pure and supported molecular metal clusters	160
5.3. XPS of molecular clusters	169
5.4. XPS of supported molecular clusters	175
5.5. Outlook for the future	179
References	180

6. R. C. THIEL, H. H. SMIT AND L. J. DE JONGH / Application of Mössbauer Effect Spectroscopy to Cluster Research	183
6.1. Introduction	183
6.2. Mössbauer Effect Spectroscopy (MES)	183
6.2.1. Elements of MES	183
6.2.2. MES specific to clusters	188
6.2.3. Previous MES measurements on gold particles	189
6.3. Our Mössbauer results	191
6.3.1. Thermal behaviour	191
6.3.2. Electronic behaviour	196
6.3.3. Effect of ligand modification	198
6.4. ^{197}Au MES on platinum clusters	201
6.5. Conclusions	206
References	207

7. H. B. BROM, J. BAAK AND L. J. DE JONGH / Specific Heat Studies on Metal Cluster Compounds	211
7.1. Introduction	211
7.2. The lattice specific heat	212
7.2.1. The elastic continuum approximation	212
7.2.2. The molecular dynamics approach	215
7.2.3. A spherical 309-atomic Pt-particle as example	216
7.3. The electronic specific heat	217
7.3.1. The degenerate electron gas. Surface effects and bulk states	218
7.3.2. The Quantum-Size limit	219

7.4. Data and discussion	220
7.4.1. Small metal particles	220
7.4.2. Metal cluster compounds	221
7.5. Summary	225
References	225

8. H. B. BROM, D. VAN DER PUTTEN AND L. J. DE JONGH / NMR in Submicron Particles

	227
8.1. Introduction	227
8.2. Surface and quantum size effects	228
8.2.1. Surface effects	229
8.2.2. Quantum size effects	229
8.3. ESR and NMR – Theory	230
8.3.1. ESR	231
8.3.2. NMR	232
8.4. Naked clusters – Experiment and discussion	234
8.4.1. ESR	235
8.4.2. NMR	235
8.5. Aggregates of metal cluster compounds – Experiment and discussion	237
8.5.1. Core-resonance	237
8.5.2. Ligand-shell resonance	242
8.5.3. Results from other experiments	245
8.6. Summary	246
References	246

9. R. E. BENFIELD / Magnetic Properties and UV-Visible Spectroscopic Studies of Metal Cluster Compounds

	249
9.1. Introduction	249
9.2. Magnetic properties	250
9.2.1. Magnetic susceptibility of low-nuclearity clusters	250
9.2.2. Magnetic susceptibility of high-nuclearity clusters	253
9.2.3. EPR studies of decanuclear osmium clusters	256
9.2.4. EPR studies of rhodium carbonyl clusters	262
9.2.5. EPR studies of other high-nuclearity clusters	263
9.3. Electronic (UV-visible-NMR) spectra	264
9.3.1. One-electron absorptions	264
9.3.2. Interband transitions	265
9.3.3. Plasma resonance absorptions	268
9.3.4. Charge transfer in the solid state	270
9.4. Conclusion	272

Acknowledgements 272
References 272

10. J. M. VAN RUITENBEEK, D. A. VAN LEEUWEN AND L. J. DE JONGH / Magnetic Properties of Metal Cluster Compounds 277

10.1. Introduction 277
10.2. Magnetic properties of atoms, metals and clusters 278
    10.2.1. Atoms 280
    10.2.2. Metals 282
    10.2.3. Clusters 285
        10.2.3.1. The spin susceptibility in metal clusters 286
        10.2.3.2. The orbital susceptibility in metal clusters 289
        10.2.3.3. Exchange and correlation effects 295
        10.2.3.4. The spin-orbit interaction 295
        10.2.3.5. Ligand coordination: Suppression and magnetisim 296
10.3. Experiments on metal cluster compounds 297
    10.3.1. The spin susceptibility 297
        10.3.1.1. The even-odd dichotomy 297
        10.3.1.2. Pauli paramagnetism 298
        10.3.1.3. Clusters containing Ni or Co 301
    10.3.2. The orbital magnetic susceptibility 304
References 305

INDEX OF CHEMICAL COMPOUNDS 307

INDEX OF SUBJECTS 315

PREFACE

On Friday, February 20, 1980, I had the pleasure to be present at the inaugural lecture of my colleague Jan Reedijk, who had just been named at the Chair of Inorganic Chemistry of Leiden University. According to tradition, the ceremony took place in the impressive Hall of the old University Academy Building. In the course of his lecture, Jan mentioned a number of recent developments in chemistry which had struck him as particularly important or interesting. Among those was the synthesis of large metal cluster compounds, and, to my luck, he showed a slide of the molecular structure of $[Pt_{19}(C))_{22}]^{4-}$. (To my luck, since at traditional Leiden University it is quite unusual to show slides at such ceremonies.) This constituted my first acquaintance with this exciting new class of materials. I became immediately fascinated by this molecule, partly because of the esthetic beauty of its fivefold symmetry, partly because as a physicist it struck me that it could be visualized as an "embryonically small" metal particle, embedded in a shell of CO ligands.

It came to me as a surprise that Nature could provide us with such a "chemical nanostructure", with its combination of metal-metal and metal-ligand bonds. However, when I later asked Jan about it, he mentioned that, to all probability the metal-ligand interactions should have washed out all potential metallic properties of the metal cluster core. Although I retorted that, still, there were two Pt atoms situated in the interior of the Pt_{19} core, and thus totally surrounded by the 17 other Pt atoms, we left the matter there at the time.

But the thought never left my mind so I brought it up from time to time and about three years later Jan passed me a copy of the memorable review paper by Muetterties entitled "Metal Clusters: Bridges between molecular and solid-state chemistry" (C & EN, Aug. 30, 1982, p. 28). From this paper I learnt that, next to the $[Pt_{19}(CO)_{22}]^{4-}$ cluster, many polynuclear metal cluster compounds had been synthesized already, some with metal cores even considerably larger, as in e.g. $[Pt_{38}(CO)_{44}H_x]^{2-}$, also synthesized by the group in Milano. I became definitively convinced that there should be a promising future in these materials as model systems for small metal particles embedded in a dielectric solid.

Since my colleagues Hans Brom and Roger Thiel in our Solid State Physics group also became enthusiastic for the idea, it was decided to try to contact a number of the chemist groups involved in the synthesis of these materials. Via Jan Reedijk, I was introduced to Jan Steggerda (Nijmegen University), and later to Günter Schmid (Universität Essen) who had just synthesized his, by now well-known, 55 metal atom clusters. Subsequently, contacts were made with Giuliano Longoni and Alessandro Ceriotti in Milano, and with Brian Johnson and Robert Benfield in England. It turned out that, at that time, the synthetic chemists involved with clusters had reached a stage where they really welcomed the arrival of a couple of physicists eager to start to study in detail the physical properties of their materials. So, a happy collaboration evolved, which, in the course of time, has proven to be very fruitful. It should be mentioned that from the very beginning the project has received the well-appreciated support from the European Community research stimulation programs, first under the Laboratory Twinning Program, then under the Science Program. More recently, Roberto Zanoni (Roma), Herman van Kempen (Nijmegen), Gianfranco Pacchioni (Milano) and Notker Rösch (München), have also joined our ranks.

The aim of this book is to present a first review of what we have learnt in the last eight years about the physical properties of these fascinating compounds. Since almost nothing was known in the beginning, we had to start from scratch, and only gradually, with ups and downs, our understanding has begun to develop. Evidently, the most important and challenging problem to unravel was the evolution with cluster size from "molecular" to "metallic" behavior. The advent of the giant clusters, containing as much as 309 Pt atoms or 561 Pd atoms (made by Günter Schmid), has greatly facilitated our task. Together with the availability of colloids of the same metal with narrow size distributions, the original dream we had at the start is now realised to a great extent: to be able to study the physical properties of metal particles of well-defined, homogeneous size as a function of increasing size. Thus the time seemed ripe to collect our present understanding of the problem in the form of a book, also with the idea to interest more physicist colleagues in these still relatively unknown materials.

As will be seen from this book, the transition from molecular to metallic behavior is probably a gradual one, with a substantial transition region in between where properties are neither molecular nor metallic. I have earlier proposed the names "meta-metallic" and/or "proto-metallic" for the behavior in this intermediate regime. In view of the latest developments in nomenclature, we might as well adopt the term "mesoscopic metallic", or "mesometallic" in short. What is important is that the onset and end of this region is certainly not universal. Even for one and the same metal it depends on the physical property one is studying and even on the temperature at which the experiment is performed.

Let me end by thanking whole-heartedly the numerous colleagues who have made invaluable contributions to this book, either directly or indirectly. I am particularly grateful, of course, to all the authors and coauthors of the various chapters. Knowing how busy lives these colleagues lead makes me all the more thankful for their willingness to spend their precious time on the realisation of this venture. I can only express my hope that they will be as satisfied with the final total product as I have been with each of their contributions. Last but not least, I wish to express my gratitude to my secretary, Mrs. Barry Cats-Houdijk for her invaluable help and patience.

L.J. DE JONGH   Leiden, Autumn 1994

1. INTRODUCTION TO METAL CLUSTER COMPOUNDS: FROM MOLECULE TO METAL!

L.J. DE JONGH
*Kamerlingh Onnes Laboratory, Leiden University,
P.O. Box 9506, 2300 RA Leiden, The Netherlands*

1.1. The Impact of Cluster-Science

The continuous quest for structures of increasingly smaller and smaller scales certainly represents one of the most fascinating scientific developments of the last decades. This trend may involve a lowering of the dimensionality of the systems studied, besides a mere reduction in size. As regards dimensionality, one may define a d-dimensional system as an object which has infinite extension only in a restricted number d of spatial directions, and finite extent in the remaining 3-d directions [1]. Thus, $d = 2$ corresponds to layered systems, which can range from a single atomic monolayer to thin films consisting of a large, albeit finite number of layers. Similarly, $d = 1$ corresponds to thin wires, (organic or inorganic) chain systems, fibers, polymers. Continuing in the same spirit, a zero-dimensional system corresponds to a small particle or cluster, i.e. a system that is finite in all directions. In the latter case the reduction in size and in dimensionality obviously amount to the same thing. Generally speaking, a reduction in dimensionality will alter drastically the physical properties of a system [1]. Size-reduction, however, may also have characteristic consequences of its own, for instance the physical properties of a system may change drastically as soon as its size becomes comparable to physical quantities like grainsize, (magnetic) domainsize, elastic or inelastic scattering length of conduction electrons, wavelength of excitations such as phonons or magnons, De Broglie wavelength of the electron, phase-coherence length in a superconductor, etc. This explains why it has proven useful to define, besides 'low-dimensional structures', a discipline such as 'submicron physics', thereby emphasizing the effects due to a mere reduction in size, as distinguished from the consequences of a lowering in dimensionality of an – otherwise infinite – system.

Many definitions of the term *'cluster'* do exist already. In this book we shall be concerned with special types of clusters, namely those occurring in metal cluster compounds. We shall use the term cluster rather loosely, meaning a small piece of matter, consisting of a few (but more than one) atoms, up to many thousands of atoms. The cluster can consist of the same kind of atoms, or of atoms of several different kinds. The general idea behind is that clusters can be viewed as forming a bridge between the microscopic world of individual atoms or molecules, and the (macroscopic) world of bulk matter [2]. Recently, the name 'mesoscopic' has become popular for systems with sizes falling in this intermediate range. There is a vivid interest in these systems, since their behaviour and properties can be completely different from both the atomic and bulk limits. Just to give an example, in many ways a cluster of a certain number of alkali-metal atoms (e.g. Na) can be seen as a 'super-atom'. Since the electrons are delocalized over the cluster as a whole, the quantum-mechanical electronic energy-level structure for the cluster can be calculated quite analogously to that of the hydrogen atom. Thus, clusters may provide in several respects a potential extension of the periodic system as we know it.

Few fields in the materials sciences are of such a rich interdisciplinary nature as cluster-science. Physicists and chemists, experimentalists and theorists, university staff members and applied industrial researchers, all can have their share, since the problems involved range from basic scientific challenges such as the properties of quantum-dots, to applications as in materials for e.g. solar cells or catalysts. At the same time, young graduate students may use it as an exciting field of exercise, in which they can put into practice most of the subjects into which they have just been initiated during their courses in physics or chemistry.

Another reason for this widespread interest and popularity is that clusters can be prepared or synthesized in many different ways. In the last fifteen years, one has witnessed a strong development in the technology of producing and probing clusters in atomic beams. In 1981 Smalley and coworkers at Rice University (Houston, Texas) introduced a new type of cluster source, in which the solid material is vaporized with the aid of a pulsed laser beam. This enabled the investigation of clusters of many materials that were hitherto inaccessible, and similar apparatus has been installed in many laboratories. The major advantage of cluster beam technology is that it enables the study of free, unsupported clusters, e.g. by spectroscopic methods. This should be contrasted by the chemical methods such as used in colloid chemistry, catalysis, or in the production of small particles on supports like zeolites, SiO_2 or Al_2O_3. On the other hand, the amount of cluster material that can be produced in a beam is evidently quite small and the number of possible physical experiments that can be preformed is also rather limited. All these methods, moreover, suffer from the fact that the clusters obtained are not monodisperse, i.e. not of the same size. The size-distribution is typically larger than 5–10%, even in the most favourable cases. Mass-selection of ionized clusters in beams is indeed possible, however, the yield is again quite small, and, after deposition, it is difficult to keep the clusters from coalescing. Therefore, experiments on mass-selected clusters have sofar been limited to *in situ* spectroscopic studies.

As physicists interested to study the evolution from molecular to bulk behaviour with increasing cluster-size, we would like to have at our disposal completely monodisperse clusters, of varying size, and furthermore in sample quantities allowing the application of all the experimental techniques available to study the properties of solids. It is for such reasons that the metal cluster compounds, which are the subject of this volume, provide a very attractive (and relatively inexpensive!) alternative 'cluster-source'. Metal cluster compounds [3–5] consist of identical macromolecules (the metal cluster molecules), which can be ionic or neutral, and which are composed of a metal core (the cluster) containing a given number (n) of metal atoms. The core is surrounded by a ligand 'shell' formed by ligand atoms (Cl, I, O, ...) or ligand molecules (CO, PPh_3, ...), which are chemically bonded to metal atoms at the surface of the metal cores. Since chemical compounds are involved, a sample of a given metal cluster compound contains only one particular type of macromolecule (provided that the compound is pure, of course), and thus presents a macroscopically large collection of identical metal clusters, mutually separated by the ligand shells (plus the counterions in case of the ionic compounds). In this way the ligand shells provide a highly effective means of 'chemical stabilization' of the small metal particles (analogous to the metal colloids). In going from one compound to the other, the type of metal atom in the clusters or the size n of the cluster can be varied. At present many hundreds of metal cluster compounds are known already, most of them with cluster sizes ranging up to $n=$ 20–30, and with many of the transition metal elements (Fe, Co, Ni, Mo, Ru, Rh, Pd, Ag, Os, Ir, Pt, Au, ...). As discussed extensively elsewhere in this volume, for some of these elements even giant metal cluster compounds are known, the metal core size becoming as large as $n = 309$ for Pt, and $n = 561$, 1415 and 2057 for Pd. Clearly, such sizes are already close or equal to those of the smallest metal particles in metal colloids ($n \simeq 10^3 - 10^4$ atoms). Most importantly, since they are chemical compounds, they can be obtained in macroscopic quantities, so that we have indeed at our disposal samples of cluster aggregates in quantities large enough even to perform experiments such as neutron scattering studies. Furthermore, many metal cluster compounds can also be obtained in the form of single crystals large enough for physical studies.

There is of course a fly in the ointment in that the same ligands which chemically stabilize the metal clusters in these materials will also alter their electronic structure. As we shall see, however, there is compelling theoretical and experimental evidence that this interaction (i.e. the ligand-metal charge-transfer) is limited to those metal atoms which are at the *surface* of the metal cluster. This implies that the remaining, *inner-core* metal atoms of the cluster can be seen as a metal particle, more or less equivalent to, say, a bare metal cluster formed in an atomic beam. Furthermore, the disadvantage of the ligands also has its interesting sides, for instance because the ligand-metal interaction is quite analogous to the chemisorption of molecules or atoms on (bulk) metal surfaces, as encountered in surface science. Since for small clusters the ratio of surface to inner-core atoms can easily become of the order of 50 to 90%, they evidently also constitute attractive model sytems for the study of such surface problems.

Most of the contents of this volume describe the results of a collaboration between chemists and physicists on cluster research, that was started about eight years ago, and was sponsored by the European Community. The purpose of this introductory chapter is to provide the reader with a 'red thread' by which he may more easily find his way through the book. Thus, in the next section we shall give a short introduction to the 'cluster-solids' in general and the metal cluster compounds in particular, as a preparation for the two chapters on the chemical synthesis by the Italian group and by Schmid. In Section 1.3 we present a brief, schematic overview of what is to be expected for the electronic structures of individual clusters, cluster solids and for ligated clusters. Thereafter, some basic experimental features found for the metal cluster compounds are summarized. This should provide a basis by which the reader may then turn to the theoretical chapter by Ellis on molecular orbital calculations for clusters, and also for the remaining chapters in which the physical measurements are described.

1.2. Structural Characteristics of Metal Clusters

As an introduction to cluster solids, in particular to the transition-*metal-cluster* compounds, let us first consider the basic structure of the majority of transition-*metal* compounds familiar to solid state physicists. Typically, in these materials a given metal atom is coordinated by a polyhedron of ligands (e.g. by an octahedron or a tetrahedron). The ligands can be nonmetallic atoms like oxygen or a halide, a chalcogenide, etc. ..., or they can be molecules like H_2O or organic molecules (pyridine, pyrazole, etc. ...). The structure of the compound is then formed by linking together these polyhedra into chains or layers, or into a 3-dimensional network. As an example Figure 1 shows the cubic perovskite structure as well as the related layered (K_2NiF_4) structure of octahedra. Oxides with these structures are well known in ferroelectrics and in high-T_c superconductivity.

Instead of a *single* metal atom, however, we may also consider a *group (cluster)* of metal atoms in a ligand shell of nonmetallic units, where the metal atoms inside the cluster are directly bound together (without intervening ligand). Even in purely inorganic compounds this situation can often be found (for reviews see e.g. [6]). An example that is well-known to solid-state physicists is formed by the Chevrel phases, the structure of which is shown in Figure 2. The building blocks in this case are octahedral clusters of six Mo (or Re, Ru) atoms, each cluster being coordinated by a cube of eight S (or Se,Te) atoms. Indeed, these cluster units are now known to generate

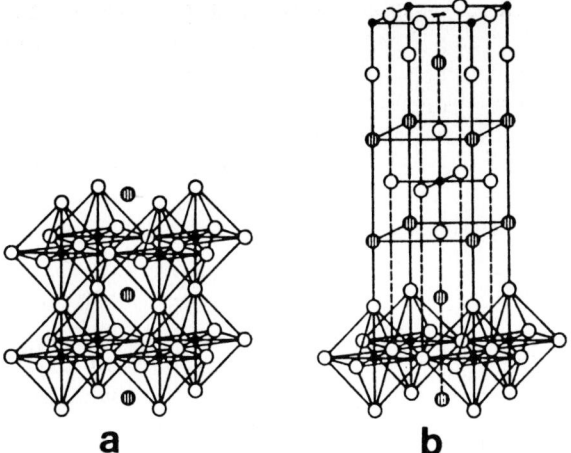

Fig. 1. Linked octahedral ML$_6$ molecules (M = metal; L = ligand), forming the cubic perovskite strucutre (a), and the layered K$_2$NiF$_4$ structure (b). Filled and open circles refer to M and L atoms, respectively.

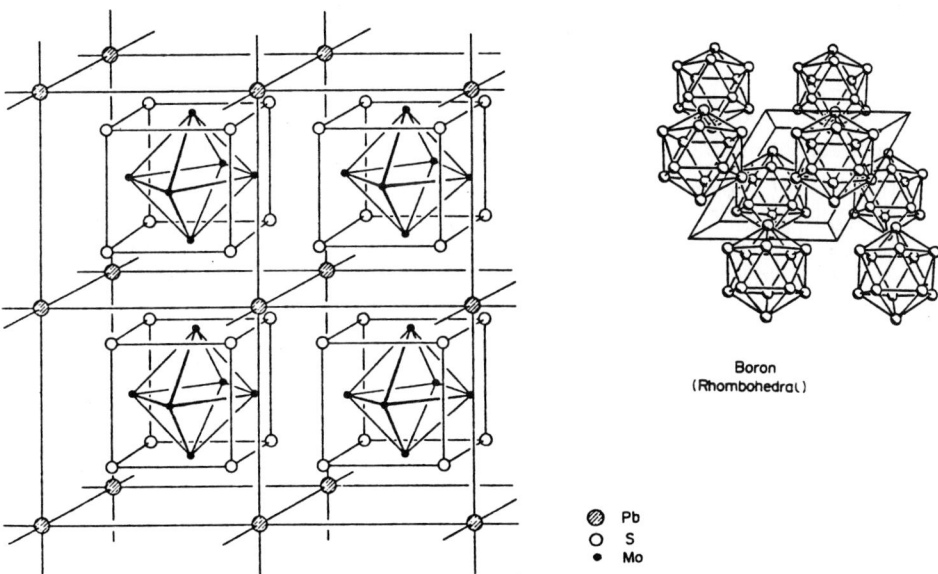

Fig. 2. Left: MO$_6$L$_8$ metal-cluster molecules found in the Chevrel phases (e.g. PbMO$_6$S$_8$). Right: Icosahedral B$_{12}$ clusters froming the rhombohedral structure of boron.

structures of different dimensionality, many of which have been found to be superconducting (see e.g.[6]).

In addition there exist compounds that can be seen as built up from 'bare' clusters, i.e. clusters of atoms without a ligand shell around them. A famous

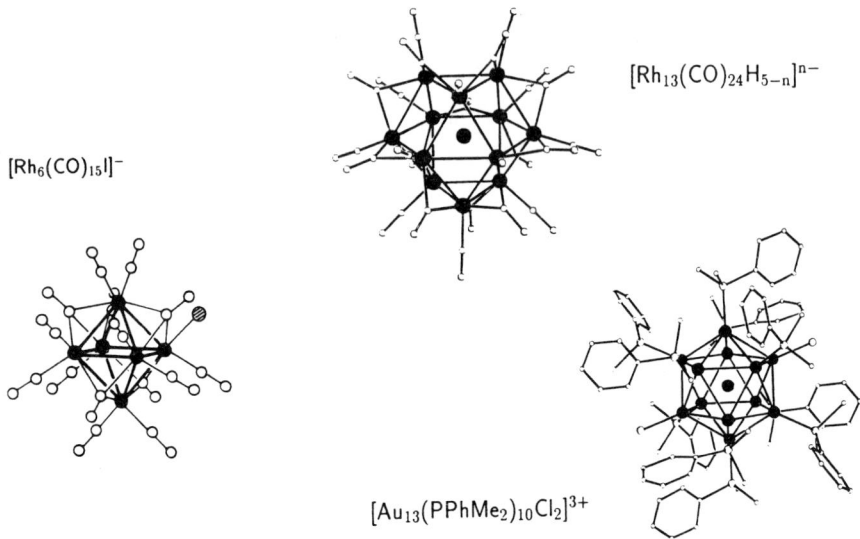

Fig. 3. Molecular structure of $[Rh_6(CO)_{14}(\eta^3\text{-}C_3H_5)]^-$, $[Rh_{13}(CO)_{24}H_{5-n}]^{n]}$ and $[Au_{13}(PPhMe_2)_{10}Cl_2]^{3+}$.

example is the structure of Boron, also shown in Figure 2, which consists of a 3-dimensional rhombohedral packing of icosahedra (of 12 B atoms each). A very recent other example is the f.c.c. packing of the C_{60} football-molecules in the solid C_{60} fullerene.

Probably less well-known is the fact that in coordination chemistry many hundreds of *metal cluster compounds* (MCC's) have been synthesized, in particular during the last two decades [3–5]. In these materials the clusterunit is a metal core consisting of a given number of metal atoms, surrounded (ligated, coordinated) by ligand atoms or molecules, as in the examples shown in Figure 3. The ligand (L) molecules can be simple, as in the large subgroup of carbonyl clusters where CO is the ligand, or also have the form of large organic molecules (e.g. phosphine). In many cases the clustermolecules are ionic, and, together with suitable counterions, build up ionic compounds. But they can also be neutral, forming crystalline solids analogous to molecular (Van der Waals) crystals, or forming dense but randomly packed structures as in the 'Schmid-clusters' which we shall now discuss.

In Figure4 it is shown how the series of magic atom-number clusters M_{13}, M_{55}, M_{147}, M_{309}, M_{561}, etc. . . . are obtained by surrounding an atom progressively with additional shells of atoms of its kind (12, 42, 92, . . .). This can be done for both icosahedral and cuboctahedral packing; the figure shows the cuboctahedral (f.c.c.-packed) particles, some of which are the metal cores of the series of giant clustermolecules synthesized by Schmid and coworkers (the Rh_{13} core in Figure 3 is an anticuboctahedron, the Au_{13} core in Figure 3 is an icosahedron). The M_{55} cores occur in the neutral cluster molecules $M_{55}L_{12}Cl_x$

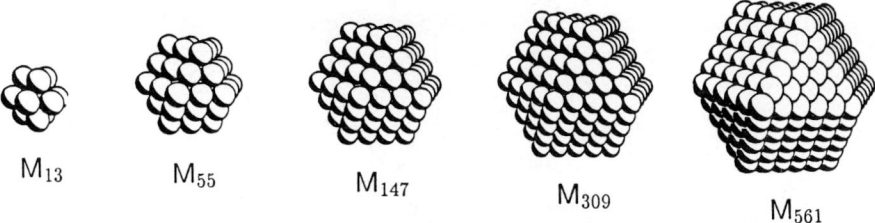

Fig. 4. Magic-number clusters M_n obtained by surrounding a given atom by successive shells of atoms (the illustration is for cuboctahedral packing).

synthesized by Schmid [4]. Here the metal atom can be either Au, Pt, Ru, Rh or Co. Depending on the metal atom, x is either 6 or 20 and the ligand L is e.g. PPh$_3$, PMe$_3$, P(t-Bu)$_3$, As(t-Bu)$_3$, or P(p-tolyl)$_3$. So far, crystalline samples of these 'Schmid-clusters' could not be obtained, the materials being either in the form of a random packing of the neutral molecules (like in a glass), or of extremely small crystallites ($< 0.1 - 0.01$ μm). The same holds for the giant cluster molecules Pd$_{570\pm30}$L$_{60\pm3}$(OAC)$_{180\pm10}$O$_{190\pm10}$ reported by Moiseev and coworkers [7] and for Pt$_{309}$Phen$^*_{36}$O$_{30\pm10}$ and Pd$_{561}$Phen$_{36}$O$_{190-200}$ synthesized likewise by Schmid and coworkers [8, 9]. Below we shall often refer to these giant Schmid-clusters just by denoting the type and number of metal atoms in the core as M_n, e.g. Pd$_{561}$ for Pd$_{561}$Phen$_{36}$O$_{190-200}$. Referring to Figure 4, we point out that the M$_{55}$, Pt$_{301}$ and Pd$_{561}$ metal cores found in these MCC's represent, respectively, the two-shell, four-shell and five-shell members of the series of multiple-shell, magic atom-number particles. Recently, Schmid has even succeeded in making samples containing a 1:1 mixture of seven- and eight-shell Pd clusters, corresponding to Pd$_{1415}$Phen$_{54}$O$_{1000}$ and Pd$_{2057}$Phen$_{78}$O$_{1600}$ [4, 10].

It should be noted that the way in which these clusters are being built up, namely metal particles stabilized by (organic) ligand molecules is quite analogous to what happens in metal colloid chemistry. In fact, Schmid has recently started to synthesize metal colloids [4, 10] using the same ligand molecules as for the metal clusters to stabilize the colloidal metal particles (which are typically much larger). All this leads to the expectation that in a very near future we shall have at our disposal cluster solids of ligand-stabilized metal cores with sizes ranging gradually from one to many nanometers!

In parallel with these developments, also new *crystalline* metal cluster compounds with increasingly larger metal cores keep being discovered. As examples we show in Figure 5 a giant Ni$_{34}$ metal cluster synthesized by Longoni and Ceriotti and coworkers [3], whereas Figure 6 depicts the largest metal cluster compound with completely solved crystal structure known until quite recently, namely the compound [Cu$_{70}$Se$_{35}$(PEt$_3$)$_{22}$], synthesized by Fenske and coworkers [11]. 'Until recently', because the same group has just succeeded in making [Cu$_{146}$Se$_{73}$(PPh$_3$)$_{30}$], which is the latest and largest member

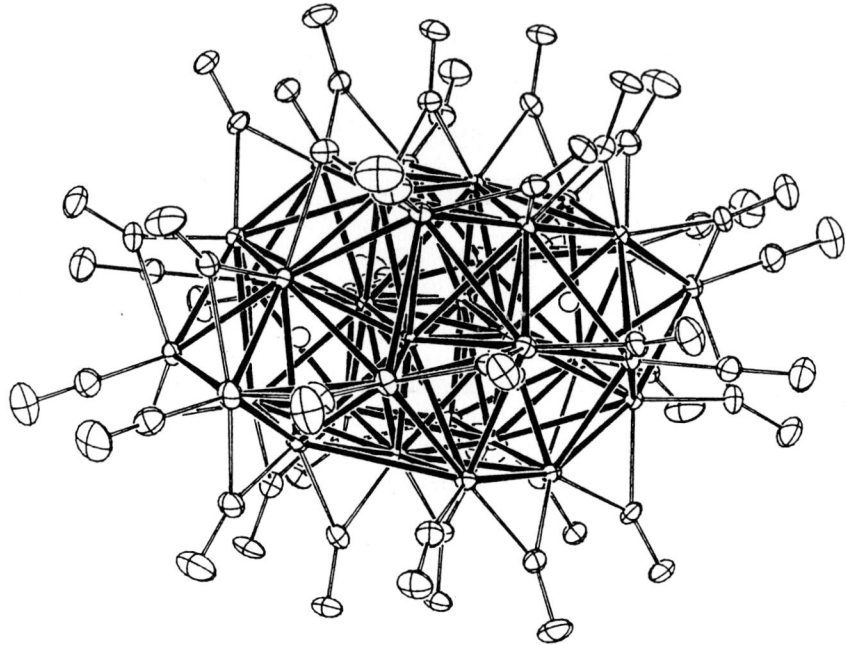

Fig. 5. Structure of the metal cluster molecule $[Ni_{34}(CO)_{38}C_4H]^{5-}$, after [3].

of a series of Cu_2Se clusters of increasing size. It is of interest to note here that the bulk material Cu_2Se is a half metal with a high conductivity. In this sense they may also be classified as metal cluster compounds.

From the above discussion it will be evident that we may indeed view the metal cluster compounds as composite materials, consisting of metal particles of a tuneable, homogeneous size in the nanometer range, which particles are embedded in the nonmetallic (dielectric) matrix formed by the ligands (and, if present, the counterions). We may thus rightly view them as *chemical nanostructures*, i.e. nanophase materials obtained by chemical synthesis [2], as opposed to the *physical nanostructures* obtained by vaporisation and subsequent deposition of materials (sputtering, molecular beam epitaxy, nanolithography, cluster formation in atomic beams, etc. . . .). Other chemical nanostructures are for instance the compounds with chain-like or layer-like structures, which are the subject of other volumes in this series on materials with low-dimensional structures. Of course it makes no sense to call every arbitrary chemical compound a chemical nanostructure! Nevertheless, there exist a multitude of, more or less complicated, chemical compounds which can be viewed as composed of a repetition of clearly recognizable (macro-) molecular units. Besides the clusters, such units may be in the form of layers or multilayers, of chains or multichains, leading respectively to quasi 2-dimensional or quasi 1-dimensional structures. Well-known examples are

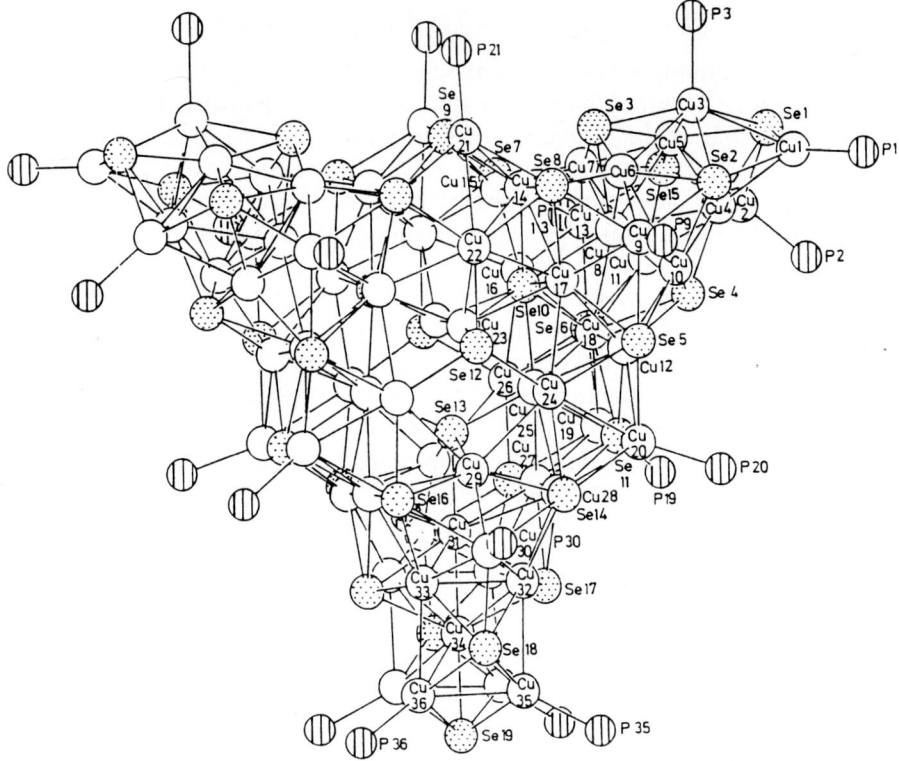

Fig. 6. Structure of the cluster molecule [$Cu_{70}Se_{35}(PEt_2Ph)_{23}$], after [10] (C-atoms are not shown).

the quasi low-dimensional magnets, the molecular (super)conductors and the high-T_c copper oxides.

Before ending this section, we wish to dwell briefly on the possible applications of the metal cluster compounds. Their importance for the field of catalysis shall be mentioned in the following chapters on the chemistry. For molecular (nano-)electronics one should imagine the many potential applications of a matrix of metal particles of tuneable size and tunable mutual electric insulation (by varying the size of the ligands). The matrix can be randomly packed, but also crystalline. Furthermore, it can be 3-dimensional, but one can also foresee the possibility of a 2-dimensional array of cluster molecules deposited on a substrate, so that individual cluster molecules in the array could be addressed e.g. with an STM tip. Last but not least, they may provide us with novel electric, magnetic, or optical properties in view of the fact that they constitute arrays of quantum wells (quantum dots), which are of tuable size, and thus of variable electronic energy-level structure.

1.3. Electronic Energy-Level Structures

The task of obtaining the energy-level scheme of a large cluster to any degree of *quantitative* accuracy represents a formidable problem, and is in fact a typical example of the difficulties met in mesoscopic physics. On the one hand, a particle of, say, 10^3 atoms, is so small that very strong size-effects are to be expected, implying that important quantum-size corrections must be applied to the theoretical approaches usually applied to bulk metals. On the other hand, such particles are generally speaking too large to be treated even by the most advanced quantum-theoretical methods presently developed for large molecules. In what follows, we shall thus only give a brief, qualitative outline of what is to be expected. We shall mention successively the Fröhlich free-electron model for a single small particle, the Kubo model for an ensemble of (noninteracting) metal particles, the Kronig-Penney model for an assembly of quantum-dots connected by δ-function shaped potential barriers, and lastly molecular-orbital calculations for small clusters. Finally, at the end of this section, the effect of the charge-transfer between ligands and metal-cores in the metal cluster molecules will be considered.

For more extensive discussions on these subjects the reader is referred to the articles mentioned below, and in particular to the valuable reviews of Perenboom, Wyder and Meijer [12], and of Halperin [13]. In addition, a very readable book on microcluster physics by Sugano has recently appeared [14].

In Fröhlich's treatment of a small particle [15] the energy levels E_n are calculated on the basis of the model for free, noninteracting electrons confined to a cubic box of side L. The Hamiltonian of this system has just the kinetic term, $\mathcal{H} = -(\hbar^2/2m)\nabla^2$, and the solutions of the Schrödinger equation depend on the boundary conditions. For a bulk metal, periodic boundary conditions are chosen, in order to enable running-wave solutions, the energy-values being

$$E_n = \frac{\hbar^2}{2m}\left(\frac{2\pi}{L}n\right)2, \quad \text{with} \quad n^2 = n_x^2 + n_y^2 + n_z^2.$$

Since the running plane waves, $\psi = A\exp(i\mathbf{k}\cdot\mathbf{r})$, are not only eigenfunctions of the Hamilton operator, but also of the momentum operator, $\hat{p} = -i\hbar\nabla$, the momentum is a good quantumnumber, so that we may associate $p/\hbar$ with the wavevector $k = n2\pi/L$.

On the other hand, to model a single small metal particle, one usually chooses for hard-walled boundary conditions, and the wavefunctions take the form of standing waves, $\psi = A\exp(i\mathbf{k}\cdot\mathbf{r}) + B\exp(-i\mathbf{k}\cdot\mathbf{r})$, which are no longer eigenfunctions of the momentum operator. Since the Hamiltonian does not depend on time, we must have stationary solutions. So, the energy values are still sharp, and are given by

$$E_n = \frac{\hbar^2}{2m}\left(\frac{\pi}{L}n\right)^2, \quad n^2 = n_x^2 + n_y^2 + n_z^2.$$

We emphasize these, perhaps trivial, details to remind the reader that for any finite system the solutions will be in the form of standing waves, so that momentum is, strictly speaking, not well-defined. Depending on the complexity of the boundary problem, the wavefunction may become an extremely complicated combination of basis functions. Nevertheless, the energy levels remain well-defined and sharp, i.e. boundary conditions cannot by themselves cause energy-level broadening. Furthermore, the uncertainty relation $\Delta x \cdot \Delta k \geq 2\pi$ is seen to be obeyed, since the electron is confined to a space $\Delta x = L$, and the associated uncertainty in the wavevector is $2\pi/L$.

Broadening may, however, originate from interactions between the quasi-particles, as in the Fermi-liquid theory. The energy then becomes a complex quantity, the imaginary part giving the relaxation time corresponding with the interaction. However, in this theory the imaginary part of the energy levels close to E_F always becomes negligible at low enough temperature, so that for $T \to 0$ the lifetime broadening will disappear, and these energy levels will become again sharply defined.

As another model for the small particle, one may consider the hard-walled sphere, i.e. the (independent) electrons are confined to a spherical volume instead of a cubic box. Then the angular momentum is conserved and is a good quantum number. The angular wave functions are the usual spherical harmonics, and the energy-levels are given by:

$$E_{nl} = \frac{\hbar^2}{2mR^2} \alpha_{nl}^2.$$

Here R is the radius of the sphere and α_{nl} is the n-th zero of the l-th order spherical Bessel function. The energy-levels are now characterized by the three well-known quantum numbers $n = 1, 2, 3, \ldots$, $l = 0, 1, 2 \ldots$, and $m = 0, \pm 1, \ldots, \pm l$. (Note however that, contrary to the hydrogen atom problem, there is no central Coulomb potential here, the spherical symmetry being imposed just by the shape of the potential well.)

These textbook examples may serve to remind us that electronic quantum-size effects are just due to the *wave-character* of the electron, combined with the fact that for *any* wave (also classical waves, as e.g. elastic waves on a string) confined to a finite volume the allowable values for the wavelength λ are quantized, that is for a cubic box of side L the ratio $2L/\lambda$ must be an integer. This implies that only discrete values for the energy are allowed, the average separation δ between adjacent energy levels becoming the larger, the smaller the box.

In a free electron model, the Fermi energy (the highest occupied molecular orbital) depends itself only on the density $\rho = N/V$ of the electrons, i.e.

$$E_F = \hbar^2 \left(\frac{3\pi^2 \rho}{2m}\right)^{2/3},$$

so that E_F is independent of particle size. Since all levels up to E_F should be occupied by the N electrons in the particle, a rough estimate of the average level spacing is obtained by

$$\delta \simeq 2E_F/N,$$

assuming non-degenerate levels with two electrons of opposite spin in each orbital. It follows from this simple argument that δ is inversely proportional to the volume $V = L^3$, of the particle, or, equivalently, $\delta \propto E_F(\lambda_F/L)^3$, with λ_F the wavelength for an electron with energy E_F. Since λ_F is of the order of interatomic distances, one finds for δ a few meV for particle sizes of a few nanometers.

For highly symmetric systems, such as the sphere, this estimate for δ will be an oversimplification, since strong degeneracies will occur, as well as bunching effects where the levels tend to group together. Such effects are in fact still the subject of intense theoretical and experimental research [16–18]. For instance, the presence of shell-structure in the electronic energy levels due to bunching effects has been beautifully demonstrated in the experiments of T.P. Martin and coworkers [16], who have studied the mass spectra of sodium clusters containing up to many thousands of atoms. Due to the bunching of levels into 'shells', there now appear two different level spacings to consider, namely the average spacing within the shells and the spacing corresponding to the periodicity of the shell-pattern [17, 18].

On the other hand, it is clear that deviations from the free electron model (e.g. spin-orbit coupling) and deviations from the regular shape of the particle, will tend to remove the degeneracies. The assumption of complete absence of degeneracies (due to a sufficient irregularity in shape) was in fact at the basis of the model of Kubo [19], who considered a large ensemble of particles for which the energy levels (near to E_F) were assumed to be randomly distributed. Then, if δ still denotes the *average* spacing between the levels, and if each energy interval has the same small probability of containing a level, it follows from elementary statistics that the values for the spacing Δ between adjacent levels will be Poisson distributed over the ensemble:

$$P(\Delta) = \frac{1}{\delta} e^{-\Delta/\delta}$$

as shown in Figure 7. Here $P(\Delta)$ is the probability for finding an energy-level at a separation Δ from a given one. It is interesting to note that $P(\Delta)$ is largest for $\Delta \to 0$. Thus, in the Poisson distribution levels 'attract' each other, leading to an accidental degeneracy. The critique on this approach has been that such a degeneracy is not likely to happen due to level repulsion effects arising from other sources. Accordingly, Gorkov and Eliashberg [20] have argued that other distributions are possibly more justified, such as the orthogonal distribution, the symplectic distribution and the unitary distribution. The applicability of a particular distribution depends on the symmetry

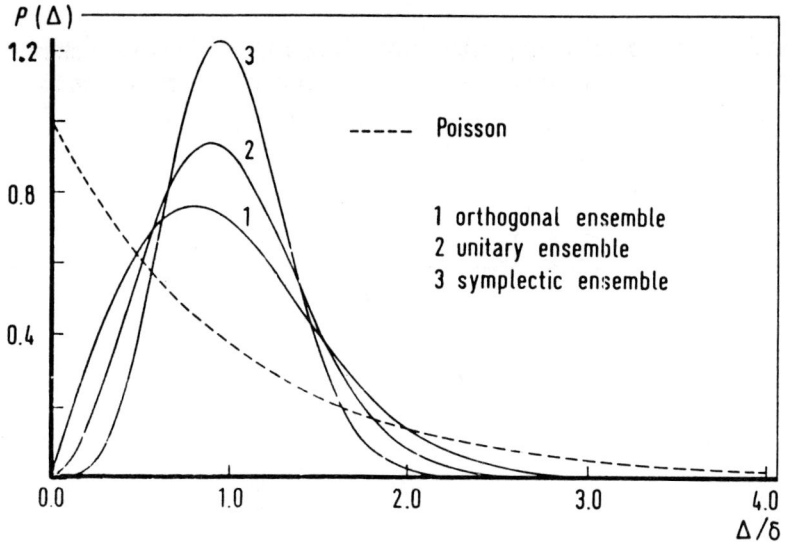

Fig. 7. Distribution functions $P(\Delta)$ versus Δ for different types of ensembles (after [12]).

properties (time-inversion, space-rotation, integer or half-integer spin) of the Hamiltonian, and we shall not discuss them here (see e.g. [12, 13] for more details. All these other ensembles have in common, however, that $P(\Delta) \to 0$ for $\Delta \to 0$, in contrast to the Poisson scheme, as is clear from the comparison given in Figure 7.

So far the discussion has been restricted to a single particle, or to an ensemble of uncoupled individual particles. In order to describe a cluster solid, we have to look for models for arrays of particles, with a finite probability for electron transfer between adjacent particles. There exists a simple model due to Kronig and Penney [21], which consists of a 1-dimensional periodic array of square wells separated from one another by potential barriers, see Figure 8. Although Kronig and Penney used this model to describe the energy levels for an electron in a periodic crystal lattice, the wells being the atoms, the model applies equally well to an array of weakly coupled mesoscopic units, as was pointed out a few years ago by Wohlleben [22] (a similar concept was also used by Mayadas and Schatzkes [23]). In particular, when the finite potential barriers are replaced by delta-functions ($V_0 \to \infty$, $b \to 0$), the problem becomes relatively simple to solve mathematically. In Figure 9 the dependence of the allowed energy values on the quantity $P/4\pi$ is shown, which quantity is a measure of the area under the potential barriers [21]. For $P = 0$ the energy spectrum is just the continuum (energy band) corresponding to non-confined, free electrons. With increasing P the electrons become more and more localized in the individual potential wells, and for $P = \infty$ the spectrum is just that mentioned above for electrons strictly confined to the hard-walled box. For intermediate values of P, the spectrum consists of a

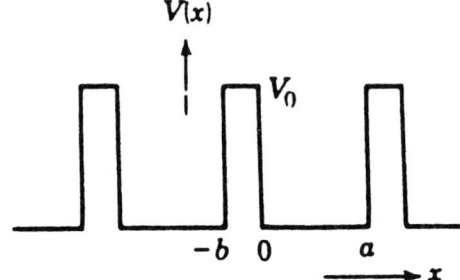

Fig. 8. One dimensional Kronig-Penney potential.

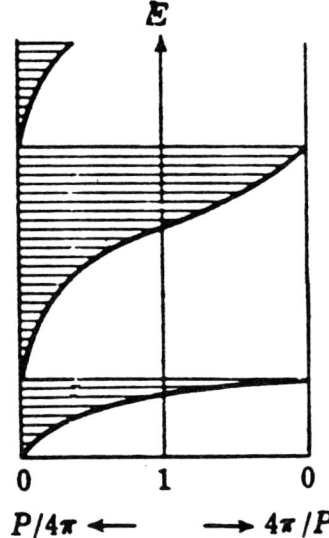

Fig. 9. Allowed and forbidden energy ranges (shaded and open respectively) as function of P. The extreme left corresponds to $P = 0$ (free electrons), the extreme right to $P = \infty$.

number of allowed energy bands, separated by forbidden regions. Thus, each of the discrete energy levels appropriate to an isolated mesoscopic unit will broaden into a miniband as soon as transfer between such mesoscopic units (in a *periodic* array) is allowed, the width of the minibands being roughly proportional to the strength of the transfer. Note that the number of minibands is given by the number of occupied levels in an isolated mesoscopic unit (molecule, cluster), and is therefore of order $2E_F/N$, with N the number of electrons in each cluster. On the other hand, the number of levels in *each* miniband is determined by the total number of allowed k-values, which equals the number of mesoscopic units in the sample.

From this simple model it can be directly concluded that conductivity only occurs when E_F lies within a miniband, otherwise the cluster solid will be a semiconductor. We mention that Kronig and Penney themselves already also presented a 3-dimensional version of their model, i.e. extension of the linear wells in Figure 8 to cubic potential wells is possible. Nevertheless, it is clear that the model is still very crude, a.o. because the internal (atomic) structure of the mesoscopic units is completely left out. It is well-known that the Kronig-Penney model can be viewed as a rudimentary precursor of the modern band theories of solids, with their rich variation in complexity and sophistication. Such calculations may in principle also be applied to cluster solids by constructing a supercell, which is in fact the cluster of atoms itself (surrounded by a region of empty space or an embedding potential). In the calculation the supercell is then repeated periodically, allowing the machinery of the bandstructure calculations for extended solids to be applied. The problem is, of course, that with increasing size of the supercell the complexity of the calculations increases very rapidly, so that so far only clusters with a small number of atoms have been treated with such methods.

The problem of incorporating the atomic structure of the cluster into the calculation is particularly acute for the transition metal clusters, since the d-bands are rather narrow so that the d-electrons can be viewed as fairly localized. For the simple s-like metals like the alkali metals the model of electrons delocalized over the cluster will evidently apply much better, which is the underlying reason why the above-mentioned shell structure is so clearly manifest in the Na clusters [16]. An important extension of the simple hard-walled potential well model that should be mentioned here is the so-called 'jellium' model. In this model the delocalized valence electrons are considered to move in a uniform background of positive charge (which neutralizes the total electronic charge). The model has been applied to calculate magnetic and electric properties of solids, cohesive energies, etc. It has also been applied to surface problems and, more recently, to clusters [24]. The merit of the model is that the computational requirements are relatively simple and that it allows the electron-electron interactions to be taken into account (Coulomb and exchange-correlation). As usual this is done within the local-density functional approximation. In this method the total energy of the interacting electron gas is expressed as a functional of the (ground-state) electron-density only. The many-body effects are treated within an effective-field approach, that is they are replaced by the problem of a single electron moving in an effective average potential in which the presence of the other electrons is incorporated as well as possible. For the exchange-correlation contribution to this local effective potential, one usually takes the Kohn-Sham exchange-correlation potential for a uniform electron gas of density ρ, which potential is proportional to $\rho^{1/3}$. Apart from a numerical factor, the Kohn-Sham exchange potential is identical to the Self-Consistent-Field $X\alpha$ potential proposed much earlier by Slater by a different method. This

$\rho^{1/3}$ approximation (with variations thereof) forms in fact the basis for most theoretical studies of solids (band-theory, molecular orbital theory). Although this one-electron approximation has the merit of rendering the many-body problem mathematically tractable, it is evident that it treats the exchange-correlation effects between the electrons only in a crude way, so that it is bound to fail as soon as such effects become really important.

Within the above jellium model then, properties such as the energy-levels (in particular the shell structure), the electronic charge densities, the ionization potentials, and the polarizabilities have been calculated as a function of size, e.g. for a sphere, and we refer to the review of De Heer et al. [24] for more information. As expected, the model is quite successful for describing general properties which do not depend too much on the precise shape or topology, and applies in particular to simple metals.

For very small clusters it is known that the electronic structure becomes quite sensitively dependent on the geometry, and it is here that quantum (theoretical) chemistry calculations become quite valuable. These have recently been reviewed by Bonačić-Koutecký et al. [25]. For clusters up to, say, about ten atoms *ab initio* calculations based on the configuration interaction method (which treats in principle all the correlation effects) are possible. These give probably the most reliable results for such detailed properties as the exact geometry, the optical transitions, etc. As the number of atoms/electrons increases, however, the required computational efforts increase dramatically. For the larger cluster-molecules, one has therefore again to take resort to one-electron approximations, such as the molecular-orbital (MO) approach within the above-mentioned Xα/Density Functional formalisms. Basically, in the MO-approach the molecular orbitals are expanded in linear combinations of atomic orbitals (e.g. Gaussian-type) centered at the atomic sites. The coefficients in these expansions are determined by solving the appropriate one-electron equations (Schrödinger/Kohn-Sham). Because the computational times do not increase exceptionally with the number of electron per atom, these methods are also well applicable to transition metal clusters. For sufficently high symmetry, clusters up to several tens of atoms can be treated.

We refer to the chapter of Ellis in this volume for further details and just show schematically in Figure 10 a typical result for the MO levels that would be obtained for a d-metal cluster with such calculations, and compare it with the density of states (DOS) for the bulk d-metal. For the cluster the quantity replacing the DOS is the degeneracy or near-degeneracy of the levels. Further, by broadening artificially each MO level in the discrete spectrum, e.g. by a Gaussian function with a width of, say, 0.1 eV, DOS-profiles can be generated which, even for relatively small clusters (13 atoms), already show many of the characteristics of the bulk spectra. This arises, of course, in part from the large number of (s and d) valence electrons for d metal atoms, implying e.g. a total number of 50–70 occupied orbitals for a 13 atom cluster already, spanning an energy width of order 5 to 8 eV, i.e. comparable to the bulk band-width.

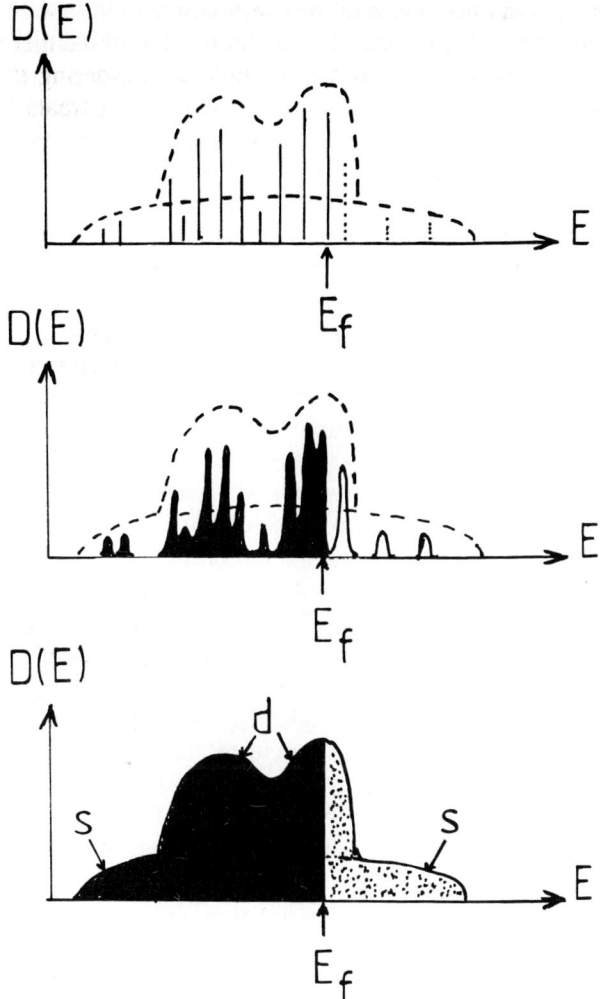

Fig. 10. Sketch of the DOS as a function of energy for a d-metal cluster as compared to the bulk Top: discrete cluster orbitals. Middle: cluster orbitals each broadened by a Gaussian distribution. Bottom: bulk metal DOS. The Fermi energy E_F separates occupied and unoccupied levels in each plot.

For magnetic metal atoms like Ni and Fe, the highest-occupied-molecular-orbitals are found to be partly occupied and nearly degenerate, leading to a magnetic moment per cluster comparable to the value expected on the basis of the bulk magnetic moment per atom (and taking into account the number of atoms in the cluster). In terms of the bulk DOS such a situation corresponds with the Fermi energy E_F lying within the d-band, the corresponding d-holes being responsible for the magnetism. On the other hand, for the nonmagnetic metals like Cu, Pd, and Pt, also the clusters are found to be nonmagnetic, since the available electrons in the cluster are found to fully occupy (spin-paired) each of the levels up to and including the highest-occupied-molecular-orbital (HOMO-level or Fermi-energy), leaving a gap between the HOMO and LUMO levels and no net unpaired spin-density. Similarly, when studying the evolution with size of the energy level structure for semiconductor clusters (like Si), it is found that the HOMO-level converges to the top of the bulk valence band, the HOMO-LUMO separation (Lowest Unoccupied Molecular Orbital) evolving into the bulk semiconducting energy gap.

Perhaps these strong resemblances between the MO schemes for clusters and bulk are not surprising, since, after all, band theory amounts to MO theory applied to the infinite solid. Nevertheless, it is important to bear in mind that the internal atomic structure in the cluster is so strongly determining the electronic properties, thereby mimicking the expected bulk behaviour even for such small clusters.

Turning once more to cluster solids, i.e. macroscopic solids built of cluster molecules, the molecular orbital calculations together with their extension to bandstructure calculations for the cluster-solid, provide a picture of the energy-level structure that is anologous to the above mentioned Kronig-Penney model, but much more detailed and accurate. Recently, such approaches have been applied for instance to molecular and solid (f.c.c.) C_{60} and the related fullerenes [26]. These calculations nicely illustrate how the discrete, molecular energy levels of the single C_{60} molecule evolve into a series of minibands in the solid. Since the intercluster transfer integral between neighbouring C_{60} molecules is small, the widths of the minibands remain small even close to E_F, in particular the HOMO- and LUMO-derived minibands remain separate from one another, as well as from other minibands. Consequently, (undoped) solid C_{60} is a semiconductor, with an energygap of about 1.5 eV. In the doped material, K_3C_{60}, the LUMO-derived band is probably half-filled, since each K-atom is expected to donate one electron to the LUMO-band, which is derived from a molecular triplet state and may thus contain upto six electrons. Indeed, the material K_3C_{60} is experimentally found to be a conductor and even a superconductor with a high T_c. Furthermore, increasing the doping to K_6C_{60} leads again to a semiconductor, since then the LUMO-derived band is full, while it is separated by another gap (of about 1 eV) from the higher-lying states.

Finally, molecular orbital theory may also be used to evaluate the effects of ligand-bonding to the metal cores in the metal cluster compounds as will also be discussed in Ellis' chapter. Evidently, the increasing amount of electrons limits the application to not too large molecules. However, recently nickel carbonyl clusters as large as $Ni_{44}(CO)_{48}$ have been successfully treated, using Gaussian-type orbitals in density functional theory [27]. Generally speaking, it is found that the ligation of the bare metal cluster core by the ligand-molecules is strongly related to the surface-chemisorption on the bulk metal surface by the same molecule. The electron transfer between ligand and metal atom may profoundly change the properties of the latter, but the ligand-metal interaction is often found to be relatively short-ranged for transition metals. In the above example of the $Ni_{44}(CO)_{48}$ cluster it was found that the interaction induces a quenching of the magnetic moment of the 38 Ni atoms which are at the surface of the cluster, whereas the magnetic moment of the 6 Ni-atoms in the interior of the metal core is largely retained.

It is of interest to mention here the basic mechanism underlying this quenching [27], since in the next section and in Chapter 10 these results shall be related to observations on Ni carbonyl clusters. In the bare metal cluster the electronic configuration of the Ni atoms is close to $3d^94s^1$. The metal-metal bonds are due primarily to the (more itinerant) $4s$ electrons, so that roughly speaking one hole is present in the (strongly localized) $3d$ shell of each Ni atom. However, by the weak $3d$–$3d$ interactions between neighbouring Ni atoms the average number of unpaired electrons is reduced, so that the moment per Ni-atom becomes about 0.6–0.7 μ_b/atom, quite similar as in the bulk. The coordination by CO now favors an intra-atomic redistribution of the Ni atom configuration from $3d^94s^1$ to $3d^{10}$, mainly due to the repulsive interaction between the CO 5σ charge density and the Ni $4s$ electrons. By the ensuing destabilization of $4s$-derived orbitals well above the HOMO-level of the bare cluster, the $4s \rightarrow 3d$ transfer takes place, the local $3d$ shell becomes filled, and the Fermi level (the chemical potential) moves *above* the $3d$-band. Effectively, the Ni-atom is 'transformed' into a Cu-atom, by which process the density of states at the Fermi-energy, $D(E_F)$, is reduced considerably (cf. Figure 10). We note that this destabilisation process, with the accompanying change in the magnetic moment, is the (macro) molecular analogue of the high-spin/low-spin transition familiar in transition metal complexes (e.g. of Mn and Fe ions) [28]. The chemical bonding of ligands to a (single) metal atom in such complexes is taken into account in the form of an electric field acting on the energy levels. This field is called crystal-field or ligand-field, and its symmetry reflects the coordination of the metal atom. In the ligand-field theory, a low-spin complex results when the crystal field is sufficiently strong to produce a splitting of the d-orbitals large enough to overcome Hund's rule. The latter favors the distribution of electrons with parallel spin in different orbitals (intra-atomic ferromagnetism!), thereby minimizing the Coulomb energy.

1.4. Brief Introduction to the Physical Properties of Metal Cluster Compounds

The physical measurements on metal cluster compounds performed sofar are discussed more extensively in the following chapters. Here, we shall attempt to provide a general background with which the reader can turn more profitably to these more detailed descriptions. On basis of the information given in the preceding paragraphs, we may arrive at a broad understanding of the observed properties of the cluster compounds of the transition metals by the following simple considerations. Due to the ligand-metal interactions, the character of the ligated metal atoms will in general be quite different from the bulk metal. These chemical bonding effects will come in addition to the usual surface effects in *bare* metal clusters or at metal surfaces. For instance, it is known that, due to the reduction in the number of neighbours, surface atoms will have considerably narrower d-bands than bulk atoms, implying that the local $D(E_F)$ for a surface atom can be quite different. On the other hand, as already discussed, the ligation effects (and also the surface effects) are short-ranged, so that in first approximation we may take them as restricted to the first layer of metal atoms. Thus we may *grosso modo* divide the metal atoms of the cluster core into *surface atoms* and *inner-core atoms*. For the magic atom number clusters shown in Figure 4, this separation is illustrated in Figure 11 (we are indebted to D.A. van Leeuwen for making this plot). For the surface atoms one expects strong deviations from bulk-metal properties, independent of the size of the metal cluster molecule. The inner-core atoms, on the other hand, should constitute a minute piece of the bulk metal, with (very strong) quantum size-effects for small clusters, evolving towards bulk properties with increasing cluster size. The rate of this evolution obviously depends also on the number of valence electrons per atom, which is rather high (about ten) for the transition metals of the second half of the $3d$, $4d$, and $5d$ series considered here (Ni, Cu; Ru, Rh, Pd; Os, Ir, Pt, Au). With Fermi-energies of the order of 5 eV, this implies average level spacings $\delta \simeq 2E_F/N$ of about 10^2 K for an inner-core of about 10^2 atoms.

Several results have already been obtained which appear to agree with this simple ansatz. For instance, as will be discussed in Chapter 6, the ^{197}Au Mössbauer spectroscopic studies on the Schmid-cluster $Au_{55}(PPh_3)_{12}Cl_6$ could be interpreted [28] in terms of four different contributions from the various identifyable Au sites, namely the 13 Au atoms forming the inner-core of the Au_{55} cuboctahedron, the 24 uncoordinated surface Au atoms, the 12 surface atoms coordinated by the PPh_3 ligands, and the 6 surface atoms ligated by Cl. It was found indeed that the Mössbauer parameters of the 13 inner-core sites are close already to the bulk metal values, the quadrupole splitting parameter being zero due to the high symmetry, and the isomer shift parameter being near to the bulk value (the isomer sift is a measure of the 6s electron density seen at the nucleus). By contrast the Mössbauer parameters for the surface sites were found to be completely different, lying in fact in the ranges of literature values known for non-conducting Au compounds. An even more dramatic confirmation of bulk-properties of the inner-core metal atoms was obtained in most recent Mössbauer studies of the giant Pt_{309} cluster, which are likewise discussed in Chapter 6.

In Chapter 10 on the magnetic properties it will be seen that the quenching of the magnetic moments of ligated surface Ni atoms, as predicted by

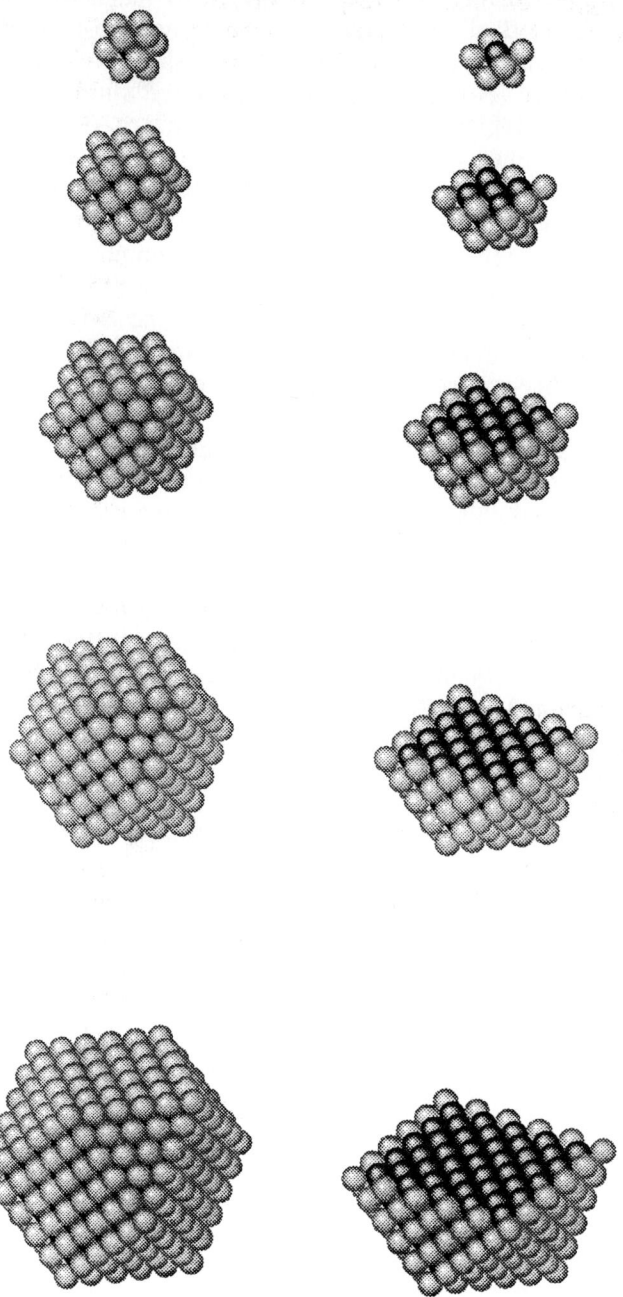

Fig. 11. Illustration of the division in surface atoms and inner-core atoms for the magic atom-number clusters.

molecular orbital theory, has been confirmed by magnetization studies on Ni carbonyl clusters of varying size [30–32]. Small cluster molecules (Ni_8, Ni_9, etc. ...) containing only surface Ni sites show no magnetic moment, as to be expected. However, also for very large cluster molecules such as $[Ni_{34}(CO)_{38}C_4H](NEt_4)_5$ and $[Ni_{38}Pt_6(CO)_{48}H_2](NEt_4)_4$, the magnetic moments of the cluster as a whole were found to amount to at most a few μ_B, and a similar conclusion could be drawn for the Schmid-cluster $Co_{55}(PMe_3)_{12}Cl_{20}$ [30, 31]. From a careful study [32] on a *single crystal* sample (18 mg) of a $[Ni_{38}Pt_6(CO)_{48}H]^{5-}$ cluster compound, it could recently be concluded that the intrinsic magnetic moment of this cluster molecule is most probably zero. Since the Ni-atoms are all on the surface, while the (presumably non-magnetic) Pt atoms form the inner-core of the metal cluster, this would agree with complete quenching of the moment of the surface Ni-atoms, as predicted by theory [27].

A dramatic reduction of the magnetic moment with respect to the bulk was also observed recently in a series of Pd clusters and colloids [33], namely for the 5-shell Schmid-cluster $Pd_{561}Phen_{36}O_{200}$, a 50/50 mixture of the 7-shell and 8-shell Schmid-clusters $Pd_{1415}Phen_{60}O_{\sim 100}$ and $Pd_{2057}Phen_{84}O_{\sim 1600}$ and a Pd colloid with an average particle size of 15 nm. As will be discussed in Chapter 10, in this series the magnetic susceptibility is found to increase only very slowly towards the bulk-value. Similar effects were also observed for the large Pt cluster $Pt_{309}Phen^*_{36}O_{30}$, and for a Pt-colloid with an average cluster diameter of about 3 nm [32].

As will be discussed in Chapter 7, specific-heat measurements may also reveal the presence of size-effects on the density-of-states (DOS) of the electronic energy spectrum. For noninteracting fermions (electrons), the contribution to the specific heat is predicted to have the form of a linear dependence on temperature, $C_{el} = \gamma T$, at temperatures much lower than the Fermi-energy, E_F. For free electrons, γ is given by $\gamma = \frac{1}{3}\pi^2 D(E_F) k_B^2$, and the proportionality with $D(E_F)$ is thought to hold also in the weakly interacting case. However, for sufficiently small particles the size-effects may become so strong that the linear term is no longer observed. This will be the case as soon as the gaps δ between the energy-levels exceed the thermal energy $k_B T$ associated with the temperature below which the linear term is expected to show up. Instead of the linear term, the electronic specific heat will then drop exponentially to zero. In this simple picture, the measurement of the linear term for a large enough metal cluster may be seen as one of the most reliable criteria for defining the onset of metallic behaviour. We note here that in what follows we shall use as a criterion for 'metallic' thermodynamic behaviour the presence of a quasi-continuous DOS near to E_F, i.e. quasi-continuous with respect to the thermal energy involved in the thermodynamic experiment in question. This means that the onset of metallic behaviour may depend on the thermodynamic quantity considered, besides on the temperature involved, i.e. a cluster may behave 'metallic' at, say, 100 K, whereas it may become non-metallic at 1 K.

Matters can become even more complicated, though, in case of Poisson-distributed energy-level structures near to E_F. Referring again to Figure 7, one observes that for the Poisson distribution (and *only* for this), the probability $P(\Delta)$ to find energy separations Δ *increases* for $\Delta \to 0$. This result may seem strange at first sight, but is merely due to the fact that for a random distribution there are much more possibilities to realize a small than a large separation, if we have to combine two levels at random within a given, fixed energy interval. As a consequence, even though the average distance δ between energy-levels may be considerable, and large compared to $k_B T$, the specific heat will still show a linear term at low enough temperatures. At a given temperature, $k_B T \ll \delta$, for a fraction of the clusters in the ensemble the nearest energy-level close to E_F will be at a separation $\Delta > k_B T$, so that their contribution to the specific heat is exponentially small ($\simeq \exp \Delta/k_B T$). However, for the remaining clusters there will be a level with $\Delta < k_B T$, and these clusters will still contribute a linear term to the total specific heat of the ensemble.

A linear term in the specific heat was first reported [34] for $Pd_{561}Phen_{36}O_{200}$ in the range T< 1 K, and later also for $Pt_{309}Phen^*_{36}O_{30}$. At first this was interpreted as a sign for bulk-like metallic behaviour (i.e. $\delta \ll k_B T$). The value of γ that was measured was about 1/3 of that known for the bulk metals, which was attributed to a strong reduction of the $D(E_F)$ due to the surface effects and ligand-bonding effects, as explained in the above. In subsequent NMR studies on the Pt_{309} cluster [35], however, strong evidence was obtained for an average $\delta \simeq 50$ K, in agreement with the value one estimates for clusters of these sizes on basis of the simple formula $\delta \simeq 2E_F/N$. Therefore, the interpretation of the specific heat in terms of bulk-like behaviour had to be revisited, and it turned out that both the NMR- and the specific heat data could be interpreted by assuming the energy-levels near E_F for the clusters to be Poisson-distributed.

As will be extensively discussed in Chapter 8 the NMR technique is a very sensitive tool since it probes the nuclei of the (metal) atoms in the cluster. The NMR-parameters, like the nuclear-spin lattice relaxation time T_1 and the NMR resonance frequency itself, are sensitive to the physical properties of the investigated material. They are particularly sensitive to the presence of delocalized electrons because of the intactions these will have with the nuclear moments. Thus, in bulk Pt the NMR frequency is considerably different from that in insulating Pt compounds. The difference is called the Knight-shift, K, and it is proportional to $D(E_F)$ similar as is the coefficient of the linear term in the specific heat. Also the behaviour of T_1 is widely different for a metal and an insulator. In a metal the coupling between the nuclear spins and the spins of the itinerant electrons provides an additional channel for nuclear-spin lattice relaxation. Accordingly T_1 is much shorter for a metal, and its temperature dependence is predicted to obey the Korringa-law, $T_1 \propto 1/T$. From these considerations it is obvious that NMR may provide a most valuable tool in establishing the presence of metallic-like properties in a metal cluster.

For the Pt_{309} cluster an extensive series of ^{195}Pt-NMR experiments has been performed [35] at temperatures between room temperature and 4.2 K. The NMR-line is quite broad and could be well-explained as being composed of different contributions coming from non-metallic surface sites and from metallic-like inner-core sites. For the surface sites the NMR intensity is found at frequencies corresponding to those known for the insulating Pt compounds, and the associated values for T_1 are quite long, as expected for nonmagnetic, nonconducting solids. By contrast, the NMR intensity associated with the inner-core sites was found to be considerably shifted into the direction of the known-value for bulk Pt. In the first experiments the temperature was 80 K or higher, and in this range the absolute values for T_1 for these inner-core sites were found to agree with those for bulk Pt, and also to show the Korringa temperature dependence. Both these results indicate metallic-like behaviour according to the definition given in the above. However, in subsequent experiments at lower temperatures these two metallic-like features were found to become lost, wich strongly indicates the presence of quantum-gaps near to E_F of the order of 50 K. As discussed by Van der Putten et al. [35], both the NMR data and the low-temperature specific heat may be analysed in terms of the Poisson-distributed energy spectrum. The question as to why for these two metal cluster compounds the cluster energy-levels near to E_F would be randomly distributed throughout the sample remains still to be answered. One possibility would be a random modification of the energy-level structures of the individual cluster molecules induced by the presence of random inter-cluster charge transfer. As is clear from the discussion of solid C_{60} and the Kronig–Penney model, a charge transfer between clusters will broaden each cluster level into a mini-band. When the clusters are randomly packed, the charge transfer and thus the degree of broadening will vary, and the ensuing variation in the HOMO-LUMO derived energy-gap around E_F might mimick the Poisson-distribution.

We shall come back to this point below, when discussing the results of the conductivity measurements in these samples.

In the heat capacity experiments on the metal cluster compounds, besides the electronic contribution at low temperatures (< 1 K), also the contributions coming from the vibrations of the atoms inside each cluster (intracluster vibrations) and the vibrations of the clusters as a whole (intercluster vibrations) could be distinguished [29, 34]. This is discussed in detail in Chapter 7. As is well-known, the vibrations of a single cluster (the intracluster vibrations) will also show size-effects, similar as for the electrons. This may be understood by realizing that in a single cluster only lattice waves with a wavelength shorter than the size of the cluster may exist. As a consequence there will be a gap on the low-energy side of the DOS of the lattice-waves (phonons). Computer calculations of the intracluster contribution to the specific heat taking into account this effect were found to agree quite well with the experimental data on a small Au cluster, $Au_{11}(PPh_3)_7(SCN)_3$, on the intermediate sized Au_{55} cluster $Au_{55}(PPh_3)_{12}Cl_6$, and on the large Pt_{309} and Pd_{561} cluster compounds. This agreement may thus be seen as an experimental confirmation of the quantum size-effect in the phonon spectra. As regards the intercluster vibrations, these correspond to the long-wavelength lattice-waves, for which the cluster compound can be viewed as an effective medium, with an elastic-stiffness constant determined by the packing of the clustermolecules in the solid. At low temperatures the intercluster vibrations thus contribute a Debye-type T^3-term to the specific heat (whereas the intracluster contribution decays exponentially due to the presence of the low-energy quantum-size-gap). The coefficient of the T^3 term is proportional to Θ^{-3}, where Θ is the Debye-temperature, which itself is proportional to $1/\sqrt{m}$ with $|M|$ the mass of the clustermolecule. The ensuing mass dependence of the coefficient is approximately found indeed. (Note, however, that also the stiffness constant of the cluster solid will vary with the clustersize.)

1.5. Conductivity Studies

Measurements of the DC conductivity and of the dielectric properties (AC conductivity in the range $10–10^6$ Hz) have sofar been mainly performed on the large $Pd_{561}Phen_{37}O_{200}$ cluster and on $Au_{55}(PPh_3)_{12}Cl_6$ and related Au_{55} clusters. Experiments were also performed on a number of small clusters like Ni_9, Au_9 and Au_{11} clusters, and on $[Ni_{38}Pt_6(CO)_{48}H_2](PPN)_4$. For these compounds, however, the resistivities were all found to be quite high, i.e. larger than 10^{10} Ωcm at room temperature. Since the resistance varies exponentially with temperature, experiments were restricted to a small range above room temperature. Except for the $Ni_{38}Pt_6$ cluster, for which a very small single crystal was available, all the other measurements were done on samples in the form of polycrystalline powders, pressed into thin pellets. The AC conductivities were measured with the impedance method, the DC conductivity with an electrometer capable of measuring resistances up to 10^{13} Ω.

Since the experiments have recently been extensively reviewed [36], we shall only give a summary of the main results here. The DC conductivity of the Au_{55} clusters and of $Pd_{561}Phen_{36}O_{200}$ was found to show a temperature dependence of the form

$$\sigma_{DC} = (\sigma_0/T)\exp[(T_0/T)^{1/2}]$$

as is illustrated in Figure 12. We remark that the deviations seen at the lowest temperatures (< 8 K) are due to non-ohmic effects. As shown in Figure 13 the conductivity becomes strongly non-ohmic below about 10 K, with σ increasing with the applied voltage. Similar effects are commonly

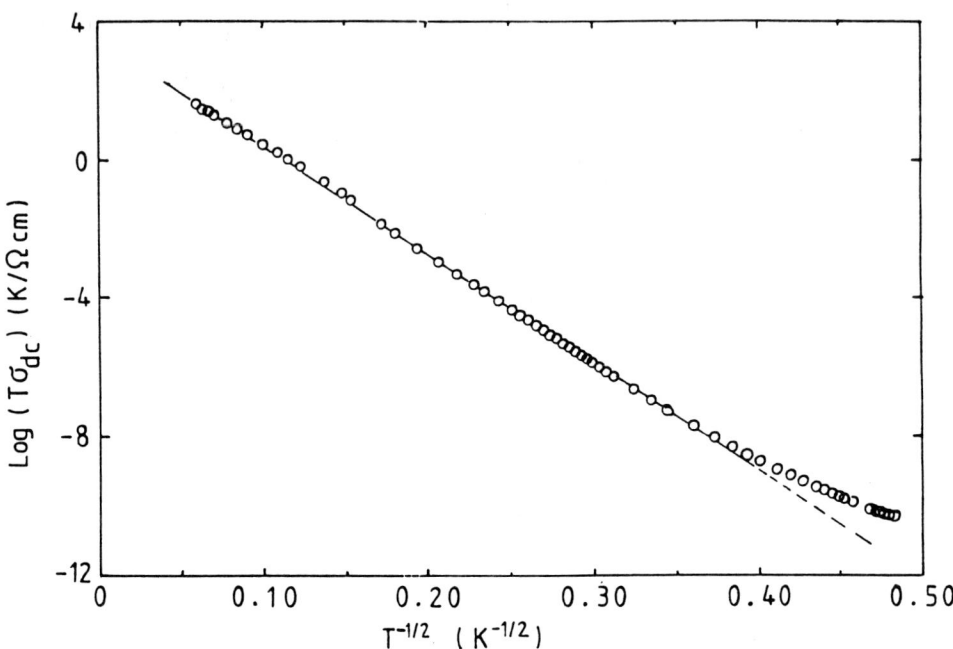

Fig. 12. $T\sigma_{dc}$ as a function of $T^{-1/2}$ for the Pd$_{561}$ compound.

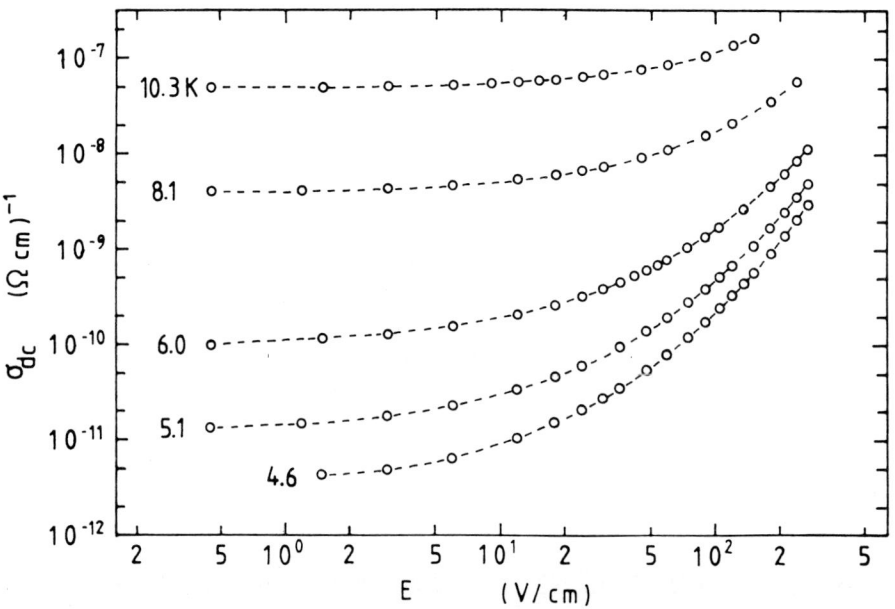

Fig. 13. DC conductivity of Pd$_{561}$Phen$_{37}$O$_{200}$ as a function of the applied electrical field at several temperatures.

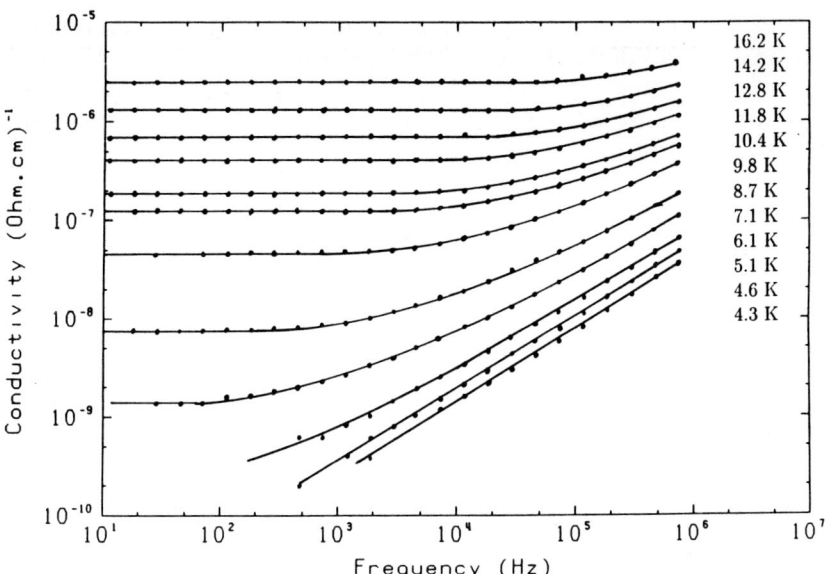

Fig. 14. AC conductivity of $Pd_{561}Phen_{37}O_{200}$.

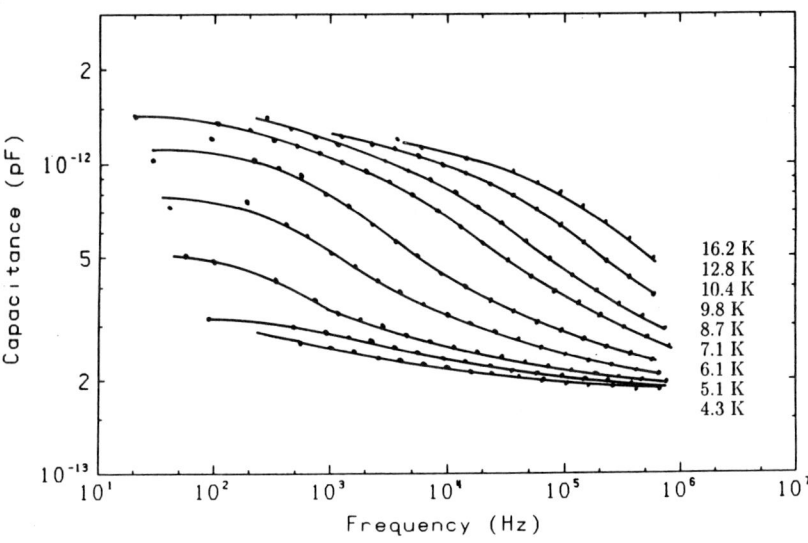

Fig. 15. AC capacitance of $Pd_{561}Phen_{37}O_{200}$.

seen in systems were the conductivity is due to tunneling (metal-insulator-metal junctions) or to hopping processes. In Figure 14 it can also be seen that, at low temperatures, the frequency dependence of the AC conductivity increases very markedly. It should be noted here that the frequency-dependent behaviour of the conductivity observed in these materials was corroborated by measurementes of the dielectric constant by the capacitive method, in the same frequency range of $10-10^6$ Hz (Figure 15). From these experiments it could be deduced that the AC-conductivity and AC dielectric constant depend on the frequency as:

$$\sigma'(\omega) - \sigma'(0) \propto \omega^{s'}$$

$$\varepsilon'(\omega) - \varepsilon'(\infty) \propto \omega^{s''-1}$$

where the exponents s' and s'' in these power-law frequency dependences are themselves also functions of the frequency, as well as of temperature. In the investigated frequency range the variation found was $0.4 \leq s' \leq 2.0$ and $0 \leq s'' \leq 1.0$.

The observed temperature dependence of the DC conductivity may be contrasted with the simply activated form, $\sigma_{DC} \propto \exp(-T_0/T)$. The square-root dependence is found in a variety of so-called variable-range-hopping models, which aim to describe the hopping conductivity in random metal-nonmetal composites, doped or amorphous semiconductors, ceramic metals ('cermets'), organic (semi)conductors, ionic conductors and other narrow-band conductors or disordered media. Lately, the underdoped high-T_c copper oxides have been added to this list. As discussed in [36], the similarities in the observed DC conductivity and dielectric behaviour of these at first sight different systems, are indeed quite remarkable. What these materials have in common is that the conductivity arises from a diffusive hopping or tunneling of charge carriers between randomly distributed sites (or through barriers of random height). Apart from the tunneling probability, it is the nature of the 'sites' that differs from one type of material to the other.

In view of the nature and structure of the metal cluster compounds, it is an obvious choice to associate the metal cluster molecular units as the sites. The origin and nature of the mobile carriers is as yet unknown. However, they might arise from a slight reduction or oxydation during the synthesis of the materials, whereby a fraction of the clustermolecules would become ionized. By intercluster charge transfer, the extra charge could then become mobile. In view of the low values of the conductivity, i.e. low mobility, one may expect the carriers to be of the polaronic type, since the carrier will stay long enough at a given cluster to establish a displacive (ionic) polarization (apart from the electronic polarization).

Thus, in the metal cluster compounds the sites between which the carriers hop are thought to be the macromolecular metal cluster units, in the cermets they are the metal particles, in carbon-polymer composites the carbon particles, in doped semiconductors the impurity atoms, in V_2O_5 gels the V atoms with a different valency, etc. Accordingly, the size of the 'sites' ranges from atomic via molecular dimensions towards the very large (500 Å) values attained in the carbon composites. Evidently, also the internal structure of the 'sites' changes drastically through this series of systems. As observed in [36], it is quite interesting that such drastic variations do not influence the *qualitative* behaviour of the dielectric properties; they appear to be only reflected in the *scaling factors* by which such external parameters as the temperature and the strength and frequency of the applied electric field have to be scaled. Also the precise nature of the charge carriers (electrons, small-polarons, hydrogenic carriers) does not appear to be very important: as long as there are fairly localized carriers, diffusing through a system of random barriers separating 'sites', the qualitative behaviour of both the AC and DC conductivity appears to be independent of the nature of 'sites' and barriers.

It is gratifying that such similarities are also found when comparing the various theoretical models that have been proposed. If one neglects on-site electron correlations, the formalism developed a.o. by Böttger and Brysksin [37] presents an attractive possibility to describe the experiments [36]. Firstly, it takes into account the polaronic effects arising from the electron-phonon interaction, which 'dresses' the charge carrier with a 'phonon cloud' arising from the atomic polarization of the atoms surrounding the charge carrier. Secondly, the formalism may treat both the quantum-mechanical (low-temperature) regime, where the carriers may move by quantum-mechanical tunneling in a band motion, as well as the high-temperature classical regime where the motion is by thermally activated, phonon-assisted hopping. Disorder may be taken into account by a distribution in the site-energies as well as in the inter-site transfer integral (jump-frequency, transition rate). Calculation of the frequency-dependent conductivity of the disordered system can be done by the density matrix method. For weak electric fields the ensuing rate equations can be linearized, and become equivalent to the Miller–Abrahams resistor-capacitive network. The Extended Pair Approximation is a mean-field solution of the Kirchoff equations for this network, and can be used to obtain the frequency-dependent conductivity. As an alternative to the density matrix method, the frequency-dependent conductivity may be calculated by interpreting the diffusive motion of the carriers as a stochastic process. Both methods lead to equivalent results, but the advantage of the stochastic interpretation is that it may yield analytical expressions in case a distribution in transition rates only is assumed (in case of a distribution in the rates as well as in the site-energies, the calculations become quite complicated in both approaches). The theory yields the frequency dependence of the conductivity in the whole range between the high-frequency limit where the pair-approximation holds, and the low-frequency limit where an infinite number of sites will be visited by a carrier during a single field cycle. The power-law frequency dependences of $\sigma'(\omega)$ and $\sigma''(\omega)$ for the metal cluster compounds agree qualitatively quite well with the predictions for multiple-site hopping. The experimentally available maximum frequency (10^6 Hz) is clearly not high enough for the pair-approximation to apply. Also

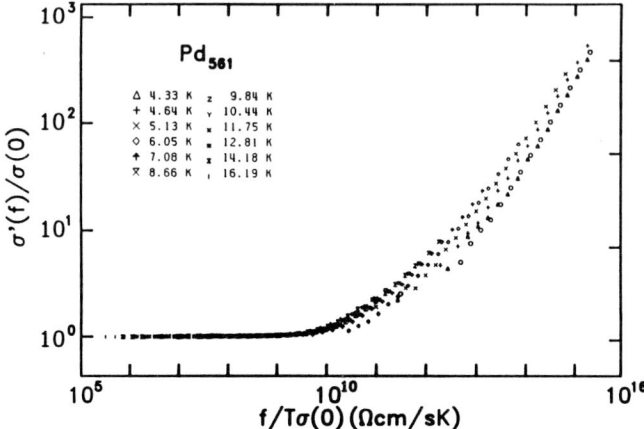

Fig. 16. Scaling of $\sigma'(\omega)$ of Pd$_{561}$.

the temperature dependences of $\sigma'(\omega)$ and $\sigma''(\omega)$ are well accounted for. Also the temperature dependences of $\sigma'(\omega)$ and $\sigma''(\omega)$ are well discribed by this theory. It predicts a scaling property, i.e. the frequency dependences of the conductivity at different temperatures should all fall unto a single curve, when plotted as $\sigma(\omega)/\sigma(0)$ versus $\omega/T\sigma(0)$. This is shown for the real part of the conductivity, $\sigma'(\omega)$, in Figure 16 (instead of $\omega = 2\pi f$, the frequency f appears in this figure). Such a detailed scaling relationship between the AC and DC conductivities, is in fact also found in many other systems showing hopping-type conduction. This may be seen as an excellent demonstration of the universality in behaviour of these materials.

For the understanding of the physics in the metal cluster compounds, two important consequences of these scaling properties should be emphasized. Firstly, it follows that the AC and DC conductivity processes are not due to different physical mechanisms (as could in principle be the case), and therefore that the AC losses cannot be attributed to spurious localized impurities. Secondly, it implies that the density of charge carriers in these systems is independent (or at most weakly dependent) of temperature. This seems to exclude the presence of a gap in the density of states around the Fermi level. The same conclusion may in fact be drawn from the experimental observation that the effect of varying the temperature on the $\sigma(\omega)$ versus ω curves (Figure 13) decreases very rapidly when increasing the frequency, as well as from the non-activated behaviour found for the DC conductivities (cf. Figure 12) down to the lowest temperatures reached (apart from non-ohmic effects). Both findings appear to be incompatible with a gap at E_F. Therefore, we should conclude that a possible intrinsic semiconducting gap in these materials (whether due to band symmetry or to electron correlations) is masked either by the effects of oxidation or reduction or else is smeared out by the effects of disorder.

Although further experimental and theoretical work is necessary, we feel that these conclusions from the conductivity data are compatible with the above-mentioned Poisson-distribution of the cluster energy-levels around E_F, which was invoked to understand the low temperature NMR (4.2 K) and specific heat data ($T < 1$ K). The presence of a finite density of states at E_F is also indicated by the photoelectron spectroscopic study on Au_{55} discussed elsewhere in this volume, although in that experiment the energy resolution is evidently very large as compared to the thermal energies involved in the conductivity studies.

As mentioned above, in the formalisms (e.g. of Böttger and Bryksin) discussed so far for the DC and AC conductivity, the correlations between the carriers are taken into account only in the sense that double-occupancy of a site is excluded. On the other hand, since the intercluster transfer integrals will in general be rather small, the metal cluster compounds will certainly fall into the class of narrow-band conductors (which in addition comprises the high-T_c superconducting oxides, the organic (semi)conductors and superconductors, and the fullerenes). For narrow-band conductors, the correlation effects will be of prime importance for the physical properties. In particular, they may be decisive in determining whether a material is a conductor or an insulator, much more than the effect of the lattice potential as considered in the band theory of solids. In view of the applicability of these concepts to the past and future experiment on the metal cluster compounds [36], we shall briefly discuss them in the last section of this introduction.

1.6. Application of the Anderson–Hubbard Approach

Localization of electronic wave functions, leading to an insulating state, can occur due to electron correlation, disorder or electron-phonon interactions. A recent review of the theoretical aspects of electron localization has been given a.o. by Ramakrishnan [39]. The essentials of correlation- or disorder-induced localization may be described by the Anderson–Hubbard Hamiltonian

$$\mathcal{H} = \sum_{m,\sigma} \varepsilon_m n_{m,\sigma} + \sum_{m \neq m,\sigma} J_{m'm} a^+_{m',\sigma} a_{m,\sigma} + U \sum_m n_{m\uparrow} n_{m\downarrow}$$

where ε_m is the energy of site m, $J_{m'm}$ is the transfer integral between sites m' and m, U is the Hubbard on-site correlation energy, $a^+_{m\sigma}$ and $a_{m\sigma}$ create, respectively annihilate an electron at site m with spin σ, and $n_{n\sigma} = a^+_{m\sigma} a_{m\sigma}$ is the number of electrons at site m with spin σ. We note that the second term in this equation would just describe a tight-binding band of electrons. The third term was introduced by Hubbard to take into account the (on-site) correlations between the electrons. In the original Hubbard model the first term was discarded, which amounts to taking ε_m to be a constant, i.e.

all sites are assumed to have the same site-energy. In this limit, it is the ratio between U and $J_{m'm}$ that drives the metal-nonmetal transition, as it determines whether the charges are localized ($U \gg J_{m'm}$), or delocalized into a band ($U \ll J_{m'm}$).

On the other hand, the situation with $U = 0$, $J_{m'm} = J_0$, and with ε_m distributed randomly over a certain range of width Γ around zero, was first investigated by Anderson [40]. If $\Gamma = 0$, all electronic states are extended, whereas if $\Gamma > 0$, the electronic states will become localized as soon as Γ/zJ_0 (where z is the number of nearest neighbours) reaches a critical value (the Anderson transition). For a given value of Γ, some states are localized and some are extended. The energy treshold between localized and extended states is called a mobility edge. States near a band edge are easiest to localize; if the disorder is large enough, then the valence- and conduction-band tails of a (small-gap) semiconductor may overlap to give a non-zero density of localized states in what would have been the energy gap it there were no disorder.

The Anderson transition is due to purely diagonal disorder ($J_{m'm} = J_0$). As already mentioned, for the case with $U > 0$, $J_{m'm} = J_0$, $\varepsilon_m = \varepsilon_0$, and exactly all sites singly occupied (half-filled band), $\mathcal{H}$ describes a system that becomes an insulator in case the ratio U/W (W is the bandwidth if $U = 0$) reaches a critical value [41]. As W increases with J_0, one may discriminate in the $U - J_0$ space between three classes of systems, namely all states very localized, very extended or intermediate.

In the field of chemistry this corresponds to the classes I, II and III of the Robin-Day classification [42], which is widely used in the field of mixed-valence compounds. These materials are built up by molecular units that can exist in different valencies (albeit with a slightly different geometry for each), and charge transfer is then associated with 'valence transfer' between molecules that are adjacent but have different valence. It should immediately be stipulated, however, that in the models for mixed-valence charge transfer the vibronic electron-lattice coupling plays an essential role. That is, since a specific geometry of the molecule is associated with a given valency, one expects the appropriate geometry to travel with the extra unit of charge, as the latter is transferred from one molecule to the other. Therefore, the mixed-valence charge transfer model is in fact equivalent to the small-polaron model, that was advanced in the theory of solid state physics for quite some time already [43, 44].

In the Hubbard–Anderson approach a rigid lattice is mostly considered, so that polaronic deformations around the moving charge are not taken into account explicitly (although they may be incorporated implicitly by a renormalization of the site-energies and transfer integrals). The charge-transfer between neighbouring sites is just due to quantum mechanical tunneling. By contrast, as soon as interaction with the phonon system is built in, charge transfer by thermally activated hopping becomes an additional possible mechanism, at least at sufficiently high temperatures. As regards the deformation around the extra charge considered in the small-polaron and mixed-valence models, the transfer of the combination of charge plus deformation (the small-polaron) is likewise assisted by the vibronic motions (phonon-assisted tunneling). It should be mentioned that the profound differences between quantum mechanical *tunneling* and thermally activated *hopping* are not always systematically distinguished in the nomenclature used in the literature. For instance, one often finds the term 'hopping' integral used for the quantum mechanical transfer integral!

Obviously, there are strong similarities between the metal cluster compounds and the small-polaron and mixed-valence charge transfer models conceived for molecular solids, if we assume that charge transfer also occurs by hopping or tunneling between the clustermolecules. The main difference would then lie in the size of the molecules considered.

It is of importance to get an idea about the relative values of the intersite transfer integral $J_{mm'}$ and the on-site interaction U, for the carriers in the metal cluster compounds. Therefore, let us start by finding upper limits, e.g. by estimating these values for an ionic transition metal semiconducting compound, such as an alkali halide or a metal oxide. In such materials, carriers can be introduced by doping with impurity ions, by deviations from stoichiometry (vacancies) or by structural defects. Yet another way of obtaining charge separation in the solid may be the disproportionation reaction of atoms or molecules, for instance $2M^{III} \rightarrow M^{II} + M^{IV}$. We note that this reaction is in many respects equivalent to the Hubbard formalism. Since U is the energy needed to put two electrons on the same site, in the atomic or molecular limit this quantity just corresponds to the difference between the ionization energy I and the electron affinity A, in other words to the energy needed for disproportionation. When the molecule or atom M is in the solid corrections to the free-ion values I and A have to be applied, so that:

$$U = I - A + E_{\text{corr}}$$

where the (often large) term E_{corr} stands for all the corrections due to screening, polarization, covalency, etc. appropriate for the solid. Evidently, the mobility of the charge carriers will enhance the degree of screening. However, even for localized carriers the polarizability leads to an important screening effect. This can be understood from the fact that the energy needed to place a point-charge in a polarizable medium is reduced by the polarization energy, $E_p = \frac{1}{2} \sum_i \alpha_i E_i^2$, where α_i and $\mathbf{E}_i$ are the polarizability and the electric field at the site of the i-th atom surrounding the point charge. Since E_p is independent of the sign of the point-charge, I and A will be reduced and increased by E_p, respectively, so that U will be reduced by $2E_p$ with respect to the free ion value $I - A$. The latter is of the order of 10–20 eV for transition metal elements. As regards the nature of the polarization, we have to distinguish between the electronic and the displacive polarizations of the medium in which the extra charge is embedded. The electronic polarization corresponds to the high-frequency dielectric constant in which all the surrounding atoms are fixed in their original positions, whereas the displacive (ionic) polarization corresponds to the low-frequency dielectric constant and represents the additional polarization obtained by allowing the surrounding atoms to relax to new equilibrium positions upon appearance of the extra charge. Here high- and low-frequency obviously have to be related to typical optical phonon frequencies characterizing the motions of the atoms. Thus, in a high-energy spectroscopic experiment (e.g. photo-ionization) the binding energy of the outcoming electron has only to be corrected for by the electronic polarization caused by the hole left behind. Furthermore in highly conducting solids, the mobility of a carrier will be so high that the displacive polaronic contribution to its total polarization cloud can be neglected. But for low conductivity (narrow-band) materials this is no longer valid. To estimate the value for σ below which the displacive contribution becomes important one may use the Einstein relation $\sigma = ne^2 \ell^2 \nu_h / k_B T$ for the diffusive hopping motion of charges. Here ℓ is the hopping distance and ν_h the hopping frequency. Equating the average time ν_h^{-1} that a carrier spends on a given site to a typical inverse optical phonon frequency (10^{-13} s), yields $\sigma = 2.5 \times 10^2 \Omega^{-1}$ cm^{-1} if we put $d = 4$ Å and adopt a (low) carrier density of $n = 10^{21}$ cm^{-3}. From this estimate it is clear that, even for relatively large values of the mobility ($\mu = \sigma/ne$) of the carriers, displacive polarization effects can still be of importance. This will certainly apply to the metal cluster compounds, so that the carriers can be considered to be polarons.

In ionic dense-packed solids both contributions to the polarization are of the order of a few eV. As a consequence, values for U are in the range of 1–10 eV. As regards the transfer integral $J_{mm'}$, this is related to the width ($= 2zJ_{mm'}$) of the metal atom band (z is the number of nearest neighbouring sites), and will be of the order of a few eV for transition metal compounds.

If we now pass to the situation in the metal cluster compounds, it is clear that both U and $J_{mm'}$ will be very much smaller. Let us consider a system of identical metallic particles embedded in a dielectric solid. A simple electrostatic model [45] then yields $I - A \simeq e^2/4\pi\epsilon R$, where R is the radius of the metal particles. For $R = 5$ Å, and a high-frequency dielectric constant ϵ for the medium of 10, this already gives $I - A \simeq 0.3$ eV. The polarization correction is difficult to estimate but would lower the resulting values for U still further. A similar reduction in U can be expected for a system of molecules (clusters) embedded in a dielectric. The point is that the larger the molecule, the more effectively it can screen an extra charge put upon it, and thus the lower will be the appropriate U. What is considered here is the *intra*molecular screening resulting from the electronic and displacive polarizations of the atoms constituting the molecule. Indeed, it is well-known that for molecules which can exist with different valences the change in valence is accompanied by changes in the molecular structure. Such changes may vary from slight differences in geometry to drastic alterations, where the molecule may even adopt completely different symmetries. These effects have been well-established also for small (bare) metal clusters, for which important changes in geometry and symmetry are predicted and observed [25] when going from the neutral to the ionized species. For metal cluster compounds too, this phenomenon is likewise well-known [46]. It is interesting in this respect that the structural changes become less pronounced the larger is the molecule, in agreement with the notion that the screening becomes more effective with increasing size of the 'site'.

In addition to this intramolecular screening will come the screening due to (electronic and displacive) polarization of the dielectric medium in which the molecule is embedded and of the other molecules (metal particles) that surround it. One may conclude from these arguments that for large metal particles or metal cluster molecules the parameter U will be quite small, i.e. of order 0.1 eV or even (much) lower for the larger particles (e.g. $R > 30$ Å).

The other important parameter to consider is the transfer integral $J_{mm'}$, which ensures the tunneling or hopping of the charges from one 'site' to the other. It is difficult to make quantitative estimates of $J_{mm'}$ for metal-insulator composites, except that it will be (very) small as compared to typical values 0.1-10 eV in the (dense-packed) transition metal compounds. In view of the exponential dependence of tunneling probabilities on intersite distances, and considering that the latter are of the order of 10 Å or more in the metal cluster compounds, one would expect $J_{mm'}$ to be less than 0.01 eV in these systems.

Nevertheless, since U will also be quite small, one may still have a ratio t/U of order unity or smaller in these materials. It is this ratio that is considered in the Hubbard model, since it determines the degree of overlap of the upper and lower Hubbard bands. In this respect we should add that our discussion of the values of t and U sofar would only apply to small metal particles, for which capacative couplings *between* neighbouring particles can be neglected compared to the self-capacitance of the particle itself. For a system of two adjacent large metal spheres, however, the energy involved in the charge transfer of an electron between them (starting from the initial neutral situation) will be determined predominantly by the capacitance of the combination. With increasing radius of the particles, their capacitive coupling will become dependent on their 'contact' area and their mutual separation. In this limit, the Hubbard-type of approach should probably be abandoned in favour of e.g. the Kronig–Penney type of Hamiltonian already discussed in the above. What this discussion shows, however, is that in approaching such a mesoscopic limit within the Hubbard formalism, the ratio U/t will decrease to below unity, so that one would have a (tight-binding) band conduction provided that there is no randomness in the parameters involved, since this will lead to localization of the states around the Fermi level. Experimentally, this localization is evidenced for the (randomly packed) Au_{55} and Pd_{561} clusters by the fact that the conductivity for these materials is described by a thermally activated hopping process, rather than by quantum mechanical band motion.

As regards the magnetic properties of the carriers expected on basis of this model, one would expect a similar behaviour as for lightly-doped semiconductors (far below the nonmetal-metal transition). For low carrier densities, the magnetism may be described by an ensemble of spins $S = \frac{1}{2}$, which are essentially localized, and which are very weakly interacting through a random-exchange Heisenberg Hamiltonian [47]:

$$\mathcal{H}_{ex} = -\sum_{i,j} K(\mathbf{r}_{mm'}) \mathbf{S}_m \cdot \mathbf{S}_{m'}.$$

The exchange interaction $K(\mathbf{r}_{mm'})$ may be assumed to be antiferromagnetic, its absolute value being a strongly decreasing function of the intersite spacing $\mathbf{r}_{mm'}$. The magnetic disorder associated with the randomness in $K(\mathbf{r}_{mm'})$ will evidently be further enhanced by the hopping of the carriers. Consequently, one expects a small paramagnetic Curie-type contribution ($\propto 1/T$) of the carriers to the magnetic susceptibility at temperatures $k_B T > K(\mathbf{r}_{mm'})$, possibly evolving into a kind of spinglass state at the lowest temperatures $k_B T < K(\mathbf{r}_{mm'})$. Reasonable estimates of $|K|$ would amount to at most 0.1 meV or so, so that such spinglass type of order may only be found at very low temperatures ($T \lesssim 1$ K).

Such predictions are not in disagreement with the observed magnetic behaviour of the metal cluster compounds in which the metal atom is an element like Au, Pt, Ru, that is nonmagnetic in the bulk-metal form. In almost all cases a weak Curie-type contribution is seen below temperatures of the order of 50 K (for a review see [31]). The magnitude of the Curie-term is found to correspond with the occupation by a spin $\frac{1}{2}$ of a fraction of the metal-cluster molecules. The presence of these spins is also seen in ESR experiments and may provide a mechanism for the observed nuclear spin-lattice relaxation for nuclear spins situated either on the metal cores themselves, or on the attached ligand molecules (e.g. P). In the above context it is also interesting to recall that such a fraction of paramagnetic centers has also been reported for other metal-nonmetal composites, a.o. for Pt-colloids (for a review, see Halperin [13]).

Summarizing, we may state that the Hubbard–Anderson Hamiltonian represents a fruitful approach by which we may understand several aspects of the conducting and localization properties of the metal cluster solids. As regards the metal cluster compounds, experiments have sofar been mainly restricted to the Au_{55} and Pd_{561} Schmid-clusters, in which the cluster molecules are randomly packed. We think this to be the basic reason for their semiconducting behaviour. At the same time this opens an attractive prospective for future studies on *crystalline* metal cluster compounds, for which such randomness should be absent. In general, such cluster solids will be characterized by small ($\simeq 0.1$–1 eV) gaps between the HOMO- and LUMO-derived 'valence' and 'conduction' bands around the Fermi level. The case of crystalline solid C_{60} serves as a good example. Provided that extra electrons or holes may be introduced, the crystalline metal cluster solids could become good conductors, similar as potassium doped C_{60}. Apart from oxydation or reduction processes, the extra carriers could be introduced in the metal cluster compounds for instance by replacing a fraction of the counterions by ions of a different valence, thereby changing the valence of part of the metal cluster molecules. It is well-known that some of the metal cluster molecules may easily adopt a scale of different valences; as an example we may mention $[Pt_{24}(CO)_{30}]^n$, $[Pt_{26}(CO)_{32}]^n$, and $[Pt_{38}(CO)_{44}]^n$, which were recently reported [48] to be able to take a charge varying between $n = 0$ to $n = 10$ electrons! As mentioned above, the intercluster transfer integrals will probably be small, smaller at least then in solid C_{60}, were the widths of the minibands is typically a few 0.1 eV. However, the metal cluster molecules offer a rich possibility of variation, and since their size can be made much larger (a few nm) than that of the C_{60} molecule (0.7 nm), the on-site Hubbard parameter U can be very much reduced. Thus there seems to be no basic factor that would prohibit to create conducting cluster solids (and perhaps also superconductors) in a near future.

References

1. L. J. de Jongh and A. R. Miedema, *Adv. Phys.* **23** (1974) 1–268.
2. L. J. de Jongh, 'Polynuclear Metal cluster Compounds: A New Chemical Submicron Structure', in *Proceedings NATO-ASI on Organic and Inorganic Low-dimensional Crystalline Materials*, Menorca, May 1987, P. Delhaes and M. Drillon (eds.), Plenum Press.
3. G. Longoni, A. Ceriotti, M. Marchionna and G. Piro, 'Large Molecular Metal Carbonyl Clusters: Models of Metal Particles', in *Surface Organometallic Chemistry: Molecular Approaches to Surface Catalysis*, J. M. Basset et al. (eds.), Kluwer Acad. Publ. (1988) pp. 157–172 (and references in this review). (See also Chapter 2 of this volume).
4. G. Schmid, 'Metal Clusters and Cluster Metals', *Polyhedron* **7** (1988) 2321–2329; *Endeavour, New Series* **14** (1990) 172. (See also Chapter 3 of this volume).
5. D. F. Shriver, H. D. Kaesz and R.D. Adams (eds.), *The Chemistry of Metal cluster Complexes*, VCH Publishers (1990).
6. A. Simon, 'Clusters of Valence Electron Poor Metals; Structure, Bonding and Properties', *Angew. Chem. Int. Ed. Engl.* **27** (1988) 159.
7. M. N. Vargaftik, V. P. Zagorodnikov, I. P. Stolyarov, I. I. Moiseev, V. A. Likholobov, D. I. Kochubey, A. I. Chuvilin, V. I. Zaikovsky, K. I. Zamaraev and G. I. Timofeeva, 'A Novel Giant Palladium Cluster', *J. Chem. Soc. Chem. Commun.* **937** (1985).
8. G. Schmid, B. Morun and J.-O Malm, '$Pt_{309}Phen^*_{36}O_{30\pm 10}$, a Four-Shell Platinum Cluster', *Angew. Chem. Int. Ed. Engl.* **28** (1989) 778–780.
9. G. Schmid, M. Harms, J.-O. Malm, J. O. Bovin, J. M. van Ruitenbeek, H. W. Zandbergen and W. T. Fu, *J. Am. Chem. Soc.* **115** (1993) 2046.
10. D. A. van Leeuwen, J. M. van Ruitenbeek, G. Schmid and L. J. de Jongh, 'Size-Dependent Magnetisation of Pd Clusters and Colloids', *Phys. Lett.* **A 170** (1992) 325–333.
11. D. Fenske and H. Krautscheid, 'New Copper Clusters Containing Se and PEt_3 as Ligands: $[Cu_{70}Se_{35}(PEt_3)_{22}]$ and $[Cu_{20}Se_{13}(PEt_3)_{12}]$', *Angew. Chem. Int. Ed. Engl.* **29** (1990) 1452–1454.

12. J. A. A. J. Perenboom, P. Wyder and F. Meier, 'Electronic Properties of Small Metallic Particles', *Phys. Rep.* **78** (1981) 173–292.
13. W. P. Halperin, 'Quantum Size Effects in Metal Particles', *Rev. Mod. Phys.* **58** (1986) 533–606.
14. S. Sugano, *Microcluster physics*, Springer Series in Materials Science, Vol. 20, Springer-Verlag (1991).
15. H. Frölich, *Physica* (Utrecht) **4** (1937) 406.
16. T. P. Martin, T. Bergmann, G. Göhlich and T. Lange, 'Electronic Shells and Shells of Atoms in Metallic Clusters', *Z. Phys.* **D 19** (1991) 25–29.
17. H. Nishioka, 'Shells and Supershells in Metal clusters', *Z. Phys.* **D 19** (1991) 19–23.
18. V. Subrahmanyam and M. Barma, 'Size-Dependent Energy Scales in the Ideal Fermi Gas', *Phase Transitions* **24–26** (1990) 419–433, and references cited therein.
19. R. Kubo, *J. Phys. Soc. Japan* **17** (1962) 975; R. Kubo, A. Kawabata and S. Kobayashi, 'Electronic Properties of Small Particles', *Ann. Rev. Mater. Sci.* **14** (1984) 49–66.
20. L. P. Gor'kov and G. M. Eliashberg, *Sov. Phys. JETP* **21** (1965) 940.
21. R. de L. Kronig and W. G. Penney, 'Quantum Mechanics of Electrons in Crystal Lattices', *Proc. Roy. Soc. London, Ser. A* **130** (1931) 499–513.
22. D. Wohlleben, 'Superconductivity by Quantum Size Effect. II Ground State Properties of a Wire with Mesoscopic Subdivisions', *J. Less Common Metals* **142** (1988) 31.
23. A. F. Mayadas and M. Schatzkes, 'Electrical-Resistivity Model for Polycrystalline Films: The Case of Arbitrary Reflection at External Surfaces', *Phys. Rev.* **B 1** (1970) 1382–1389.
24. For a review of the application to clusters, see W. A. de Heer, W. D. Knight, M. Y. Chou and M. L. Cohen, 'Electronic Shell Structure and Metal clusters', *Solid State Physics* **40** (1987), 94–181.
25. V. Bonačić-Koutecký, P. Fantucci and J. Koutecký, 'Quantum Chemistry of Small Clusters of Elements of Groups Ia, Ib, and IIa: Fundamental Concepts, Predictions, and Interpretation of Experiments', *Chem. Rev.* **91** (1991) 1035–1108; see also J. Koutecký and P. Fantucci, 'Theoretical Aspects of Metal Atom Clusters', *Chem. Rev.* **86** (1986) 539–587.
26. For an example see S. Saito and A. Oshiyama, 'Cohesive Mechanism and Energy Bands of Solid C_{60}', *Phys. Rev. Lett.* **66** (1991) 2637–2640 (in particular Figure 4).
27. N. Rösch, L. Ackermann, G. Pacchioni and B. I. Dunlap, 'Paramagnetism of High Nuclearity Metal cluster Compounds as Derived from Local Density Functional Theory', *J. Chem. Phys.* **95** (1991) 7004–7007; see also *Chem. Phys. Lett.* **199** (1992) 275 and *Int. J. Quant. Chem: Quantum Chem. Symp.* **26** (1992) 605, by the same group.
28. L. J. de Jongh, H. B. Brom, G. Longoni, P. R. Nugteren, B. J. Pronk, G. Schmid, H. H. A. Smit, M. P. J. van Staveren and R. C. Thiel, 'Physical Properties of High-Nuclearity Metal cluster Compounds', in P. Jena *et al.* (eds.), *Physics and Chemistry of Small Clusters*, NATO-ASI Series B, Vol. 158, Plenum (1986), pp. 807–812; ibidem, *J. Chem. Research* (S) (1987) 150–151.
29. H. H. A. Smit, P. R. Nugteren, R. C. Thiel and L. J. de Jongh, 'Mössbauer and Specific Heat Studies of the Vibrations of Metal Core Atoms in Polynuclear Gold Cluster Compounds', *Physica* **B 153** (1988) 33–52; F. M. Mulder, E. A. van de Zeeuw, R. C. Thiel and G. Schmid, 'Physical Properties of Metal Cluster Compounds VI: The Influence of the Ligands on the ^{197}Au Mössbauer Spectra of Three Different Au_{55} Molecular Clusters', *Solid State Commun.* **85** (1993) 93–97.
30. B. J. Pronk, H. B. Brom, L. J. de Jongh, G. Longoni and A. Ceriotti, 'Physical Properties of Metal cluster Compounds: Magnetic Measurements on High-Nuclearity Nickel and Platinum Carbonyl Clusters', *Solid State Commun.* **59** (1986) 349–354; D. A. van Leeuwen, J. M. van Ruitenbeek and L. J. de Jongh (to be published).
31. L. J. de Jongh, 'Magnetic Measurements on Polynuclear Metal cluster Compounds: Between Molecule and Metal', *Physica* **B 155** (1989) 289–295.
32. D. A. van Leeuwen, Thesis, University of Leiden, September 1993; D. A. van Leeuwen, J. M. van Ruitenbeek and L. J. de Jongh (to be published).

33. D. A. van Leeuwen, J. M. van Ruitenbeek, G. Schmid and L. J. de Jongh, 'Size-Dependent Magnetisation of Pd Clusters and Colloids', *Physics Lett.* **A 170** (1992) 325–333.
34. J. Baak, H. B. Brom, L. J. de Jongh and G. Schmid, 'The Electronic Contribution to the Low-Temperature Specific Heat in the Metal cluster Compounds $Au_{55}(PPh_3)_{12}Cl_6$, $Pt_{309}Phen^*_{36}O_{30}$ and $Pd_{561}Phen_{36}O_{200}$ – A Comparison', *Suppl. Z. Phys.* **D 26** (1993) S30–32; J. Baak, H. B. Brom, L. J. de Jongh and G. Schmid, 'Vibrational and Electronic Contributions to the Specific Heat of $Pd_{561}Phen_{36}O_{200}$', in P. Jena *et al.* (eds.), *Physics and Chemistry of Finite Systems: From Clusters to Crystals*, NATO-ASI Series C, Vol. 374, Kluwer Acad. Publ. (1992), pp. 339–344.
35. D. van der Putten, H. B. Brom, L. J. de Jongh and G. Schmid, '^{195}Pt NMR of $Pt_{309}Phen^*_{36}O_{30}$ Metallic Clusters', in P. Jena *et al.* (eds.), *Physics and Chemistry of Finite Systems: From Clusters to Crystals*, NATO-ASI Series C, Vol. 374, Kluwer Acad. Publ. (1992), pp. 1007–1012; D. van der Putten, H. B. Brom, J. Witteveen, L. J. de Jongh and G. Schmid, 'The Electronic Quantum Size Effect Observed by ^{195}Pt NMR in the Metal cluster Compound $Pt_{309}Phen^*_{36}O_{30}$', *Suppl. Z. Phys.* **D 26** (1993) S21–23.
36. M. P. J. van Staveren, H. B. Brom and L. J. de Jongh, 'Metal Cluster Compounds and Universal Features of the Hopping Conductivity of Solids', *Phys. Rep.* **208** (1991) 1–96.
37. A valuable account is given by H. Böttger and V. V. Bryksin in *Hopping Conduction in Solids*, VCH Verlag, Weinheim (1985).
38. A. Miller and E. Abrahams, *Phys. Rev.* **120** (1960) 745.
39. T. V. Ramakrishnan, in P. P. Edwards and C. N. R. Rao (eds.), *The Metallic and Nonmetallic States of Matter*, Taylor and Francis (1985), Ch. 2, pp. 23–64.
40. P. W. Anderson, *Phys. Rev.* **109** (1958) 1492.
41. J. Hubbard, *Proc. Roy. Soc.* **A277** (1964) 237, **A 281** (1964) 401.
42. M. B. Robin and P. Day, *Adv. Inorg. Chem. Radiochem.* **10** (1967) 247.
43. T. Holstein, *Ann. Phys.* **8** (1959) pp. 325 and 343.
44. I. G. Austin and N. F. Mott, *Adv. Phys.* **18** (1969) 41.
45. M. P. J. van Staveren, H. B. Brom, L. J. de Jongh and I. Ishii, *Phys. Rev.* **B 35** (1987) 7749.
46. P. R. Raithby, in B. F. G. Johnson (ed.), *Transition Metal Clusters*, Wiley, New York (19??), Ch. II (and other chapters in this book).
47. R. N. Bhatt, M. A. Paalanen, and S. Sachdev, *J. de Phys.* **49** Coll. C8, suppl. nr. **12** (1988) 1179.
48. J. D. Roth, G. J. Lewis, L. K. Safford, X. Jiang, L. F. Dahl and M. J. Weaver, *J. Am. Chem. Soc.* **114** (1992) 6159–6169.

2. HIGH-NUCLEARITY CARBONYL METAL CLUSTERS

ALESSANDRO CERIOTTI, ROBERTO DELLA PERGOLA and
LUIGI GARLASCHELLI
*Dipartimento di Chimica Inorganica, Metallorganica e Analitica,
via G. Venezian 21, 20133 Milano, Italy*

Glossary of Abbreviations

acac	acetylacetonato
AFM	Atomic Force Microscopy
b.c.c.	body centered cubic
Bu	butyl
Bu^t	*tert*-butyl
c.c.p.	cubic close packing
cod	1,5-cyclooctadiene
c.p.	close packing
cp	cyclopentadienyl
cp^*	pentamethylcyclopentadienyl
CVE	cluster valence electrons
Cy	cyclohexyl
dmso	dimethylsulfoxide
EELS	Electron Energy Loss Spectroscopy
ESR	Electron Spin Resonance
Et	ethyl
FAB-MS	Fast Atom Bombardment – Mass Spectroscopy
FTIR	Fourier Transform Infrared Spectroscopy
h.c.p.	hexagonal close packing
HOMO	highest occupied molecular orbital
HREM	High Resolution Electron Microscopy
LUMO	lowest unoccupied molecular orbital
Me	methyl
NMR	Nuclear Magnetic Resonance
Ph	phenyl
phen	1,10-phenanthroline
$phen^*$	4,7-p-phenylsulfonate–1,10-phenanthroline disodium salt
Pr	propyl
R	alkyl group
STM	Scanning Tunneling Microscopy
thf	tetrahydrofuran
UPS	UV-Photoelectron Spectroscopy
XPS	X-ray-Photoelectron Spectroscopy

2.1. Introduction

The field of metal clusters nowadays represents a large area of inorganic chemistry. Relevant and unexpected achievements have been obtained and the existence of cluster compounds has been observed for many elements in very different environments. This class of compounds can be present in alkaline-metal oxides, in Chevrel phases, in biological enzymes and in metal vapour beams [1–3].

Cluster compounds consisting of either rings, cages or extended chains are formed by main group elements [4, 5]. Cage-like molecular clusters are present in metallo-boranes [5] and in metallo-carbohedrenes [6]. The recently characterized fullerenes, gigantic molecules formed by up to 500 carbon atoms, can be considered as clusters of this element [7].

According to the classical definition, a metal cluster is a compound formed by at least three metal atoms, stabilized by small molecules – acting as ligands – and characterized by metal-metal bonds, either localized or delocalized over the metal framework [8, 9]. Most common metal clusters are formed by early transition metals, stabilized by halide and oxygen atoms, by metals of groups 6–10, stabilized by CO molecules, or by transition or post-transition metals, having halogen atoms, phosphine groups, and main group elements as ligands. Thus, molecular metal clusters can be considered as stable fragments of metal aggregates, electronically saturated and stabilized by surrounding ligands and/or free charges. In this sense, they are different from small, electronically unsaturated, naked metal particles, which are extremely reactive unless stabilized by a support or an inert matrix, or trapped in zeolites [10, 11].

Originally, the interest in molecular metal clusters was limited to the metal-metal interactions. More recently, since molecular metal clusters are in a privileged position between the atomic and the metallic state, such compounds have been considered as excellent model systems for understanding important events occurring on small metal particles, such as the chemi- and physisorption of small molecules onto metallic surfaces (e.g. coordination modes at high coverage, fluxionality, etc.) [12–15], the mechanism of hydrogen diffusion inside a metallic network [16], the distortions in a metal close-packing induced by interstitial heteroatoms [17], and the nucleation processes involving different metal particles [18–20].

Furthermore, the chemical reactivity of coordinated ligands in molecular metal clusters has been widely used for studying the reaction mechanisms at the metal centres [21], and for mimicking elementary processes extremely relevant for homogeneous and heterogeneous catalysis [22].

Recent developments in metal cluster chemistry have allowed the isolation and characterization of molecular species having a metal nuclearity (circa 560 metal atoms) approaching the size of small metal colloids [23], ultra small amorphous metal particles and crystallites [24, 25], resembling either miniature bulk metal packings (viz. c.c.p., h.c.p. or b.c.c.) [26], or quasicrystalline phases (viz. amorphous metals or metallic glasses) [27], and showing the same physical properties (viz. electronic, magnetic, optical) [28, 29] as bulk metals.

New results in cluster chemistry have increased the interest for their possible applications and, at the same time, have stimulated the use of new solid-state techniques for their study and characterization (HREM, STM, AFM, XPS, UPS, EELS, etc.) and improved the techniques normally used for their investigation in solution at molecular level (FTIR, NMR, ESR, etc.) [30].

Molecular metal clusters have been tested in catalysis either homogeneous, due to their good solubility, or heterogeneous, anchored on resin supports or deposited on inert oxides [30, 31]. An exciting potential application for molecular metal clusters might be their use as starting compounds for preparing new materials in unconventional situations that do not take place in extended metallic lattices, when using traditional metallurgical methods.

The full elucidation of the structure of $Rh_6(CO)_{16}$, in 1963 by L. Dahl, may be regarded as the birth of the chemistry of molecular metal carbonyl clusters [32]. Since then, several hundreds if not a thousand compounds of almost every transition element of the periodic table have been successfully synthesized and several reviews and books concerning carbonyl metal clusters have been published [33–42].

A large number of small metal carbonyl clusters possessing up to 6 metal atoms have been synthesized, with a variety of metal cages and a great number of ligands other than CO, such as cyclopentadiene, phosphines, halides, unsaturated organic molecules, etc. Isolation and purification of the products requires typical chemical procedures such as chromatography, fractional crystallization, sublimation, etc., their identification can be performed with very sophisticated spectroscopic techniques. Preliminary structural information can be inferred on the basis of the stoichiometry of the reaction, elemental analysis and spectroscopic characterization, however, X-ray analysis is required for the complete structure determination.

In the last decade highly intriguing, giant-sized transition metal carbonyl clusters have been isolated and characterized. Only a few laboratories in the world are dedicated to this investigation: this fact reflects the experimental difficulties which are present in the preparation, purification and crystallization of this class of compounds. In spite of continous efforts, their number remains relatively small and a very large gap exists between the largest known molecular carbonyl cluster (52 metal atoms) and the smallest metal particle.

In this field, the molecular nature of the particles is still a fundamental feature: a high-nuclearity metal cluster is still a chemical compound and a sample, no matter how large, is ideally composed of one single species, with rigourously identical molecules, thus possessing homogeneous properties and composition.

The synthesis of high-nuclearity carbonyl metal clusters usually requires unselective conditions. Mixtures of products with very similar properties are normally obtained and either separation or isolation of such products are frequently a matter of luck: very few compounds can be purified to a sufficient extent and even less can be characterized, not necessarily the most abundant, nor the most stable. Due to the small quantities of product available, the very high molecular weight, and the relative low solubility, spectroscopic techniques frequently fail to give relevant information about the formula and the structure of the products. FT-IR, high-resolution multinuclear magnetic resonance and mass spectrometry are the most powerful methods for characterization. The shape of the IR spectra in the ν-CO stretching region can be considered as a fingerprint for carbonyl compounds, but, for large clusters, the spectra consist of very few bands, attributable to terminal and bridging CO's. The ν-CO stretching frequency gives information only about the ratio between the number of negative charges and the number of metal atoms. The NMR signals are frequently too weak to be detected and, when spectra are available, consist of too many signals to be assigned, if the structure is unknown. The recent development of FAB-MS, which can vaporize and detect ions of very large molecular weight, can give some precious information about the metal cluster nuclearity only when the compound is stable under the experimental conditions. Thus, X-ray crystallographic analysis is almost the only way to characterize a high-nuclearity metal cluster. Unfortunately, only a small percentage of the observed clusters form reasonably stable, good quality crystals, suitable for X-ray analysis. Thus, computing programs have been developed taking into account the large number of variables to be determined, and the copresence of light and heavy elements. Even so it has not been possible to characterize the majority of compounds already isolated. Most high-nuclearity carbonyl clusters are neutral or negatively-charged, forming part of a salt. Bulky organic cations normally act as positive counterions. In the solid state, such polynuclear metal compounds form a macroscopic ordered array of identical metal particles, embedded in a dielectric matrix (ligands and/or counterions). Both in the solution and in the solid state, very weak interactions exist between ions of different electric charge and it is the general belief that the structure, the spectroscopic properties and the reactivity of the clusters are not affected by the counterion. In a few cases, the cation can play a crucial role in modifying the solubility of the salt, and therefore the solution equilibria to which it is subject. More often, the correct choice of the cation can modify the crystal packing and help in stabilizing the solid state, giving prolonged life to the crystal under the X-ray beam. Moreover, a problem frequently occurring with large carbonyl metal clusters is the loss of crystallization solvent which causes the loss of the X-ray diffraction pattern, either during data collection or indeed even before data collection is initiated.

The highest nuclearity transition metal carbonyl clusters generally contain electron-rich metals (e.g. Ni, Pd, Pt) because large aggregates of such metal atoms may be stabilized with relatively few carbon monoxide ligands, minimising the crucial steric hindrances. As a matter of fact, due to the progressive flattening of the polyhedral surfaces, the steric requirements of the ligands increase exponentially with the cluster size [43]. In other cases (e.g. Co), the presence of interstitial atoms of main group elements may reduce the number of external ligands which are necessary for stabilizing the cluster molecule, by supplying electrons from the inside. High-nuclearity carbonyl metal-clusters may also be obtained with metals having high metal-metal bond energy (e.g. Ru, Os, Rh, Ir) which can contribute to support a big metal skeleton.

Past investigations have been concerned with either homo- or heterometallic carbonyl clusters. Nowadays, a great interest in cluster chemistry is centred on heterometallic systems. The presence of two metals having different electronic and steric requirements allows the study of synergistic effects, the chemical reactivity at different sites and the effect of alloying on the structural properties. The structural behaviour of mixed-metal clusters is useful when compared with the structural situation found in the bulk, where spontaneous or induced phenomena of partial or total segregation have been predicted and experimentally proved ('cherry' phases, 'chemisorption induced' surface segregations, etc.) [44–46].

Recent results in mixed-metal cluster systems have shown unexpected skeletal arrangements which are unexplainable by using solid-state data for their interpretation, and have strengthened the idea that bulk situations may not be crudely related to relatively small molecular systems such as high-nuclearity metal carbonyl clusters. Molecular metal carbonyl clusters must find a compromise between optimizing metal-metal and metal-carbon monoxide bond energies simply, either by means of small skeletal rearrangements or by minor changes in metal nuclearity. On the other hand, small metallic crystallites having a relatively small number of surfacial atoms interacting with the adsorbed molecules compared to the number of interior metal atoms, should suffer dramatic structural rearrangements for the above optimization. In other words, the major difference between finite clusters and infinite lattices stems from the different degree of freedom of arrangements around a central metal atom. As a consequence of the greater role played by the ligands in determining the cluster geometry, high-nuclearity metal clusters could adopt very different arrangements from those present in metal crystallites.

It is actual opinion that cluster chemistry can afford data and models for a solid-state situation, referred to as meta-metallic [47]. The chemical compounds and the particles belonging to this state seem to have particular chemical and physical properties, being in a situation where the molecular properties are lost but the bulk properties have not yet started to show up. This 'quantum-size regime' is a size-dependent region resulting from the presence of a particular energy gap between the highest-occupied (HOMO) and the lowest-unoccupied (LUMO) molecular orbitals [48].

TABLE 1
Homometallic carbonyl clusters

Index No.	CLUSTER	C.V.E.*	Metal frame**	Method of synthesis***	Ref.
	Ruthenium				
1	$[\{Ru_6C(CO)_{16}\}_2Tl]^-$	182	c	d	49
	Osmium				
2	$[Os_{17}(CO)_{36}]^{2-}$	210	a	a	50
3	$[Os_{20}(CO)_{40}]^{2-}$	242	a	a	51
	Cobalt				
4	$[Co_{13}C_2(CO)_{24}]^{4-}$	177	d	a	52
5	$[Co_{13}C_2(CO)_{24}]^{3-}$	176	d	g	53
6	$[Co_{14}N_3(CO)_{26}]^{3-}$	196	d	a	54
7	$[Co_{14}P_2(CO)_{27}]^{4-}$	194	e	c	55
8	$\{Co_8As_2(CO)_{16}(AsPh)_2\}_2$	244	e	a	56
9	$Co_4O\{O_2CCCo_3(CO)_9\}_6$	348	c	d	57
	Rhodium				
10	$[\{Rh_6(CO)_{15}\}_2]^{2-}$	170	c	c	58
11	$H_2Rh_{12}(CO)_{25}$	160	a	g	59
12	$Rh_{12}C_2(CO)_{25}$	166	d	d; g	60
13	$[Rh_{12}C_2(CO)_{24}]^{2-}$	166	d	d+a	61
14	$[Rh_{12}C_2(CO)_{23}]^{4-}$	166	d	g	62
15	$[Rh_{12}C_2(CO)_{23}]^{3-}$	165	d	g	63
16	$[HRh_{12}N_2(CO)_{23}]^{3-}$	168	d	c+a	64
17	$[Rh_{12}Sb(CO)_{27}]^{3-}$	170	b	c+a	65
18	$[HRh_{13}(CO)_{24}]^{4-}$	170	a	h	66, 70b
19	$[H_2Rh_{13}(CO)_{24}]^{3-}$	170	a	h	67, 70b
20	$[H_3Rh_{13}(CO)_{24}]^{2-}$	170	a	c+a	68, 70b
21	$[Rh_{14}(CO)_{25}]^{4-}$	180	a	c+a; f	69, 70
22	$[HRh_{14}(CO)_{25}]^{3-}$	180	a	h; e	70b, 71
23	$[Rh_{14}(CO)_{26}]^{2-}$	180	a	c+a; d	72
24	$[\{Rh_6(CO)_{14}(CN)_2\}_2\{Rh(CO)_2\}_2]^{2-}$	202	c	c	73
25	$[\{Rh_6C(CO)_{15}\}_2\{Rh_2(CO)_3\}]^{2-}$	202	c	e	74
26	$[Rh_{14}N_2(CO)_{25}]^{2-}$	188	a	d+a	75
27	$[Rh_{15}(CO)_{27}]^{3-}$	192	a	c+a	70, 76
28	$[Rh_{15}(CO)_{30}]^{3-}$	198	a	c+a	76
29	$[Rh_{15}C_2(CO)_{28}]^-$	200	b	d	77
30	$[Rh_{17}(CO)_{30}]^{3-}$	216	a	c+a	78
31	$[Rh_{17}S_2(CO)_{32}]^{3-}$	232	d	c+a	79
32	$[Rh_{22}(CO)_{37}]^{4-}$	276	a	c+a	80
33	$[H_xRh_{22}(CO)_{35}]^{5-(\lozenge)}$	276	a	c+a	81
34	$[H_{x+1}Rh_{22}(CO)_{35}]^{4-(\lozenge)}$	276	a	c+a	81
35	$[Rh_{23}N_4(CO)_{38}]^{3-}$	306	e	a	82

TABLE 1
(Continued)

Index No.	CLUSTER	C.V.E.*	Metal frame**	Method of synthesis***	Ref.
	Iridium				
36	$[Ir_{12}(CO)_{24}]^{2-}$	158	a	c+a	83
37	$[Ir_{12}(CO)_{26}]^{2-}$	162	a	d	84
38	$[Ir_{14}(CO)_{27}]^{-}$	181	a	d	85
	Nickel				
39	$[Ni_{12}(CO)_{21}]^{4-}$	166	a	c; h	86
40	$[HNi_{12}(CO)_{21}]^{3-}$	166	a	d; h	86, 87
41	$[H_2Ni_{12}(CO)_{21}]^{2-}$	166	a	d	86, 87
42	$[Ni_{12}C_2(CO)_{16}]^{4-}$	164	d	f	88
43	$[Ni_{12}Ge(CO)_{22}]^{2-}$	170	b	d	89
44	$[Ni_{12}Sn(CO)_{22}]^{2-}$	170	b	d	89
45	$[Ni_{11}Sb_2(CO)_{18}\{Ni(CO)_3\}_2]^{3-}$	191	b	d	90
46	$[Ni_{11}Sb_2(CO)_{18}\{Ni(CO)_3\}_2]^{2-}$	190	b	d	90
47	$Ni_{15}Se_{10}(CO)_3(Cp^*)_8$	226	e	f	91
48	$Ni_{15}Se_{10}(CO)Cl_2(Cp^*)_8$	226	e	f	91
49	$[Ni_{16}(C_2)_2(CO)_{23}]^{4-}$	226	d	f	92
50	$[HNi_{34}C_4(CO)_{38}]^{5-}$	438	e	d	93
51	$[Ni_{35}C_4(CO)_{39}]^{6-}$	450	e	d	93
52	$[HNi_{38}C_6(CO)_{42}]^{5-}$	494	e	d	94
	Palladium				
53	$Pd_{16}(CO)_{13}(PEt_3)_9$	204	b	f	95
54	$Pd_{23}(CO)_{22}(PEt_3)_{10}$	294	a	f	96
55	$Pd_{23}(CO)_{20}(PEt_3)_8$	286	a	f	97
56	$Pd_{38}(CO)_{28}(PEt_3)_{12}$	460	e	f	98
	Platinum				
57	$[Pt_{12}(CO)_{24}]^{2-}$	170	d	c	99
58	$[Pt_{15}(CO)_{30}]^{2-}$	212	d	c	99
59	$H_xPt_{15}(CO)_8(PBu^t_3)_6^{(\Diamond)}$	186	a	f	100
60	$Pt_{17}(CO)_{12}(PEt_3)_8$	210	b	a	101
61	$[Pt_{18}(CO)_{36}]^{2-}$	254	d	d	99
62	$[Pt_{19}(CO)_{22}]^{4-}$	238	b	a	102
63	$[Pt_{24}(CO)_{30}]^{2-}$	302	a	a	103
64	$[Pt_{26}(CO)_{32}]^{2-}$	326	a	a	104
65	$[Pt_{38}(CO)_x]^{2-(\Diamond)}$	470	a	d	104
66	$[Pt_{52}(CO)_x]^{2-(\Diamond)}$	642	a	d	105

$^{(\Diamond)}$ for **33** and **34** ($x \approx 3$); for **59** ($x \approx 8$); for **65** ($x \approx 44$); for **66** ($x \approx 60$). * observed Cluster Valence Electrons. ** as from Section 3, this review: (a) polyhedra as bulk metal fragments, (b) polyhedra having 5-fold symmetry, (c) polyhedra linked through metal-metal or metal-non-metal bonds, (d) vertex-, edge-, face-shared polyhedra, (e) complex nets or polyhedra. *** as from Section 2, this review: (a) thermal activation, (b) photochemical activation, (c) reductive condensation, (d) oxidative condensation or coupling, (e) redox condensation, (f) removal of metal fragments, (g) oxidation-reduction reactions, (h) acid-base equilibria, (i) electrochemical activation.

TABLE 2
Heterometallic carbonyl clusters

Index No.	CLUSTER	C.V.E.*	Metal frame**	Method of synthesis***	Ref.
	Molybdenum				
67	$Mo_2\{O_2CCCo_3(CO)_9\}_4\{HO_2CCCo_3(CO)_9\}_2$	310	c	d	106
68	$Mo_4Hg_4\{Mo(CO)_3Cp\}_4$	140	c	c	107
	Rhenium				
69	$[\{Re_7AgC(CO)_{21}\}_2Br]^{5-}$	220	c	d	108
	Iron				
70	$[HFe_6Pd_6(CO)_{24}]^{3-}$	160	a	d	109
71	$[Fe_8Ag_{13}(CO)_{32}]^{4-}$	275	d	d	110
	Ruthenium				
72	$[Ru_6Pd_6(CO)_{24}]^{2-}$	158	a	d	111
73	$[Ru_9Pt_6(CO)_{28}]^{4-}$	192	a	d+a	112
74	$\{Ru_5C(CO)_{14}Cl\}_2(HgCl)_2$	180	c	d	113
75	$[\{Ru_6C(CO)_{16}\}_2Hg]^{2-}$	182	c	d	113
76	$[\{Ru_9C(CO)_{21}\}_2Hg_3]^{2-}$	274	c	d	114
	Osmium				
77	$Os_6Pt_7(CO)_{21}(cod)_2$	168	d	d	115
78	$\{Os_3Hg(CO)_{11}\}_3$	174	c	d	116
79	$Os_{10}C(CO)_{24}\{Au(AuPcy_3)_3\}$	182	c	d	117
80	$[Os_{11}C(CO)_{27}\{Cu(NCCH_3)\}]^-$	160	d	d	118
81	$[\{Os_9C(CO)_{21}\}_2Hg_2]^{2-}$	262	c	b	119
82	$[\{Os_9C(CO)_{21}\}_2Hg_2]^{4-}$	264	c	g	120
83	$[\{Os_9C(CO)_{21}\}_2Hg_3]^{2-}$	274	c	d	121
84	$[\{Os_{10}C(CO)_{24}\}_2Hg]^{2-}$	278	c	d	122
	Cobalt				
85	$[Co_2Ni_{10}C(CO)_{20}]^{2-}$	164	d	d	123
86	$[Co_3Ni_9C(CO)_{20}]^{3-}$	164	d	d	124
87	$[Co_3Ni_9C(CO)_{20}]^{2-}$	163	d	g	123
88	$\{Co_3Hg_3(CO)_9\}_2Hg_3$	198	c	c	125
89	$Zn_4O\{O_2CCCo_3(CO)_9\}_6$	354	c	d	126
	Rhodium				
90	$Rh_{10}C_2(CO)_{18}(AuPPh_3)_4$	186	d	g	127
91	$Rh_{10}C_2(CO)_{20}(AuPPh_3)_4$	190	d	g	127
92	$[Rh_{11}Pt_2(CO)_{24}]^{3-}$	170	a	c+a	128
93	$[Rh_{12}Pt(CO)_{24}]^{4-}$	170	a	c+a	128
94	$[Rh_{12}Pt_2(CO)_{26}]^{2-}$	182	a	e	129
95	$[\{Rh_6C(CO)_{15}\}_2Ag]^{3-}$	190	c	d	130
96	$[Rh_{12}C_2(CO)_{23}(AuPPh_3)]^-$	176	d	d	131
97	$[Rh_{13}Pt(CO)_{25}]^{3-}$	180	a	d	129
98	$[Rh_{18}Pt_4(CO)_{35}]^{4-}$	276	a	c+a; d	132

HIGH-NUCLEARITY CARBONYL METAL CLUSTERS

TABLE 2
(Continued)

Index No.	CLUSTER	C.V.E.*	Metal frame**	Method of synthesis***	Ref.
	Nickel				
99	$[Ni_6Pt_6(CO)_{21}]^{4-}$	166	a	e	133
100	$[Ni_9Pt_3(CO)_{21}]^{4-}$	166	a	f; h	134
101	$[HNi_9Pt_3(CO)_{21}]^{3-}$	166	a	d	135
102	$[Au_6Ni_{12}(CO)_{24}]^{2-}$	236	d	d	136
103	$[Ni_{36}Pt_4(CO)_x]^{6-(\lozenge)}$	494	a	d	137
104	$[HNi_{38}Pt_6(CO)_{48}]^{5-}$	542	a	d	138
105	$[H_2Ni_{38}Pt_6(CO)_{48}]^{4-}$	542	a	h	138

Palladium: see index No. 70, 72
Platinum: see index No. 73, 77, 92, 93, 94, 97, 98, 99, 100, 101, 103, 104, 105
Copper: see index No. 80
Silver: see index No. 69, 71, 95
Gold: see index No. 79, 90, 91, 96, 102
Zinc: see index No. 89
Mercury: see index No. 68, 74, 75, 76, 78, 81, 82, 83, 84, 88

$^{(\lozenge)}$ for **103** ($x \approx 44$). * observed Cluster Valence Electrons. ** as from Section 3, this review: (a) polyhedra as bulk metal fragments, (b) polyhedra having 5-fold symmetry, (c) polyhedra linked through metal-metal or metal-non-metal bonds, (d) vertex-, edge-, face-shared polyhedra, (e) complex nets or polyhedra. *** as from Section 2, this review: (a) thermal activation, (b) photochemical activation, (c) reductive condensation, (d) oxidative condensation or coupling, (e) redox condensation, (f) removal of metal fragments, (g) oxidation-reduction reactions, (h) acid-base equilibria, (i) electrochemical activation.

The present chapter will be restricted to high-nuclearity metal carbonyl clusters, containing twelve or more metal atoms of d-elements in their metal framework, as a lower limit for metal nuclearity arbitrarily fixed for our description.

A comprehensive compilation of the known homo- and hetero-metallic carbonyl clusters, structurally characterized to date, belonging to the above class is presented in Tables 1 and 2, respectively.

2.2. Synthesis and Reactivity

Intense activity has been addressed so far to the synthesis rather than to the reactivity of high-nuclearity carbonyl metal-clusters. The main reason for this stems from the difficulties in the separation and characterization of the products of synthetic reactions. Thus, reactivity studies are typically restricted to oxidation-reduction reactions, acid-base equilibria, degradation reactions, and thermal decomposition. Little information is available about the chemical behaviour towards coordination of organic ligands, following the unsupported belief that small and large clusters behave similarly. Even less information is available about skeletal and ligand isomerism in solution (viz. fluxionality).

The border-line between synthesis and reactivity in high-nuclearity carbonyl metal-clusters is ill-defined and unclear, so much so that the efforts devoted in the past to rationalizing the carbonyl cluster reactivity resulted in some cases in the preparation of new compounds through tailored syntheses. For such a reason, in this review, the synthesis will be discussed together with the reactivity and the latter will be emphasized only when representing a specific synthetic pathway.

Knowledge of the mechanisms of cluster growth is generally still too limited to devise a specific synthesis for new compounds only on the basis of the reagents and the experimental conditions. The synthesis of high-nuclearity carbonyl metal clusters normally occurs via a condensation of preformed carbonyl metal clusters or carbonyl complexes, after generation of unsaturated fragments. Here the discussion has been organized according to the type of the reactions, irrespective of wheter heteroatoms of transition metals or of main group elements are present or not.

The preparation methods may be classified as follows:

(I) reactions not involving redox conditions:
   a) *thermal activation.*
   b) *photochemical activation.*

(II) reactions requiring redox conditions:
   c) *reductive condensation*, from mononuclear complexes or preformed carbonyl clusters, including reductive carbonylation.
   d) *oxidative condensation or coupling.*
   e) *redox condensation.*
   f) *removal of metal fragments*, inducing fragmentation or build-up.
   g) *oxidation-reduction reactions*, including demolition.
   h) *acid-base equilibria.*
   i) *electrochemical activation.*

Moreover, the synthetic procedure is frequently a miscellanea of methods and the same compound may often be prepared by using completely different approaches. The following discussion is merely a presentation of the above methods, exemplified by some representative compounds.

(I) REACTIONS NOT INVOLVING REDOX CONDITIONS

a) *Thermal Activation*

Unlike small metal carbonyl clusters, large clusters are characterized by a high number of metal-metal interactions and by a relatively small number of ligands. Their formation from small carbonyl clusters, because of different average bond energies ($E_{M-CO} > E_{M-M}$) and of CO loss, is an endothermic process, associated with a positive change of entropy, and is therefore favoured by high temperature. On the contrary, metal cluster degradation by CO addition is favoured by low temperature [33, 139]. From the experimental data, the difference $E_{M-CO} - E_{M-M}$ decreases on descending down a group in the periodic table. Thus, the pyrolytic activation of preformed carbonyl clusters is probably the most used method for the preparation of high-nuclearity carbonyl metal clusters, for compounds containing second- or third-row transition metals. Normally the experimental conditions are severe and unselective, thus affording reaction mixtures containing different compounds which require a long time to be separated and purified. The isolation of the components is carried out, in some cases, by means of a very delicate fractional precipitation or, in other cases, by means of a selective extraction with different organic solvents. The positively charged counterion plays a very important role here inducing differences in the solubility between the various reaction products.

Thermal condensation has been widely used for (i) the preparation of homo- or hetero-metallic clusters, containing in some cases main group elements, such as interstitial atoms, and (ii) the generation of carbide atoms by CO splitting, as indicated by Equation 1:

$$2CO \rightarrow C + CO_2. \tag{1}$$

Method (ii) has lead to the preparation of several carbido carbonyl clusters, such as $[Re_7C(CO)_{21}]^{3-}$ [140], $[Ru_{10}C_2(CO)_{24}]^{2-}$ [141], $[Ru_{10}C(CO)_{24}]^{2-}$ [142], $[Os_{10}C(CO)_{24}]^{2-}$ [143].

Thermal condensation (i) is a general synthetic route for the preparation of the largest binary carbonyl clusters in the iron triad, e.g. $[Os_{17}(CO)_{36}]^{2-}$ **2** [50] and $[Os_{20}(CO)_{40}]^{2-}$ **3** [51], which have been obtained from the vacuum pyrolysis (up to 300°C, for several hours) of the chemically activated $Os_3(CO)_{10}(NCCH_3)_2$ species. The final mixture contains a large number of products, their amounts and nuclearities depending on chosen time and temperature. The preparation affords, as by-products, several low-nuclearity carbonyl clusters containing interstitial carbon atoms. Compounds **2** and **3** have been separated from the reaction mixture by means of thin-layer chromatography.

The pyrolytic activation under more controlled conditions, for example in refluxing solvent, has been used for the preparation of several high-nuclearity Pt carbonyl clusters, by heating the anions of the series $[Pt_3(CO)_6]_n^{2-}$ (n = 2–6) [99]. Thus, a mixture of $[Pt_9(CO)_{18}]^{2-}$ and $[Pt_{12}(CO)_{24}]^{2-}$ **57** in the molar ratio 5:1, in refluxing acetonitrile, affords the $[Pt_{19}(CO)_{22}]^{4-}$ **62** [102] cluster in about 80% yield, according to Equation 2:

$$5[Pt_9(CO)_{18}]^{2-} + [Pt_{12}(CO)_{24}]^{2-} \rightarrow 3[Pt_{19}(CO)_{22}]^{4-} + 48\,CO. \tag{2}$$

The $[Pt_{24}(CO)_{30}]^{2-}$ **63** [103] and $[Pt_{26}(CO)_{32}]^{2-}$ **64** [104] dianions have been prepared and isolated from the reaction mixtures, by refluxing the

$[Pt_{15}(CO)_{30}]^{2-}$ **58** and $[Pt_{18}(CO)_{36}]^{2-}$ **61** dianions, respectively, in THF. In this case, the yields of the reaction are low (ca. 15%) owing to the formation of uncharacterized products which may account for the different metal to negative charge ratios between reagents and products. A better separation of the final mixture has been obtained working with bulky counterions, e.g. $[PPh_4]^+$, $[AsPh_4]^+$, which decrease the solubility of the reaction side-products.

The neutral carbonyl phosphine cluster, $Pt_{17}(CO)_{12}(PEt_3)_8$ **60** [101], has been prepared from the pentanuclear $Pt_5(CO)_6(PEt_3)_4$, by thermally activating the Pt-C and Pt-P bonds.

In some cases, transition metal carbonyl clusters, having a metal skeleton stabilized by interstitial atoms of main group elements, have been used as starting materials for the preparation of high-nuclearity carbonyl metal clusters. Relevant examples are the synthesis of the dicarbido $[Co_{13}C_2(CO)_{24}]^{4-}$ **4** [52], the trinitrido $[Co_{14}N_3(CO)_{26}]^{3-}$ **6** [54] and the tetranitrido $[Rh_{23}N_4(CO)_{38}]^{3-}$ **35** [82] carbonyl clusters, which have been prepared by thermal condensation from the corresponding isoelectronic $[Co_6C(CO)_{15}]^{2-}$, $[Co_6N(CO)_{15}]^-$ and $[Rh_6N(CO)_{15}]^-$ anions, respectively.

b) *Photochemical Activation*

The only known example of high-nuclearity carbonyl cluster synthesis by means of a photochemical activation is the light-induced loss of Hg from $[\{Os_9C(CO)_{21}\}_2Hg_3]^{2-}$ **83** to form $[\{Os_9C(CO)_{21}\}_2Hg_2]^{2-}$ **81** [119]. The reaction can be reversed by adding Hg metal in the dark and so far, this process is unique in cluster chemistry.

II) REACTIONS REQUIRING REDOX CONDITIONS

c) *Reductive Condensation*

The salts of noble metals (second and third-row transition elements) are reduced by CO even at atmospheric pressure. The presence of carbon monoxide as a complexing agent avoids the formation of powdered metal, thus forming carbonyl complexes or clusters. This reaction, called reductive carbonylation, has been used for the preparation of most binary carbonyl species [33]. The reductive ability of CO enhances as the basicity of the reaction medium increases. Thus, the progressive reduction of a Pt (IV) salt in basic conditions, under a carbon monoxide atmosphere, affords high-nuclearity platinum carbonyl clusters of general formula, $[Pt_3(CO)_6]_n^{2-}$ ($n = 2-6$). The reductive carbonylation proceeds following the sequence:

$$[PtCl_6]^{2-} \rightarrow [Pt(CO)Cl_3]^- \rightarrow [Pt_3(CO)_6]_n^{2-}$$

$$(n = 5\ \mathbf{58}, n = 4\ \mathbf{57}, n = 3).$$

The final product depends on the experimental conditions [99], and the overall process is indicated by Equation 3:

$$3n[PtCl_6]^{2-} + (12n+1)CO + (12n+1)OH^-$$
$$\rightarrow [Pt_3(CO)_6]_n^{2-} + 18Cl^- + (6n+1)CO_2 + (6n+1)H_2O. \quad (3)$$

Further reactions with oxidizing Pt(IV) or reducing Li agents can be used for a cluster build-up or demolition respectively, as shown in Equations 4 and 5:

$$3[Pt_3(CO)_6]_4^{2-} + [PtCl_6]^{2-} + CO$$
$$\rightarrow 2[Pt_3(CO)_6]_6^{2-} + [Pt(CO)Cl_3]^- + 3\,Cl^- \quad (4)$$

$$[Pt_3(CO)_6]_4^{2-} + 2Li \rightarrow 2[Pt_3(CO)_6]_2^{2-} + 2Li^+. \quad (5)$$

Reactions (4) and (5) represent a good synthetic pathway for the synthesis of $[Pt_3(CO)_6]_n^{2-}$ ($n = 6$ **61**, $n = 2$), which are not easily prepared from direct reductive carbonylation [99].

Reductive carbonylation has also been used for the synthesis of high-nuclearity mixed metal carbonyl clusters. Thus, the $[Rh_{11}Pt_2(CO)_{24}]^{3-}$ **92** cluster has been prepared starting from $RhCl_3 \cdot 3H_2O$ and Na_2PtCl_6 salts in the presence of Na_2CO_3, under a carbon monoxide atmosphere [128].

In several cases, a preformed carbonyl complex is subjected to a condensation process induced either by the presence of reducing agents (viz. H_2, alkali metals, alkali hydroxides, carbonates, acetates, etc.) or by thermal activation, or even by the combination of both conditions. Most high-nuclearity Rh carbonyl clusters have been prepared in this way. Thermal activation of $Rh_4(CO)_{12}$ with sodium hydroxide in isopropanol, under a nitrogen atmosphere, affords a mixture of products, i.e., $[Rh_{14}(CO)_{26}]^{2-}$ **23** [72a], $[Rh_{15}(CO)_{27}]^{3-}$ **27** [70], $[Rh_{17}(CO)_{30}]^{3-}$ **30** [78] and $[Rh_{22}(CO)_{37}]^{4-}$ **32** [80]. The anion $[H_3Rh_{13}(CO)_{24}]^{2-}$ **20** has been selectively synthesized in ca. 50% yields, performing the reaction under hydrogen instead of a nitrogen atmosphere [70b]. The reaction of $Rh_4(CO)_{12}$ with sodium acetate, at room temperature, proceeds very quickly to a violet solution containing the dimeric species $[\{Rh_6(CO)_{15}\}_2]^{2-}$ **10** [58a, c].

Reduction of $Rh(CO)_2(acac)$ with cesium benzoate at high CO/H_2 pressure, and at high temperature, in glyme yields several large carbonyl anions, e.g. $[Rh_{14}(CO)_{25}]^{4-}$ **21** [69a], $[Rh_{14}(CO)_{26}]^{2-}$ **23** [72b], $[Rh_{15}(CO)_{27}]^{3-}$ **27** [76], $[Rh_{15}(CO)_{30}]^{3-}$ **28** [76], $[H_xRh_{22}(CO)_{35}]^{5-}$ **33** [81] and $[H_{x+1}Rh_{22}(CO)_{35}]^{4-}$ **34** [81] ($x \approx 3$). Some of these clusters have the same metal nuclearity but a different number of ligands, and different metal frames than those prepared starting from $Rh_4(CO)_{12}$. The comparison of the product illustrates the difficulties of rationalizing the cluster synthesis and understanding the relevance of different parameters.

Under similar conditions, but in the presence of an external source of heteroatoms, e.g. PPh_3, $AsPh_3$, $SbPh_3$, H_2S, the same reactants can afford clusters containing interstitial atoms of main group elements. This approach has been used to prepare $[Rh_{12}Sb(CO)_{27}]^{3-}$ **17** [65] and $[Rh_{17}S_2(CO)_{32}]^{3-}$ **31** [79].

The preparation of the nitrido species, $[HRh_{12}N_2(CO)_{23}]^{3-}$ **16**, containing two interstitial nitrogen atoms [64], is an example of thermally-assisted reductive condensation. The dianion has been obtained by pyrolysis of the $[Rh_6N(CO)_{15}]^-$ cluster, in refluxing methanol, in the presence of NaOH. In this case the starting hexanuclear anion already contains an interstitial atom and the reaction looks like a simple reductive dimerization.

The analogous reduction of $Ir_4(CO)_{12}$ with KOH in ethanol, performed under hydrogen atmosphere at 80°C, has been used for the preparation, in low yields, of the $[Ir_{12}(CO)_{24}]^{2-}$ **36** [83] cluster, the main product of the reaction, however being the hexanuclear dianion $[Ir_6(CO)_{15}]^{2-}$.

Reductive condensation has been applied to synthesize heterometallic carbonyl clusters, belonging to second- and third-row transition element systems. The two cluster anions, $[Rh_{11}Pt_2(CO)_{24}]^{3-}$ **92** [128] and $[Rh_{12}Pt(CO)_{24}]^{4-}$ **93** [128], have been obtained by controlled pyrolysis of the mixed-metal cluster $[Rh_5Pt(CO)_{15}]^-$ in the presence of $NaHCO_3$. Both anions are isoelectronic with the $[H_{5-n}Rh_{13}(CO)_{24}]^{n-}$ ($n = 2, 3, 4$) family, a Pt atom being electronically equivalent to a 'RhH' group.

In the first-row transition elements (Fe, Co, Ni), the direct carbonylation of metal salts cannot be performed with standard laboratory techniques, owing to the high toxicity of the corresponding binary mono- or bi-nuclear carbonyl compounds and to the high pressure required. Thus, due to the difficulty of condensation when excess carbon monoxide is present (high E_{M-CO}), the carbonylation step is carried out before the condensation step.

An example of a reductive condensation is the reaction of $Ni(CO)_4$ with alkali hydroxides in DMF or in methanol, which generally affords low-nuclearity carbonyl anions; a remarkable exception is the formation of $[Ni_{12}(CO)_{21}]^{4-}$ **39** [86], which has been obtained, in low yields, performing the reduction in a saturated-KOH methanolic solution.

d) *Oxidative Condensation or Coupling*
More frequently, oxidizing agents can labilize the metal-CO bond and promote either dimerization or self-assembling of activated fragments into new, more stable, metal frames. The coupling is often accompanied by the incorporation of the oxidant inside the metal skeleton. The oxidative condensation has also been widely used for the preparation of heterometallic clusters or clusters containing interstitial atoms. A dimerization induced by acids is often observed when a first protonation step gives rise to unstable hydride derivatives. The dimerization product can be either a simple oligomeric species or a more complex derivative. Several cases belong to this section and a further subdivision is convenient in order to understand the following presentation better. We can roughly distinguish the oxidative condensation processes into:

i) dimerization (coupling) and oligomerization reactions with retention of the basic unit or with metal frame rearrangement, *ii*) condensation reactions without incorporation of the oxidizing agent, *iii*) condensation reactions with incorporation of the oxidizing agent, and *iv*) addition of mononuclear complexes.

(i) Dimerization and oligomerization reactions with retention of the basic unit or with metal frame rearrangement. A dimerization of $[Rh_6(CO)_{15}]^{2-}$ and $[Rh_6C(CO)_{15}]^{2-}$ into $[\{Rh_6(CO)_{15}\}_2]^{2-}$ **10** [144] and $[Rh_{12}C_2(CO)_{24}]^{2-}$ **13** [61] respectively, has been observed when treating the two hexanuclear clusters with acids. NMR has established [145] that the protonation of both reagents proceeds with the formation of the unstable intermediate species $[HRh_6(CO)_{15}]^-$ and $[HRh_6C(CO)_{15}]^-$, which then transform into the dodecanuclear clusters, with contemporary evolution of molecular hydrogen. A similar pathway is probably involved in the reaction of $[Rh_7(CO)_{16}]^{3-}$ with trifluorosulfonic acid, which results in a selective oxidative coupling to give $[Rh_{14}(CO)_{26}]^{2-}$ **23** [72b].

Stable hydride derivatives have been found during the reaction of $[Ni_6(CO)_{12}]^{2-}$ with acids. In this case the condensation process occurs through a sequence of reactions to give hydridic anions of general formula $[H_{4-n}Ni_{12}(CO)_{21}]^{n-}$ ($n = 2$ **41**, $n = 3$ **40**) [86], as represented by Equations 6–8:

$$3[Ni_6(CO)_{12}]^{2-} + 2 H^+ \rightarrow 2[Ni_9(CO)_{18}]^{2-} + H_2 \qquad (6)$$

$$2[Ni_9(CO)_{18}]^{2-} + H^+$$
$$\rightarrow [HNi_{12}(CO)_{21}]^{3-} + 3 Ni(CO)_4 + 3 Ni + 3CO \qquad (7)$$

$$[HNi_{12}(CO)_{21}]^{3-} + H^+ \rightarrow [H_2Ni_{12}(CO)_{21}]^{2-}. \qquad (8)$$

A different case is the dimerization of $[Ir_6(CO)_{15}]^{2-}$ induced by electrophiles, e.g. $[Cu(NCCH_3)_4]^+$, partially affording the $[Ir_{12}(CO)_{26}]^{2-}$ **37** [84] cluster.

Particular examples of dimerizations have been found when the oxidizing agent bridges two or more cluster units. These coupling reactions are induced by salts of Ag(I), Au(I), Hg(I), Hg(II), Tl(I), Tl(III). This is the case of the formation of $[\{Ru_6C(CO)_{16}\}_2Tl]^-$ **1** [49a] shown by Equation 9:

$$2[Ru_6C(CO)_{16}]^{2-} + Tl^{3+} \rightarrow [\{Ru_6C(CO)_{16}\}_2Tl]^- \qquad (9)$$

or the preparation of $[\{Rh_6C(CO)_{15}\}_2Ag]^{3-}$ **95** [130] starting from $[Rh_6C(CO)_{15}]^{2-}$ with a stoichiometric amount of $AgBF_4$. This reaction yields a series of inorganic oligomers containing the Rh_6 prismatic units bridged by silver atoms. At a ratio Ag(I):$[Rh_6C(CO)_{15}]^{2-} = 1$, a polymeric species of high molecular weight is formed, as evidenced by NMR studies.

Condensation of triangular units has been found during the reaction of $[HOs_3(CO)_{11}]^-$ with $HgBr_2$ to yield the raft-like cluster, $\{Os_3Hg(CO)_{11}\}_3$ **78** [116]. Formation of the $[\{Os_{10}C(CO)_{24}\}_2Hg]^{2-}$ **84** [122] dianion from the reaction of $[Os_{10}C(CO)_{24}]^{2-}$ with RHgCl (R = C_6F_5, C_6H_5, Cl) has been observed to occur via the formation of the $[Os_{10}C(CO)_{24}(HgR)]^-$ intermediate. A certain number of high-nuclearity Ru-Hg carbonyl clusters have been recently synthesized by the condensation of the hexanuclear $[Ru_6C(CO)_{16}]^{2-}$ with $HgCl_2$. The dodecanuclear $[\{Ru_6C(CO)_{16}\}_2Hg]^{2-}$ **75** [113] dianion contains two octahedral Ru_6C units linked through a mercury atom. The reaction of $Ru_5C(CO)_{15}$ with $HgCl_2$ yielded the compound $\{Ru_5C(CO)_{14}Cl\}_2(HgCl)_2$ **74** [113] which showed two Ru_5C units joined through a Hg_2Cl_2 bridging unit.

A very special case is the dimerization reaction of $[Re_7AgC(CO)_{21}]^{2-}$ induced by bromide. The unexpected product, $[\{Re_7AgC(CO)_{21}\}_2Br]^{5-}$ **69** [108], is formed on the addition of halides such as $[N(C_4H_9)_4]Br$. The dimerization, represented by Equation 10:

$$2[Re_7AgC(CO)_{21}]^{2-} + Br^- \rightarrow [\{Re_7AgC(CO)_{21}\}_2Br]^{5-} \quad (10)$$

is probably a consequence of the high E_{Ag-Br}.

Oligomerization reactions induced either by strong acids, e.g. H_2SO_4, or by strong oxidizing agents, e.g. Fe^{3+}, transform the $[Pt_{19}(CO)_{22}]^{4-}$ **62** tetranion and the $[Pt_{26}(CO)_{32}]^{2-}$ **64** dianion into $[Pt_{38}(CO)_x]^{2-}$ **65** [104] and $[Pt_{52}(CO)_x]^{2-}$ **66** [105], respectively. Such large platinum carbonyl clusters, i.e. **65** and **66**, have been partially characterized by X-ray diffraction analysis. Unfortunately, the scattering from the many heavy atoms obscures that of the few carbonyl ligands, therefore, the number and the disposition of the ligands in Pt_{38} and Pt_{52} is still unknown. Elemental analysis suggests 44 CO's for compound **65**, in agreement with the predictions of some electron-counting theories [146]. On the same electron-counting basis, 60 CO's are predicted for compound **66**.

(ii) Condensation reactions without incorporation of the oxidizing agent. In some cases the oxidation reaction results only in a cluster expansion. For example, reaction of $[Rh_6C(CO)_{15}]^{2-}$ with Fe^{3+} instead of H_2SO_4, gives a brown intermediate product from which it has been possible to isolate $Rh_8C(CO)_{19}$ or $[Rh_{15}C_2(CO)_{28}]^-$ **29** [77b] depending on the reaction conditions. The pentadecanuclear complex is quantitatively converted into $Rh_{12}C_2(CO)_{25}$ **12** in the presence of CH_2Cl_2 [60]. The $[Rh_{14}N_2(CO)_{25}]^{2-}$ **26** [75] anion has been synthesized by reacting $[Rh_6N(CO)_{15}]^-$ with acetic acid, in prolonged pyrolytic conditions. Treatment of the mixed metal cluster, $[Rh_{11}Pt_2(CO)_{24}]^{3-}$ **92**, with acids affords $[Rh_{13}Pt(CO)_{25}]^{3-}$ **97** via a complicated oxidative condensation reaction [129].

Very recently, the largest known iridium carbonyl cluster, the paramagnetic $[Ir_{14}(CO)_{27}]^-$ **38** anion, has been recognized among other products when treating $[Ir_6(CO)_{15}]^{2-}$ with $[Fe(Cp)_2]^+$ [85].

(iii) Condensation reactions with incorporation of the oxidizing agent. A very important class of reactions is the oxidative condensation in which the oxidizing agent remains in the final product, being either a transition metal or a main group element. Several high-nuclearity heterometallic carbonyl clusters have been synthesized in this way.

Thus, reacting $[Fe_4(CO)_{13}]^{2-}$ with K_2PdCl_4, affords a mixture of products containing small amounts (5–10%) of $[HFe_6Pd_6(CO)_{24}]^{3-}$ **70** [109], with low-nuclearity homo- and hetero-metallic clusters as the main products. An analogous reaction carried out starting from a preformed Ru carbonyl anion, viz. $[HRu_3(CO)_{11}]^-$, and some Pd(II) or Pt(II) salts resulted in the preparation of the $[Ru_6Pd_6(CO)_{24}]^{2-}$ **72** [111] and $[Ru_9Pt_6(CO)_{28}]^{4-}$ **73** [112] anions, respectively. The preparation of **72** is performed at low temperature, $-78°C$, to reduce the unwanted formation of Pd metal; on the contrary, the synthesis of **73** is assisted by a thermal activation.

A series of mixed Ni-Pt metal clusters have been obtained by a similar synthetic method. Indeed the reaction of $[Ni_6(CO)_{12}]^{2-}$ with $PtCl_2$ or K_2PtCl_4 in acetonitrile, results in the synthesis of cluster anions having general formulae $[H_{6-n}Ni_{38}Pt_6(CO)_{48}]^{n-}$ ($n = 4$ **105**, $n = 5$ **104**) or $[H_{4-n}Ni_9Pt_3(CO)_{21}]^{n-}$ ($n = 3$ **101**, $n = 4$ **100**). Compound **104** is the largest mixed metal carbonyl cluster fully characterized to date. During the synthesis another high-nuclearity carbonyl anion, viz. $[Ni_{36}Pt_4(CO)_x]^{6-}$ **103** [137], was isolated but only partially characterized. On the electron-counting basis [147], 44 CO's are predicted for compound **103**. The distribution of the final products depends either on the initial $[Ni_6(CO)_{12}]^{2-}$:Pt(II) molar ratio or on the acidity of the reaction medium [138].

Recent achievements are the syntheses of $[Au_6Ni_{12}(CO)_{24}]^{2-}$ **102** [136], from the reaction between $[Ni_6(CO)_{12}]^{2-}$ and $(Ph_3P)AuCl$, and of $[Fe_8Ag_{13}(CO)_{32}]^{4-}$ **71** [110], from the reaction between $[Fe(CO)_4]^{2-}$ and $AgBF_4$. Clusters **102** and **71** are examples of molecular Au-Ni and Ag-Fe bimetallic species.

High-nuclearity Ru or Os carbonyl clusters, containing interstitial carbon atoms, have been prepared by exploiting the ability of mercury to bridge small metal aggregates. Two remarkable examples are represented by the $[\{M_9C(CO)_{21}\}_2Hg_3]^{2-}$ (M = Ru **76** [114], M = Os **83** [121]) anions. These two clusters have been synthesized by reacting $[M_{10}C(CO)_{24}]^{2-}$ with mercury(II) trifluoroacetate.

The preparation of a broad class of clusters containing interstitial atoms of main group elements (C, Si, Ge, Sn, P, As, Sb, Bi, S, Se, Te) can be described as oxidative condensation.

Interstitial nickel carbide clusters have been obtained from the reaction of $[Ni_6(CO)_{12}]^{2-}$ with halocarbons. The use of CCl_4 afforded the monocarbide derivative, $[Ni_9C(CO)_{17}]^{2-}$ [148], whereas the dicarbide derivative, $[Ni_{10}C_2(CO)_{16}]^{2-}$ [149], has been synthesized by using hexachloroethane as a source of carbon atoms. During the $[Ni_{10}C_2(CO)_{16}]^{2-}$ preparation other derivatives having general formula $[H_{6-n}Ni_{34}C_4(CO)_{38}]^{n-}$ ($n = 5$ **50**, $n = 6$) were isolated [93]. The hexacarbide cluster, $[HNi_{38}C_6(CO)_{42}]^{5-}$ **52** [94], was obtained by reaction of $[Ni_6(CO)_{12}]^{2-}$ with hexachloropropene, probably through the condensation of an uncharacterized tricarbido derivative. The synthesis of $[Ni_{35}C_4(CO)_{39}]^{6-}$ **51** [93], obtained by refluxing the $[Ni_6(CO)_{12}]^{2-}$ dianion in CH_2Cl_2, is also reported. In this case the dichloromethane behaves as the source of carbon atoms.

Nickel carbonyl clusters containing interstitial carbon-congener atoms, viz. Ge, Sn, have been isolated from the reaction of $GeCl_4$ or $SnCl_2 \cdot 2H_2O$ with $[Ni_6(CO)_{12}]^{2-}$. The isoelectronic compounds, $[Ni_{12}Ge(CO)_{22}]^{2-}$ **43** [89] and $[Ni_{12}Sn(CO)_{22}]^{2-}$ **44** [89] showed an interstitial germanium and tin atom in an icosahedral cavity.

The new $[Ni_{13}Sb_2(CO)_{24}]^{n-}$ ($n = 2$ **46**, $n = 3$ **45**) [90] clusters have been obtained by addition of $SbCl_3$ to a tetrahydrofuran solution of $[Ni_6(CO)_{12}]^{2-}$. The interconversion between compounds **46** and **45** has been selectively obtained by means of clean one-electron chemical and electrochemical oxidation-reduction reactions.

The chloromethynyl derivative $Co_3(CO)_9CCl$ has been used as a source of $Co_3(CO)_9C$ fragments for the synthesis of large mixed-metal carbido clusters. Thus the reaction of $Co_3(CO)_9CCl$ with $[Ni_6(CO)_{12}]^{2-}$ results in a mixture of clusters, among which the $[Co_3Ni_9C(CO)_{20}]^{3-}$ **86** [124] trianion has been isolated and fully characterized. The condensation may be represented by Equation 11:

$$2[Ni_6(CO)_{12}]^{2-} + Co_3(CO)_9CCl$$
$$\rightarrow [Co_3Ni_9C(CO)_{20}]^{3-} + Cl^- + 3\, Ni(CO)_4 + CO. \qquad (11)$$

However, the reaction stoichiometry is more complicated than shown in Equation 11, since another dodecanuclear $[Co_2Ni_{10}C(CO)_{20}]^{2-}$ **85** monocarbido cluster is obtained in the same yield under these reaction conditions [123].

(iv) Addition of mononuclear complexes. Several heterometallic carbonyl clusters have been synthesized by condensation of mononuclear cationic fragments into anionic carbonyl clusters.

The species $[M(NCCH_3)_4]X$ (M = Cu, Ag; X = Cl, PF_6) and $[M(PPh_3)]X$ (M = Cu, Ag, Au; X = Cl, Br, I) are frequently added to a cluster solution, in the presence of halide acceptors, to form the derived cationic fragments, $[M(NCCH_3)]^+$ and $[M(PPh_3)]^+$. These cationic species, behaving as 12 electron donors, normally act as capping groups without altering the basic metal skeleton (see Section 2.4).

High-nuclearity carbonyl clusters containing the above fragments are almost exclusively restricted to Ru and Os, and more recently to Rh as well. The mono- and di-derivatives of both Au and Cu, viz. $[Os_{11}C(CO)_{27}(AuPPh_3)]^-$ [118b], $Os_{11}C(CO)_{27}(AuPPh_3)_2$ [118b], $[Os_{11}C(CO)_{27}\{Cu(NCCH_3)\}]^-$ **80** [118a, b] and $Os_{11}C(CO)_{27}\{Cu(NCCH_3)\}_2$ [118b] have been produced from $[Os_{11}C(CO)_{27}]^{2-}$ and $[AuPPh_3]^+$ or $[Cu(NCCH_3)_4]^+$, respectively. Only the undecaosmium monocopper derivative **80** has been characterized by X-ray diffraction analysis.

An alternative source for the $[AuPPh_3]^+$ fragment is represented by the gold oxonium salt, $[O(AuPPh_3)_3]BF_4$. The preparation of the tetra-gold decaosmium cluster, $Os_{10}C(CO)_{24}\{Au(AuPCy_3)_3\}$ **79** [117], belongs to this type of synthesis.

Recently, mixed Au-Rh metal clusters have been prepared starting from the high-nuclearity dicarbido cluster $[Rh_{12}C_2(CO)_{24}]^{2-}$ **13**. Reaction of the dianion with stoichiometric amounts of $Au(PPh_3)Cl$ affords the monoanionic derivative, $[Rh_{12}C_2(CO)_{23}(AuPPh_3)]^-$ **96** [131], while an excess of $Au(PPh_3)Cl$ affords the neutral compound, $Rh_{10}C_2(CO)_{18}(AuPPh_3)_4$ **90** [127], through a partial oxidative demolition and rearrangement of the Rh cluster. This compound is in reversible equilibrium with $Rh_{10}C_2(CO)_{20}(AuPPh_3)_4$ **91** through association-dissociation of two carbon monoxide ligands [127].

$Me_3NO \cdot 2H_2O$, has also been used (at 0°C) as an activating agent, for the preparation of the mixed Os-Pt carbonyl cluster, i.e., $Os_6Pt_7(CO)_{21}(cod)_2$ **77** [115], starting from $Os_6Pt_4(CO)_{22}(cod)$ and $Pt(cod)_2$.

e) Redox Condensation

The addition of a mononuclear carbonyl complex to a preformed cluster in a different oxidation state was defined as redox condensation reaction [33]. In a wider meaning, every condensation between two carbonyl species in different oxidation states, fusing into a single product, can be classified in this section. Thus, the redox condensation allows a well-controlled growth of the cluster in nearly quantitative yields. Very often redox condensation occurs via evolution of carbon monoxide deriving from the breaking of metal-carbon monoxide bonds and the formation of new metal-metal bonds.

This synthetic approach has been used most widely for the preparation of small or medium-sized metal clusters, being quite rarely applied to the case of high-nuclearity metal clusters. The reason for this probably stems from the scarce availability of suitable reagents. A step-by-step growth is well exemplified by the preparation of the $[HRh_{14}(CO)_{25}]^{3-}$ **22** and $[Rh_{15}(CO)_{27}]^{3-}$ **27** anions [70b], with the following sequence of reactions (12–14), which implies a cluster build-up via incorporation of $[Rh(CO)_2]^+$ fragments, and an interposed deprotonation step:

$$[HRh_{13}(CO)_{24}]^{4-} + [Rh(CO)_2(NCCH_3)_2]^+$$

$$\rightarrow [HRh_{14}(CO)_{25}]^{3-} + CO + 2CH_3CN \qquad (12)$$

$$[HRh_{14}(CO)_{25}]^{3-} + OH^- \rightarrow [Rh_{14}(CO)_{25}]^{4-} + H_2O \qquad (13)$$

$$[Rh_{14}(CO)_{25}]^{4-} + [Rh(CO)_2(NCCH_3)_2]^+$$
$$\rightarrow [Rh_{15}(CO)_{27}]^{3-} + 2\,CH_3CN. \qquad (14)$$

Redox condensations of carbonyl clusters of different metals provides a selective synthetic route for the synthesis of heteronuclear high-nuclearity cluster anions. Thus, the undecanuclear $[Rh_9Pt_2(CO)_{22}]^{3-}$ [150] cluster has been obtained by a redox condensation reaction, according to Equation 15:

$$[Rh_5Pt(CO)_{15}]^- + [Rh_4Pt(CO)_{12}]^{2-} \rightarrow [Rh_9Pt_2(CO)_{22}]^{3-} + 5\,CO. \qquad (15)$$

The reaction pathway is more complex than represented, owing to the presence, in the reaction mixture, of a larger Pt-Rh carbonyl cluster, viz. $[Rh_{18}Pt_4(CO)_{35}]^{4-}$ **98** [132], probably formed through a further oxidative condensation. Reaction of $[Rh_{11}Pt_2(CO)_{24}]^{3-}$ **92** with $[Rh(CO)_2(NCCH_3)_2]^+$ to give $[Rh_{12}Pt_2(CO)_{26}]^{2-}$ **94** [129] reproduces the situation found in homometallic Rh carbonyl clusters, when a $[Rh(CO)_2]^+$ fragment is added. This reaction is represented by Equation 16:

$$[Rh_{11}Pt_2(CO)_{24}]^{3-} + [Rh(CO)_2(NCCH_3)_2]^+$$
$$\rightarrow [Rh_{12}Pt_2(CO)_{26}]^{2-} + 2\,CH_3CN. \qquad (16)$$

In some cases, probably due to instability of the intermediate product, further condensation occurs. The preparation of $[\{Rh_6C(CO)_{15}\}_2\{Rh_2(CO)_3\}]^{2-}$ **25** [74] exemplifies this situation, as shown by Equation 17:

$$2[Rh_6C(CO)_{15}]^{2-} + 2[Rh(CO)_2(NCCH_3)_2]^+$$
$$\rightarrow [\{Rh_6C(CO)_{15}\}_2\{Rh_2(CO)_3\}]^{2-} + CO + 4\,CH_3CN. \qquad (17)$$

A similar situation has been found in mixed metal systems. The unexpected formation of $[Ni_6Pt_6(CO)_{21}]^{4-}$ **99** [133] is attributable to a dimerization process following a redox condensation between $[Ni_6(CO)_{12}]^{2-}$ and $[Pt_6(CO)_{12}]^{2-}$. The reaction proceeds via the formation of an intermediate species, formulated as $[Ni_3Pt_3(CO)_{12}]^{2-}$ on the basis of ^{195}Pt NMR and derived from statistical exchange of $M_3(CO)_6$ moieties [151]. The supposed $[Ni_3Pt_3(CO)_{12}]^{2-}$ is unstable during crystallization and the dodecanuclear **99** separates out, due to the self-condensation reaction, as in Equation 18:

$$2\,[Ni_3Pt_3(CO)_{12}]^{2-} \rightarrow [Ni_6Pt_6(CO)_{21}]^{4-} + 3CO. \qquad (18)$$

f) *Removal of Metal Fragments*
Addition of mononuclear fragments affording a cluster build-up has been discussed in the previous section (Equations 12, 14, 16 and 17).

Reaction of high nuclearity carbonyl clusters with soft nucleophiles such as CO, PR_3, NO_2^- and halides, results in addition, substitution products, cluster demolition and, sometimes, cluster build-up. The addition of nucleophiles fills the antibonding molecular orbitals, thus affording a metal-metal bond breaking and an opening of the cluster framework. The resulting compound can be stabilized either by finding a new situation in the metal-metal and metal-carbon monoxide interactions, or by losing a small fragment or even by inducing condensation towards more stable clusters of higher metal nuclearity.

The first case is exemplified by the addition of CO to the octahedral, 86 CVE, $[Rh_6C(CO)_{13}]^{2-}$ anion [152]. The resulting, 90 CVE, $[Rh_6C(CO)_{15}]^{2-}$ anion assumes a trigonal prismatic array of metal atoms by breaking 3 M-M bonds and forming 2 extra M-CO bonds.

Degradation of preformed clusters induced by carbon monoxide has been observed in the Ni carbonyl cluster system owing to the formation of the very highly stable $Ni(CO)_4$, as a degradation product. The capped square-antiprismatic $[Ni_9C(CO)_{17}]^{2-}$ is indeed decapped by CO to give the square-antiprismatic $[Ni_8C(CO)_{16}]^{2-}$, which cannot be obtained by a different synthetic way [148]. A similar decapping reaction has been observed in the transformation of the icosahedral array of $[Ni_{12}Ge(CO)_{22}]^{2-}$ **43** into the pentagonal antiprismatic array of $[Ni_{10}Ge(CO)_{20}]^{2-}$ [89], by exposing the former to a carbon monoxide atmosphere.

An example of halide-induced degradation is the transformation of $[Rh_{15}(CO)_{27}]^{3-}$ **27** into $[Rh_{14}(CO)_{25}]^{4-}$ **21** [70] which occurs by dissolving the former in acetonitrile containing bromide ions, as represented in Equation 19:

$$[Rh_{15}(CO)_{27}]^{3-} + 2\,Br^- \rightarrow [Rh_{14}(CO)_{25}]^{4-} + [Rh(CO)_2Br_2]^- \quad (19)$$

Reactions with triphenylphosphine usually result either in a carbonyl substitution or in a disproportionation-induced degradation. Examples of this type of reaction are present in the inorganic oligomeric series, $[Pt_3(CO)_6]_n^{2-}$ (n = 3, 4 and 5) [99]. The reaction proceeds through the formation of Pt(0) species, according to Equation 20:

$$[Pt_3(CO)_6]_5^{2-} + 6\,PPh_3 \rightarrow [Pt_3(CO)_6]_4^{2-} + 3Pt(CO)_2(PPh_3)_2. \quad (20)$$

An unusual disproportionation-induced condensation has been found in the reaction of $[Ni_{10}C_2(CO)_{16}]^{2-}$ with 4 equivalents of PPh_3 in THF, to give $[Ni_{16}(C_2)_2(CO)_{23}]^{4-}$ **49** [92] according to Equation 21:

$$2[Ni_{10}C_2(CO)_{16}]^{2-} + 8PPh_3$$
$$\rightarrow [Ni_{16}(C_2)_2(CO)_{23}]^{4-} + CO + 4Ni(CO)_2(PPh_3)_2. \quad (21)$$

This behaviour is probably due to the unstable nature of the degradation product of $[Ni_{10}C_2(CO)_{16}]^{2-}$. A similar reaction has been used for the preparation of the dicarbido cluster, $[Ni_{12}C_2(CO)_{16}]^{4-}$ **42** [88] starting from $[Ni_9C(CO)_{17}]^{2-}$. A stepwise addition of triphenylphosphine to a $[Ni_9C(CO)_{17}]^{2-}$ solution gives rise to $[Ni_7C(CO)_{12}]^{2-}$ [88] as in Equation 22:

$$[Ni_9C(CO)_{17}]^{2-} + 4\,PPh_3$$
$$\rightarrow [Ni_7C(CO)_{12}]^{2-} + 2Ni(CO)_2(PPh_3)_2 + CO. \qquad (22)$$

The further degradation of $[Ni_7C(CO)_{12}]^{2-}$ with PPh_3 affords an unstable uncharacterized species which transforms into $[Ni_{12}C_2(CO)_{16}]^{4-}$ **42** during crystallization. The net result of the reaction is a disproportionation-induced condensation with removal of the mononuclear $Ni(CO)_2(PPh_3)_2$ fragment.

The syntheses of large neutral Pd carbonyl phosphine clusters, viz. $Pd_{23}(CO)_{22}(PEt_3)_{10}$ **54** [96], $Pd_{23}(CO)_{20}(PEt_3)_8$ **55** [97] and $Pd_{38}(CO)_{28}(PEt_3)_{12}$ **56** [98], which have been obtained by a general phosphine-abstracting route performed in mild conditions using $Pd(OAc)_2$ and $Pd_{10}(CO)_{12}(PEt_3)_6$, belong to an even more complicated type of reaction. The role of the Pd(II) salt is not clear, however, probably an oxidation of CO to CO_2, a concomitant formation of acetic acid, and a reduction of Pd(II) to Pd(0) are necessary in order to subtract phosphine ligands as mononuclear metal complexes such as $Pd(CO)(PEt_3)_3$ or $Pd(CO)_2(PEt_3)_2$ and to induce condensation of the resulting unstable intermediate cluster. The preparation of $Pd_{16}(CO)_{13}(PEt_3)_9$ **53** [95] has recently been reported using the same procedure, this time adding $Me_3NO \cdot 2H_2O$ to activate the metal-carbon monoxide bond.

g) *Oxidation-Reduction Reactions*

Oxidation and reduction reactions can change the free negative charge on the cluster molecule. When the metal frame is not able to support such changes, the above reactions can be complicated either by oxidative and redox condensations or by cluster demolition.

Only recently a systematic investigation into the possibility of preparing oxidized or reduced species of unchanged metal nuclearity has been undertaken thanks to the rapidly expanding field of electrochemical techniques which have undoubtedly improved the study of cluster reactivity.

Very few examples of clean oxidation reaction occurring without rearrangements are known to date. The paramagnetic dicarbido derivative, $[Co_{13}C_2(CO)_{24}]^{4-}$ **4**, has been used as the starting material for the preparation of the oxidized derivative, $[Co_{13}C_2(CO)_{24}]^{3-}$ **5** [53], by the addition of iodine, according to Equation 23:

$$[Co_{13}C_2(CO)_{24}]^{4-} + 1/2\,I_2 \rightarrow [Co_{13}C_2(CO)_{24}]^{3-} + I^-. \qquad (23)$$

The other reported example is the preparation of the mixed Co-Ni metal cluster, $[Co_3Ni_9C(CO)_{20}]^{2-}$ **87** [123], quantitatively obtained by reaction of the previously quoted $[Co_3Ni_9C(CO)_{20}]^{3-}$ **86** with protonic acids. The reaction, performed in acetone, provides the evolution of molecular hydrogen, as simply described in Equation 24:

$$[Co_3Ni_9C(CO)_{20}]^{3-} + H^+ \rightarrow [Co_3Ni_9C(CO)_{20}]^{2-} + 1/2\ H_2. \tag{24}$$

In some cases, the oxidation reaction takes place with partial cluster demolition and is used for preparative purposes. For example, the neutral species $H_2Rh_{12}(CO)_{25}$ **11** [59] has been nicely obtained by reaction of $[Rh_{14}(CO)_{25}]^{4-}$ **21** with trifluoroacetic acid in the presence of CO and chloride ions. It has been postulated that the role of the chloride ions in this reaction is the stabilization of the monomeric fragment $[Rh(CO)_2Cl_2]^-$ deriving from the decomposition, as shown in Equation 25:

$$[Rh_{14}(CO)_{25}]^{4-} + 6\ H^+ + 4\ Cl^- + 4CO$$
$$\rightarrow H_2Rh_{12}(CO)_{25} + 2H_2 + 2[Rh(CO)_2Cl_2]^-. \tag{25}$$

As in oxidation, the reduction processes may be complicated by side reactions such as condensation or degradation reactions. In this case, the degradation is induced by the increased number of electrons present in antibonding molecular orbitals, which afford a metal-metal bond-breaking and an opening of the cluster metal frame. This possibility is responsible for the reduction of $[Pt_{12}(CO)_{24}]^{2-}$ **57** to $[Pt_9(CO)_{18}]^{2-}$ [99] and of $[\{Rh_6(CO)_{15}\}_2]^{2-}$ **10** to $[HRh_6(CO)_{15}]^-$ [34, 153], with molecular hydrogen, as represented in the Equations 26 and 27, respectively:

$$3[Pt_{12}(CO)_{24}]^{2-} + H_2 + 2H_2O \rightarrow 4[Pt_9(CO)_{18}]^{2-} + 2H_3O^+ \tag{26}$$

$$[\{Rh_6(CO)_{15}\}_2]^{2-} + H_2 \rightarrow 2[HRh_6(CO)_{15}]^-. \tag{27}$$

When strong nucleophiles such as OH^-, OR^-, H^- are used as reducing agents and the reduction reaction affords evolution of carbon monoxide, the further reaction of carbon monoxide with the reduction product is avoided since CO is readily transformed into CO_2, HCO_3^-, CO_3^{2-}, $HCOO^-$ depending on the basicity of the solution.

The preparations of the anionic species, e.g. $[Rh_{12}C_2(CO)_{23}]^{n-}$ ($n = 3$ **15** [63], $n = 4$ **14** [62]), from $[Rh_{12}C_2(CO)_{24}]^{2-}$ **13** [61], are relevant examples of this situation. Reduction of the dicarbido dianion proceeds via the formation of the paramagnetic trianion and the diamagnetic tetraanion, by adding NaOH in MeOH solution. A very large excess of reducing agent is necessary to complete the reduction to the final product. The second step of the reduction is easily reversed by I_2, while the dianion cannot be regenerated under these reaction conditions.

The recent preparation of the $[\{Os_9C(CO)_{21}\}_2Hg_2]^{4-}$ **82** tetraanion is another example of controlled reduction [120]. The tetraanion has been prepared starting from $[\{Os_9C(CO)_{21}\}_2Hg_3]^{2-}$ **83** by careful addition of cobaltocene in THF solution. The reduction affords the simultaneous elimination of one Hg atom from the inner Hg_3 unit. Compound **83** is obtained by treating compound **82** with Hg^{2+} as oxidazing agent.

h) *Acid-Base Equilibria*

As discussed before, addition of acids or bases to clusters may afford complicated chemical transformations when unstable derivatives are formed. Thus, reactions of oxidative coupling, reductive condensation, or demolition can occur. In some cases, however, protonation or deprotonation reactions afford the formation of new derivatives, without any change in metal nuclearity. In suitable conditions of acidity and solvent medium, sequential protonation or deprotonation reactions have been used for preparative purposes, as shown, for example, by Equation 28:

$$[H_3Rh_{13}(CO)_{24}]^{2-} \underset{H^+}{\overset{MeOH/CO_3^{2-}}{\rightleftarrows}} [H_2Rh_{13}(CO)_{24}]^{3-}$$

$$\underset{H^+}{\overset{CH_3CN/Bu^tO^-}{\rightleftarrows}} [HRh_{13}(CO)_{24}]^{4-} \underset{H^+}{\overset{DMSO/Bu^tO^-}{\rightleftarrows}} [Rh_{13}(CO)_{24}]^{5-} \quad (28)$$

The trihydrido dianion **20** [70b] can be selectively deprotonated to the dihydrido trianion **19** [70b] or to the monohydrido tetraanion **18** [70b], whereas the pentaanion may be obtained in very strong deprotonating conditions, but has never been crystallized owing to its extremely high reactivity. The stereochemistry of the ligands remains unchanged and only minor changes in the metal-metal bond lengths have been detected in the solid state. A different situation has been found for $[HRh_{14}(CO)_{25}]^{3-}$ **22** [70b, 71], which has been obtained through a protonation reaction of the parent $[Rh_{14}(CO)_{25}]^{4-}$ **21**. In this case, the protonation of the tetranion leads to distortion in either the cluster geometry or in the carbonyl stereochemistry.

Another relevant example is the preparation of the clusters of general formula $[H_{4-n}Ni_{12}(CO)_{21}]^{n-}$ ($n = 2$ **41**, $n = 3$ **40**, $n = 4$ **39**). Dissolving $[Ni_6(CO)_{12}]^{2-}$ in a buffered acqueous solution (pH 5), brings about the slow formation of **40** [86]. The reaction is extremely rapid on lowering the pH to 2–3. In such acidic conditions, the **41** derivative was the only product obtained [86]. The two clusters **41** and **40** can be interconverted by using typical protonating-deprotonating agents, as shown in Equation 29:

$$[H_2Ni_{12}(CO)_{21}]^{2-} \underset{H^+}{\overset{CH_3OH/OH^-}{\rightleftarrows}} [HNi_{12}(CO)_{21}]^{3-}$$

$$\underset{H_2O}{\overset{DMSO/OH^-}{\rightleftarrows}} [Ni_{12}(CO)_{21}]^{4-}. \tag{29}$$

The second deprotonation step is incomplete and the tetranion **39** is not easily prepared by this way. Examples of protonation-deprotonation reactions are also present in the heterometallic Ni-Pt system. The cluster $[H_{6-n}Ni_{38}Pt_6(CO)_{48}]^{n-}$ ($n = 3$, $n = 4$ **105**, $n = 5$ **104**, $n = 6$) series, exists as an equilibrium mixture, which can be easily interconverted by controlled acid-base addition [138], as shown in Equation 30:

$$[H_3Ni_{38}Pt_6(CO)_{48}]^{3-} \underset{H^+}{\overset{CH_3CN}{\rightleftarrows}} [H_2Ni_{38}Pt_6(CO)_{48}]^{4-}$$

$$\underset{H^+}{\overset{CO_3^{2-}}{\rightleftarrows}} [HNi_{38}Pt_6(CO)_{48}]^{5-} \underset{H_2O}{\overset{OH^-}{\rightleftarrows}} [Ni_{38}Pt_6(CO)_{48}]^{6-}. \tag{30}$$

The tetranion **105** [138] and pentanion **104** [138] derivatives have been isolated in a crystalline state and characterized by X-ray analysis, showing identical metal arrangements. On the contrary, isolation of the trianion and hexanion has been hampered by their instability, when dissolved in common organic solvents.

As for the $[H_{4-n}Ni_{12}(CO)_{21}]^{n-}$ series, the dodecanuclear isoelectronic Ni-Pt series $[H_{4-n}Ni_9Pt_3(CO)_{21}]^{n-}$ ($n = 2$, $n = 3$ **101**, $n = 4$ **100**) undergoes reversible protonation-deprotonation equilibria [135]. Thus, the best synthetic procedure for **100** was found to be the deprotonation reaction of the corresponding monohydrido derivative **101** in DMSO with alkaline reagents [134].

i) *Electrochemical Activation*
In a typical cyclovoltammetric experiment, the electric potential E between two electrodes is scanned back and forth (triangular potential ramp) at different rates (usually in the range 1.0–0.01 V/s), while measuring the current passing through the solution to be analyzed. If the species under investigation reacts at a given potential, a sudden increase of the current is measured and, if the product is stable enough, in the returning scan it undergoes the reverse reaction, producing an opposite current. On the other hand, if the electron transfer is followed by a chemical transformation, the reverse wave is missing.

Thus, in a single experiment, several pieces of information can be inferred: the number of redox states available to the compound, the potentials E required to generate them and their life-time. Owing to the rapidity of the experiment, the absence of chemical reaction associated with the electron transfer and the fact that potentials are controlled at every moment, all these data can be obtained even for very reactive species, which cannot be observed under normal redox conditions. Paramagnetic radical anionic species can be detected, and their nature confirmed by electron spin resonance (ESR) spectroscopy, when the species is sufficiently stable. If two microelectrodes are immersed in a spectrophotometric cell, the transient products of each step can be characterized *in situ*, avoiding any loss of time and accidental exposure to air.

The most impressive results of this technique have been obtained with $[Pt_{24}(CO)_{30}]^{2-}$ **63**: seven different oxidation states of the cluster, with the charge ranging from 0 to -6, were electrochemically generated in a reversible manner. The IR spectra of the generated species were recorded (each spectrum requiring about 3 s), so that the ν-CO strecthing frequencies could be associated to the 'surface charge density' [154]. A striking feature of the voltammogram, common to other very large clusters, is that the electric potential E are spaced with small and large values alternately, suggesting that the electrons occupy successive cluster orbitals in a pairwise fashion.

The same experiment, carried out for $[Os_{10}C(CO)_{24}]^{2-}$, showed five different oxidation states from 0 to -4, most of them characterized by their IR spectrum [155]. Also in this case, the electrons are transferred as couples, and the two reduction processes can be separated at low temperature only. The entire process is reversible.

The electrochemistry of high-nuclearity metal carbonyl clusters has been systematically undertaken. Preliminary experimental data about the redox behaviour of several compounds, viz. $[Rh_{12}C_2(CO)_{24}]^{2-}$ **13**, $[Rh_{12}C_2(CO)_{23}]^{4-}$ **14**, $[Co_{13}C_2(CO)_{24}]^{4-}$ **4** $[Pt_{19}(CO)_{22}]^{4-}$ **62** [156] and the recent results on $[Pt_{26}(CO)_{32}]^{2-}$ **64** and $[Pt_{38}(CO)_{44}]^{2-}$ **65** [157], show the intrinsic ability of these clusters to add/lose electrons reversibily.

Unfortunately, cyclovoltammetry itself does not have a synthetic utility, since only a very thin layer of solution around the electrode is examined. However, when stable oxidation states are evidenced by cyclovoltammetry, a macroelectrolysis on a larger scale can be performed in order to generate new compounds not easily prepared with standard chemical methods, as has been reported for several electrochemically generated organometallic complexes [158–160]. Nevertheless, cyclovoltammetry gives important information about the frontier orbitals of the clusters.

The $[Ni_{11}Bi_2(CO)_{18}]^{3-}$ cluster, which is the most stable of the three $[Ni_{11}Bi_2(CO)_{18}]^{n-}$ (n = 2–4) [161] derivatives in solution, and which is obtained from the corresponding dianion by controlled reduction with sodium diphenylketyl, has been tested in a cyclovoltammetric experiment. The compound undergoes both a reduction and an oxidation process each displaying a directly associated response in the reverse scan and involving a one-electron transfer. Macroelectrolysis tests, unlike chemical reactions, showed that both dianion and tetranion redox clusters are quite stable.

Analogously, the $[Ni_{13}Sb_2(CO)_{24}]^{n-}$ (n = 2 **46**, 3 **45**) [90b] congeners displayed reversible redox changes. The dianion undergoes two further reduction processes, whereas the trianion undergoes both an oxidation and a reduction process. Coulometric tests indicated that redox changes involved one-electron step. Compound **46** showed further irreversible oxidation processes indicative of declustering situations. The substantial reversibility of the redox changes -2/-3 and -3/-4 suggests that no important geometrical rearrangements of the metal core occur during the electrochemical studies, as has also been suggested for the previous $[Ni_{11}Bi_2(CO)_{18}]^{n-}$ series.

On this ground, high-nuclearity clusters can be viewed as electron reservoirs, which can accumulate and release electrons, in some cases without structural transformations: this might have some relevance both in catalytic cycles or, hopefully, in constructing new doped materials, having isostructural metal particles with different electronic configuration in the same lattices. A recent paper reported the preparation of electrode surfaces, modified by cluster-derived metal particles, which have been used in electrochemical organic synthesis [162].

2.3. Structural Aspects

A small metallic crystallite can be considered to be a very large metal cluster or, *vice versa*, a high-nuclearity carbonyl metal-cluster can be described as a fragment of bulk lattices (viz. hexagonal or cubic close packed, body centered cubic, hexagonal, tetragonal packings, etc.) and, sometimes, as a fragment of particular bulk pseudolattices (viz. icosahedral, pentagonal packings, etc.). Generally, the adopted metal frameworks do not support the theoretical approach in which the growth of a metal particle should occur through an icosahedral rather than a close-packing sequence [18–20]. The first difference between small crystallites and molecular clusters is that a molecular cluster is coated with covalently bound CO molecules, which are essential for stabilization of the cluster. The second difference stems from the surfacial metal connectivity, which is normally lower in molecular clusters than in bulk metals. As a general consequence, in the cluster molecule, the metal-metal interactions are the most easily broken and/or deformed ones, and small changes in the metal polyhedron (viz. icosahedron – cuboctahedron interconversion), are energetically compensated by either the formation of strong M-CO interactions or a more sterically favoured distribution of the CO's on the cluster surface. Thus, a prerogative of high-nuclearity molecular metal-clusters, strictly related to the presence of ligands, is the possibility for a given metal of adopting different metal packings or even combined packings, independent of the metal packing adopted in the bulk. For example, high-nuclearity Rh or Pt metal carbonyl clusters show a high variety of metal packing arrangements (viz. h.c.p., c.c.p., b.c.c., 5-fold packings, etc.). In addition, mainly when atoms of main group elements are encapsulated in interstitial sites, more complicated metal frameworks are present in the solid state, arising from the condensation of small stable different polyhedra.

Such a fact gives rise to chained metal structures, or, in some cases, to concavities on the cluster surface. This latter case mimics the situation of steps, and unevennesses present in a carbidized metal particle [17]. In conclusion, the molecular nature of the high-nuclearity carbonyl metal-clusters plays an important role in stabilizing unusual structures forbidden in the bulk, where requirements of infinite lattices are present, and the occurrence of any particular metal polyhedron is probably due to a delicate balance between electronic needs, internal and external steric factors (viz. interstitial atoms and ligands), and crystal lattice requirements (viz. counterions, clathrate solvent), etc.

Classification of high-nuclearity carbonyl metal-clusters based on structure is quite difficult and sometimes artful, however, an initial segregation based on the overall metal geometry can be made. The following categories will be discussed:

a) *polyhedra as bulk metal fragments* (h.c.p., c.c.p., c.p., b.c.c., b.c.c./c.p.).

b) *polyhedra having five-fold symmetry* (icosahedral, pentagonal packings).

c) *polyhedra linked through metal-metal (M-M) or metal-non-metal (M-E) bonds* (chained structures).

d) *vertex-, edge-, face-shared polyhedra* (layered, condensed structures).

e) *complex nets or polyhedra*.

a) *Polyhedra as Bulk Metal Fragments*

This very large class is well represented in both homo- and hetero-metallic clusters. Compounds of this class have played an important role in strengthening the analogies between molecular clusters, small metal particles and metal surfaces [12–15]. In several cases, hydrogen atoms have been found interstitially in octahedral or square pyramidal cavities, giving rise to a slight expansion of the metal skeleton [16].

(i) Hexagonal close packing (h.c.p.). The $[H_3Rh_{13}(CO)_{24}]^{2-}$ **20** dianion was the first reported example of a cluster containing an interstitial metal atom, coordinated with twelve surrounding rhodium atoms, reproducing the coordination of a bulk atom by its nearest neighbours in a hexagonal close packed lattice [68]. The metallic skeleton can be described as a twinned-cuboctahedron, (Figure 1) consisting of a 3-layer polyhedron (3–7–3) (number of metal atoms for each layer). Such a metal polyhedron has also been found in the series of compounds having the general formula $[H_{5-n}Rh_{13}(CO)_{24}]^{n-}$ ($n = 3$ **19** [67], $n = 4$ **18** [66]). In this series the Rh-Rh bond distances decrease on going from the tri-hydrido to the di-hydrido and to the mono-hydrido derivative. The $[Rh_{17}(CO)_{30}]^{3-}$ **30** [78] trianion has a related metal skeleton, with four of the six square faces of the twinned-cuboctahedron capped by Rh atoms. The whole metallic array is a 3-layer (4–7–6) h.c.p. metal fragment (Figure 2).

The metal frameworks of $H_2Rh_{12}(CO)_{25}$ **11** [59] and $[Ir_{12}(CO)_{26}]^{2-}$ **37** [84] can be described as formed by stacking four triangles in an ABAB

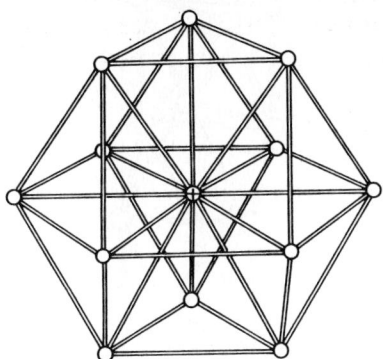

Fig. 1. The metal skeleton of $[H_3Rh_{13}(CO)_{24}]^{2-}$ **20**. ⊕ twelve coordinated Rh atom.

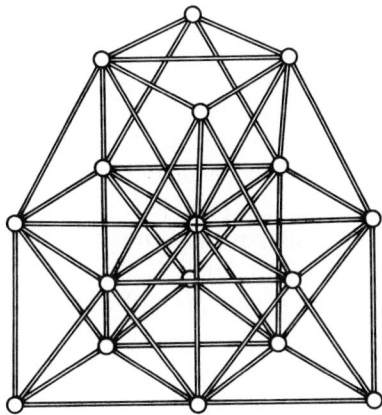

Fig. 2. The metal skeleton of $[Rh_{17}(CO)_{30}]^{3-}$ **30**. ⊕ twelve coordinated Rh atom.

sequence; the metal skeletons can be seen as deriving from the condensation of two octahedral units by connecting the six vertices of two faces with six intertriangular metal-metal bonds, generating a third octahedral unit (Figure 3). Interestingly, although they are isostructural, the compounds **11** and **37** possess a different number of cluster valence electrons (C.V.E.) (see Table 1).

The metal skeletons of the clusters with general formula $[H_{4-n}Ni_{12}(CO)_{21}]^{n-}$, ($n$ = 2 **41** [87], n = 3 **40** [87], n = 4 **39** [86]) can be described as a truncated ν_2 trigonal bipyramid elongated along the 3-fold axis, and formed by a 3-layer (3–6–3) metal frame. The interlayer Ni-Ni bond distances are longer than the intralayer Ni-Ni bond distances (2.8 Å vs. 2.4 Å). In the metal polyhedron there are two octahedral holes lodging one or two hydrogen atoms, as shown by neutron diffraction analysis. The longer

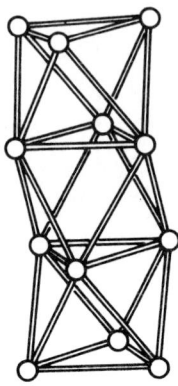

Fig. 3. The metal skeleton of $[Ir_{12}(CO)_{26}]^{2-}$ **37**.

interlayer distance has been associated with a metal-metal bond weakened by the presence of interstitial hydrides.

The largest cluster possessing a h.c.p. arrangement of metal atoms is the $[Pt_{26}(CO)_{32}]^{2-}$ **64** dianion, whose structure is composed of a 3-layer (7–12–7) metal frame, containing a fully encapsulated Pt_3 triangle [104].

The $[Rh_{11}Pt_2(CO)_{24}]^{3-}$ **92** [128] and $[Rh_{12}Pt(CO)_{24}]^{4-}$ **93** [128] clusters present a metal skeleton similar to that found in the previously discussed isoelectronic series, $[H_{5-n}Rh_{13}(CO)_{24}]^{n-}$ (n = 2–4), but with a different ligand distribution. In agreement with the bulk behaviour (viz. 'cherry' phase), a kernel of Pt atoms, *i.e.* the metal with the higher atomisation energy, is completely surrounded by a skin of Rh atoms. Thus, in **93** the single Pt atom lies at the centre of the twinned-cuboctahedron, while in **92** one Pt atom is in the centre of the twinned-cuboctahedron and a second Pt atom is disorderly located on the polyhedral surface.

An analogous situation has been found in the $[H_{4-n}Ni_9Pt_3(CO)_{21}]^{n-}$ series (n = 3 **101** [135], n = 4 **100** [134]) and in the $[Ni_6Pt_6(CO)_{21}]^{4-}$ **99** [133] anion. Compounds **99–101** are isoelectronic and isostructural with the compounds of the $[H_{4-n}Ni_{12}(CO)_{21}]^{n-}$ (n = 2–4) series, with three Pt atoms lying at the vertices of the inner triangle of the central Ni_3Pt_3 layer, in the positions of highest connectivity. Compound **99** presents three additional Pt atoms statistically disordered over the remaining positions. This latter situation is reminiscent of disordered alloys belonging to solid solutions.

(ii) Cubic close packing (c.c.p.). A centered Pt_{13} cuboctahedron with two exopolyhedral $Pt(CO)_2(PBu_3^t)$ groups, bridging diametrically related edges of the cuboctahedron, is the basic metal skeleton of the $H_xPt_{15}(CO)_8(PBu_3^t)_6$ **59** cluster [100], where $x \approx 8$ is calculated on the basis of electron counting theories for this type of polyhedron (Figure 4).

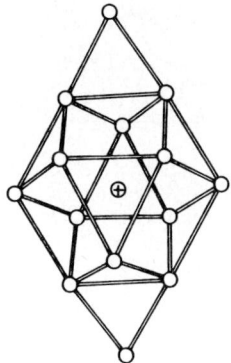

Fig. 4. The metal skeleton of $H_xPt_{15}(CO)_8(PBu_3^t)_6$ **59**. ⊕ twelve coordinated Pt atom.

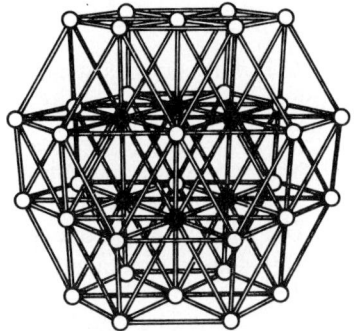

Fig. 5. The metal skeleton of $[Pt_{38}(CO)_x]^{2-}$ **65**. • interstitial octahedron of Pt atoms.

The $[Pt_{24}(CO)_{30}]^{2-}$ **63**, $[Pt_{38}(CO)_x]^{2-}$ **65** ($x \approx 44$) and $[Pt_{52}(CO)_x]^{2-}$ **66** ($x \approx 60$) dianions are closely related species, all possessing metal arrangements which correspond to c.c.p. fragments of the bulk metal. As previously reported in Section 2.2, the number and the location of the carbonyl ligands, in **65** and **66**, are still unknown. Moreover, for the time being, the presence of hydride atoms in the $[Pt_{38}(CO)_x]^{2-}$ anion cannot definitively be ruled out. The Pt_{38} core of the $[Pt_{38}(CO)_x]^{2-}$ dianion is a truncated ν_3 octahedron containing a fully encapsulated Pt_6 octahedron for an overall O_h symmetry [104]. (A metal ν_n polyhedron is defined as a polyhedron having $n + 1$ metal atoms on each edge [163]). In the Pt_{38} core, (Figure 5) four layers (7–12–12–7) of metal atoms can be recognized along the $\langle 111 \rangle$ direction, giving rise to six Pt_4 $\langle 100 \rangle$ and eight Pt_7 $\langle 111 \rangle$ faces. The whole metal framework contains 13 octahedral holes and 32 tetrahedral holes. The Pt_{38} core of the $[Pt_{38}(CO)_x]^{2-}$ cluster can be converted into the cuneane-like Pt_{24} core of the $[Pt_{24}(CO)_{30}]^{2-}$ cluster by simply removing two opposite Pt_7 $\langle 001 \rangle$ faces, symmetrically related by a C_4 axis [103]. The resulting metal arrangement

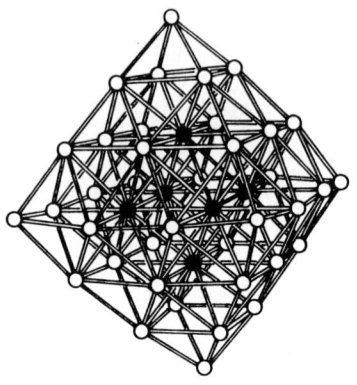

Fig. 6. The metal skeleton of $[HNi_{38}Pt_6(CO)_{48}]^{5-}$ **104**. • interstitial octahedron of Pt atoms.

shows two miniature Pt_{10} $\langle 111 \rangle$ flat surfaces which may provide a realistic model for the chemisorption, at high coverage, of carbon monoxide onto a bulk Pt single-crystal surface.

The partially characterized $[Pt_{52}(CO)_x]^{2-}$ dianion has a geometry consisting of four layers (10–16–16–10) of metal atoms, and may be considered as a mirror-twinned Pt_{38} core. The whole framework contains 10 fully encapsulated Pt atoms forming an edge-shared bioctahedron which is covered by a monolayer of 42 Pt atoms [105].

The molecular structures of the anions having general formula $[H_{6-n}Ni_{38}Pt_6(CO)_{48}]^{n-}$ ($n = 4$ **105**, $n = 5$ **104**) consist of a Pt_6 octahedral kernel fully encapsulated by a Ni_{38} ν_3 octahedron (Figure 6), thus possessing a c.c.p. metal frame [138]. In this sense, compounds **104** and **105** have a metal geometry closely related to that of the homometallic $[Pt_{38}(CO)_x]^{2-}$ cluster. As for compounds **104** and **105**, the $[Ni_{36}Pt_4(CO)_x]^{6-}$ **103** ($x \approx 44$) cluster [137] is another interesting molecular model for the 'cherry' metallic crystallites. Compound **103** has a metal skeleton formed by a truncated ν_5 tetrahedron having a tetrahedron of Pt atoms completely covered by a skin of Ni atoms. As a unique situation in cluster chemistry, a very large, almost flat, Ni_{18} $\langle 111 \rangle$ surface is present in this compound. Unfortunately the lack of good quality crystals for X-ray analysis has prevented the location of the carbon monoxide ligands.

The metal framework of the high-nuclearity carbonyl cluster $Pd_{23}(CO)_{22}(PEt_3)_{10}$ **54** can be regarded as a ν_2 centered octahedron formed by 19 Pd atoms with four additional $Pd(CO)_2(PEt_3)$ edge bridging groups (Figure 7) [96].

The largest transition metal clusters, structurally characterized, have a c.c.p. metal arrangement and, interestingly, present more $\langle 111 \rangle$ than $\langle 100 \rangle$ surfaces. As suggested by Dahl [25], probably these smooth $\langle 111 \rangle$ surfaces are

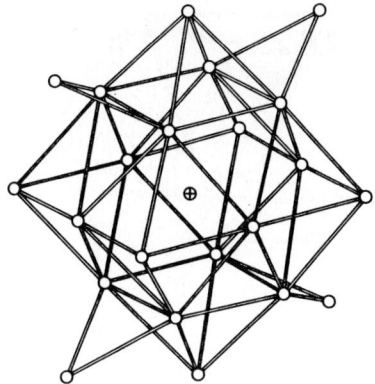

Fig. 7. The metal skeleton of $Pd_{23}(CO)_{22}(PEt_3)_{10}$ **54**. ⊕ interstitial Pd atom.

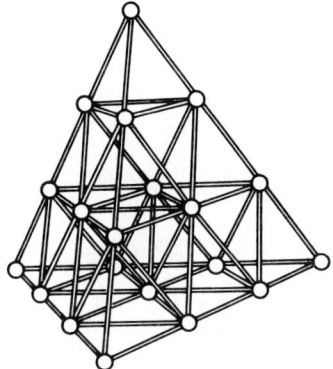

Fig. 8. The metal skeleton of $[Os_{20}(CO)_{40}]^{2-}$ **3**.

the key factor for stabilizing very large molecular structures. More recently an example of a c.c.p. metal framework has been found in Os cluster chemistry.

An overall T_d symmetry is present in the $[Os_{20}(CO)_{40}]^{2-}$ **3** dianion (Figure 8). Four layers (1–3–6–10) of metal atoms give rise to a beautiful ν_3 tetrahedron [51].

A quite interesting situation has been discovered lately in the interstitial $[Rh_{14}N_2(CO)_{25}]^{2-}$ **26** cluster. Surprisingly, the encapsulated nitrides, which often lodge in unusual and irregular sites, in this case are contained in an only slightly distorted c.c.p. metal arrangement helping to stabilize the molecule [75]. Indeed the Rh_{14} metallic array can be viewed as a 3 layer (3–7–4) metal frame which represents a centered and elongated monocapped cuboctahedron with one broken edge (Figure 9). The longer metal-metal bonds are associated with the presence of interstitial nitrogen atoms. The almost regular array is an exception in the species containing heteroatoms which normally afford distortion due to the steric requirements of the encapsulated atom.

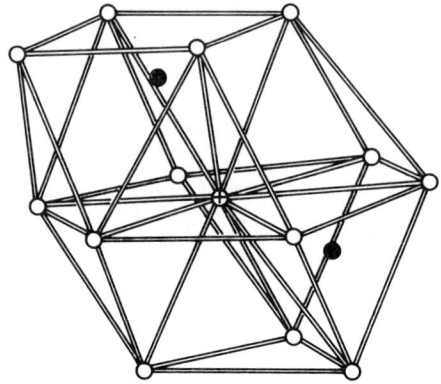

Fig. 9. The metal skeleton of $[Rh_{14}N_2(CO)_{25}]^{2-}$ **26**. • interstitial N atoms. ⊕ interstitial Rh atom.

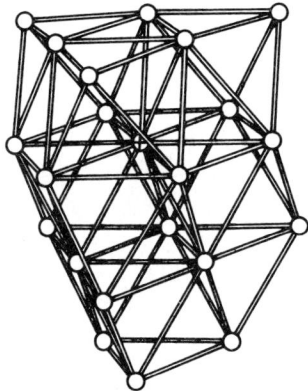

Fig. 10. The metal skeleton of $[Rh_{22}(CO)_{37}]^{4-}$ **34**. ⊕ interstitial Rh atom.

(iii) Close packing (c.p.). A mixed close packing situation, that is neither an ABAB (h.c.p.) nor an ABCABC (c.c.p.) sequence of layers, was discovered for the first time at the molecular level in the structure of the $[Rh_{22}(CO)_{37}]^{4-}$ **32** anion [80]. The metal skeleton of this cluster is formed by the superimposition of four compact layers (3–6–7–6) with an ABCB sequence, thus resulting in a mixed h.c.p./c.c.p. array of metal atoms (Figure 10). The metallic polyhedron can be described as a twinned-cuboctahedron capped on all square faces by six Rh atoms and condensed with an additional layer of three Rh atoms. Such a particular metal array is present in the lattices of lanthanum, neodymium and praseodymium bulk metals.

The metal skeleton shown by $[Ir_{14}(CO)_{27}]^-$ **38**, the largest Ir carbonyl cluster so far structurally characterized, is another remarkable example of mixed close packing [85]. Its metal frame consists of a perfect ν_2 trigonal

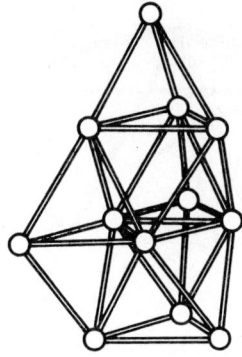

Fig. 11. The metal skeleton of $[Ir_{12}(CO)_{24}]^{2-}$ **36**.

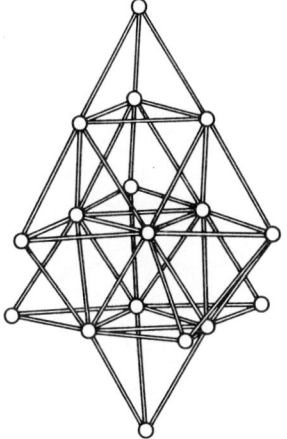

Fig. 12. The metal skeleton of $[Os_{17}(CO)_{36}]^{2-}$ **2**.

bypiramid formed by an ABCBA sequence of five close-packed layers (1–3–6–3–1) of Ir atoms. A closely related metal skeleton has been previously found in the $[Ir_{12}(CO)_{24}]^{2-}$ **36** dianion [83]. The Ir_{12} core (Figure 11) may be obtained from the Ir_{14} core by simply removing two metal atoms, one from an apical position and another from the central Ir_6 layer.

A slightly distorted tricapped ν_2 trigonal bypiramid is the basic metal framework of the $[Os_{17}(CO)_{36}]^{2-}$ **2** cluster [50]. In this case the stacking sequence is also a mixed packing (Figure 12).

The two isostructural mixed-metal clusters $[HFe_6Pd_6(CO)_{24}]^{3-}$ **70** [109] and $[Ru_6Pd_6(CO)_{24}]^{2-}$ **72** [111], display two nearly planar M_3Pd_3 layers stacked in a close packed manner. This type of metal skeleton may be alternatively seen as a hexacapped octahedron (Figure 13). The relative disposition of the metal atoms in **70** and **72** is reminiscent of the ordered superstructure

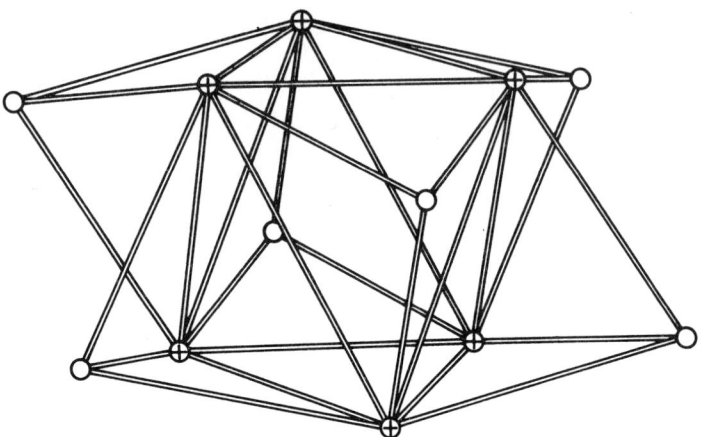

Fig. 13. The metal skeleton of $[HFe_6Pd_6(CO)_{24}]^{3-}$ **70**. ⊕ Pd atoms.

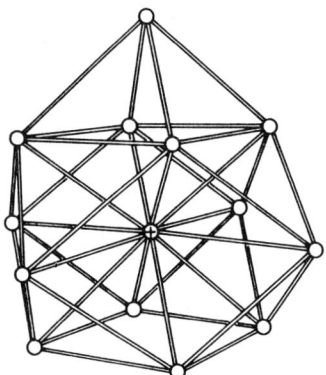

Fig. 14. The metal skeleton of $[Rh_{14}(CO)_{25}]^{4-}$ **21**. ⊕ interstitial Rh atom.

shown by bulk bimetallic M$_3$M' alloys such as Cu$_3$Au (c.c.p.) and Ni$_3$Sn (h.c.p.). The metal stereochemistry, presenting Pd atoms at the core and Fe or Ru atoms at the periphery, is only apparently in disagreement with the previously mentioned trend for the 'cherry' phases ($E_{Fe-Fe} > E_{Pd-Pd}$; $E_{Ru-Ru} > E_{Pd-Pd}$). In this case, the presence of carbon monoxide as the ligand, which forms stronger chemical bonds with Fe or Ru rather than with Pd, determines the reversal of the metal disposition. Indeed, this apparently anomalous situation is well known in bimetallic crystallites and the phenomenon is called 'chemisorption induced surface segregation' [46].

(iv) Body centered cubic (b.c.c.). A molecular geometry which approximately conforms to a body centered cubic packing of metal atoms is present in

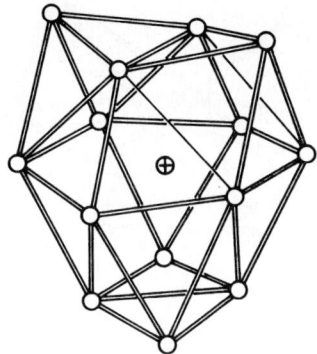

Fig. 15. The metal skeleton of $[Rh_{15}(CO)_{27}]^{3-}$ **27**. ⊕ interstitial Rh atom. Two long edges are drawn as single lines.

the $[Rh_{15}(CO)_{30}]^{3-}$ **28** trianion [76]. The metal core of this cluster can be described either as a centered cube, capped on all faces, or, alternatively, as a regular rhombic dodecahedron with eight short and six long metal-metal contacts between the central and the surface metal atoms. A similar metal array has also been found in the mixed Ru-Pt carbonyl cluster, $[Ru_9Pt_6(CO)_{28}]^{4-}$ **73** [112]. This compound is not isoelectronic with **28**, having six fewer valence electrons. Such a difference has some influence on the observed distortion of the metal skeleton, which, for **73**, may be better described as a contracted hexacapped cube. In **73**, the ratio between the two sets of distances (eight short and six long metal-metal contacts), is 1.27 instead of 1.15, as required for a regular b.c.c. packing. The resulting elongation along the six caps gives rise to a concavity for the twelve rhombic faces of the metal polyhedron.

The first reported molecular model for a b.c.c. packing was the $[Rh_{14}(CO)_{25}]^{4-}$ **21** cluster, which has a metal framework formed by an incomplete rhombic dodecahedron elongated along the four-fold axes and obtained by capping five of the six faces of a centered cube with metal atoms (Figure 14) [69b].

The $Pd_{23}(CO)_{20}(PEt_3)_8$ **55** cluster is another remarkable example of a b.c.c. metal fragment, with a nucleus consisting of a centered cube of metal atoms capped on each face by Pd atoms and on eight of the twelve edges by a $Pd(CO)_2(PEt_3)$ group [97].

(v) Mixed body centered cubic/close packing (b.c.c./c.p.). The Rh system is peculiar in showing several examples of high-nuclearity metal carbonyl clusters having a metal packing intermediate between b.c.c. and c.p.

The metal skeletons of the $[HRh_{14}(CO)_{25}]^{3-}$ **22** [71], $[Rh_{14}(CO)_{26}]^{2-}$ **23** [72a] and $[Rh_{15}(CO)_{27}]^{3-}$ **27** (Figure 15) [70a] anions may all be derived from that of $[Rh_{14}(CO)_{25}]^{4-}$ **21** [69b] by slight distortions of the metal-metal

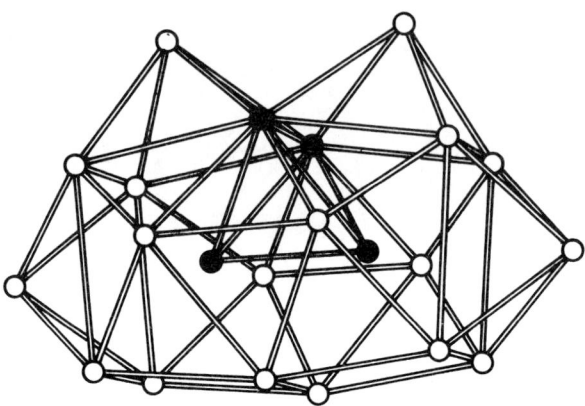

Fig. 16. The metal skeleton of $[Rh_{18}Pt_4(CO)_{35}]^{4-}$ **98**. • encapsulated tetrahedron of Pt atoms.

bond distances. Two different subunits of metal packing can be recognized in the metal frameworks of these clusters. In the b.c.c. part, a central metal atom lies in an incomplete capped cube, while in the c.p. part the central metal atom has nine or ten short metal-metal interactions in **23** and **22**, respectively. It is worth noting that the compounds **21–23** are isoelectronic. A simple substitution of a double negative charge with one carbonyl groups, in the case of $[Rh_{14}(CO)_{26}]^{2-}$, and the insertion of a hydride ligand, in the case of $[HRh_{14}(CO)_{25}]^{3-}$, affords a detectable structural change, resulting in a b.c.c.-c.p. interconversion. This phenomenon has been found in binary hydrides of transition metals [164]. The structures already discussed are all derived from the twinned-cuboctahedron present in the isoelectronic series $[H_{5-n}Rh_{13}(CO)_{24}]^{n-}$ ($n = 2$–4), by systematically capping square faces of the metal framework. Since the Rh_{14} anions are isoelectronic, the simple relationship between electron counting and structure is lost and probably the stereochemistry of the ligands plays an important role in determining the whole metal geometry.

It has been suggested that the metal cores of $[H_xRh_{22}(CO)_{35}]^{5-}$ **33** and $[H_{x+1}Rh_{22}(CO)_{35}]^{4-}$ **34** ($x \approx 3$) could derive from a mixed packing situation [81]. The metal cluster framework consist of two centered rectangular prisms sharing a common face with all except two adjacent and coplanar rectangular faces capped by eight μ_4-Rh atoms. Thus, the metal frame may be described as b.c.c. and c.p. in two different parts of the cluster. However the slight distortions observed in the rectangular prisms do not completely justify this description, and an idealized form for the metal skeleton is a b.c.c. fragment.

An analogous situation has been found in the Pt-Rh system. The $[Rh_{12}Pt_2(CO)_{26}]^{2-}$ **94** and $[Rh_{13}Pt(CO)_{25}]^{3-}$ **97** compounds, the latter being isoelectronic with the previously mentioned $[Rh_{14}(CO)_{26}]^{2-}$, have a metallic skeleton intermediate between close packing and b.c.c. [129]. Furthermore, the metal skeleton of the recently reported $[Rh_{18}Pt_4(CO)_{35}]^{4-}$ **98** is a fragment of a b.c.c. metal lattice with a semi-exposed distorted Pt_4 core (Figure 16) [132]. This cherry-like cluster has a metal cage very similar to that previously found in the homometallic $[H_xRh_{22}(CO)_{35}]^{5-}$ **33** anion, confirming once more the structural correspondence between the Rh and Pt-Rh isoelectronic derivatives, when Rh is the major component of the cluster molecule.

b) *Polyhedra Having Five-Fold Symmetry*
Five-fold symmetry is explicitly excluded in periodic Bravais lattices, due to the impossibility of filling space with pentagons, and thus it is not allowed from a crystallographic point of view [165]. Nevertheless, packing models of rigid equal spheres having an overall five-fold symmetry have been proposed by Bagley [166] and Mackay [167]. In Bagley's models a continuous systematic packing around a pentagonal bypiramidal nucleus results in an infinite array of concentric pentagons, giving rise to a three-dimensional aggregate which is not rigorously a lattice, since the spheres occupy three distinct types of place [168]: in the Bagley packing the packing efficiency is 73.42%, *i.e.* only slightly lower than in close-packing (c.c.p. or h.c.p. have 74.05%), but higher than in body-centered cubic (b.c.c. has 68.02%). In the Mackay model, an icosahedral packing of rigid equal spheres, around a central one, gives rise to concentric shells; in this case, the packing efficiency decreases on increasing the number of shells from 72.58% down to a limiting value of 68.82%. Molecular carbonyl metal-clusters can optimize these particular situations by filling the interstitial cavities with atoms having suitable radius. In some cases, the correct size for the central atom is simply obtained from its positive or negative polarization, which can be enhanced by the intrinsic acidity or basicity of the external ligands. Interestingly, any icosahedral shell has the same number of spheres as the corresponding shell of c.c.p. or h.c.p. packing. Thus, metal clusters with different arrays but identical nuclearities can exist: a ready transformation of an icosahedron into a cuboctahedron or a twinned-cuboctahedron is supposed to occur via a simple mechanism [167]. In molecular metal-clusters the type and the number of ligands are probably decisive for the choice of a given metal frame. Both close-packed and icosahedral structural models, based on concentric shells of metal atoms according to 'magic numbers', have been proposed for molecular compounds such as $Au_{55}(PPh_3)_{12}Cl_6$ [169], $Pt_{309}phen^*_{36}O_{30\pm 10}$ [170] and $Pd_{561}phen_{36}O_{200\pm 20}$ [171].

The presence of five-fold symmetry has been well known for a long time in some pseudolattices of small metallic crystallites, e.g. Fe, Co, Ni, Cu, Pt, Au in the form of vapour- or electro-deposited films [172, 173], in some metastable quasicrystalline metal alloys, e.g. Mn-Al [174], and in some structural units of amorphous metal alloys or metallic glasses [175]. At the molecular level, five-fold symmetry has been found in borane chemistry [5], in Au or Au-Ag phosphine clusters [26] and, more recently, in some carbonyl metal-clusters.

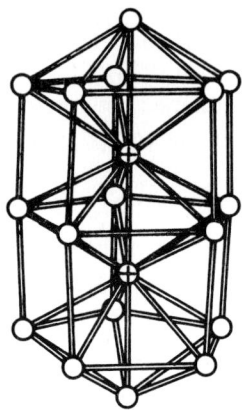

Fig. 17. The metal skeleton of $[Pt_{19}(CO)_{22}]^{4-}$ **62**. ⊕ interstitial Pt atoms.

The $[Rh_{15}C_2(CO)_{28}]^-$ **29** [77a] monoanion is the first example known in cluster chemistry adopting a metal frame based on five-fold symmetry. It has a Rh-centered pentagonal prism, with both pentagonal faces and two non-adjacent square faces capped by Rh atoms. The metal atoms capping the square faces identify two octahedral cavities which contain two carbide atoms out of range of bonding to each other. Two basal edges of the pentagonal prism are sketched with long Rh-Rh interactions.

A closely related metal framework is present in $[Pt_{19}(CO)_{22}]^{4-}$ **62** [102]. This anion has an idealized D_{5h} symmetry with a metal skeleton corresponding to a double pentagonal prism bicapped on both pentagonal faces with two interstitial Pt atoms sandwiched between the pentagonal units (Figure 17). A striking feature of this cluster is the small number of carbonyl ligands per surfacial metal atom, a situation reminiscent of the low-coverage chemisorption of carbon monoxide on a metal crystallite surface.

Several examples of icosahedral metal clusters are present in Au and Au-Ag phosphine molecular systems [26]. A certain number of metal carbonyl clusters showing either a non-centered or a centered icosahedral arrangement of metal atoms have recently been synthesized and structurally characterized [176]. Examples of this second class are the $[Ni_{12}Ge(CO)_{22}]^{2-}$ **43** and $[Ni_{12}Sn(CO)_{22}]^{2-}$ **44** dianions, which present a Ni_{12} icosahedral core centered by Ge or Sn, more or less elongated along the C_5 axis [89]. An icosahedral cavity is necessary to lodge atoms larger than carbon, and in the case of the $[Ni_{12}Sn(CO)_{22}]^{2-}$ derivative, a further elongation of the interlayer Ni-Ni bonds is unavoidable in order to accommodate the bulky Sn atom. The structure of the $[Rh_{12}Sb(CO)_{27}]^{3-}$ **17** anion is another example of five-fold symmetry, consisting of a distorted, Sb-centred, Rh_{12} icosahedron [65].

A very interesting situation is exhibited by the Ni-Sb carbonyl clusters having general formula $[Ni_{13}Sb_2(CO)_{24}]^{n-}$ ($n = 2$ **46**, $n = 3$ **45**, $n = 4$), where

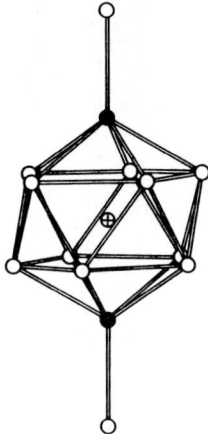

Fig. 18. The metal skeleton of $[Ni_{13}Sb_2(CO)_{24}]^{2-}$ **46**. ⊕ interstitial Ni atom. • Sb atoms.

a transition metal atom is encapsulated in an icosahedral cavity. The dianion and trianion are isostructural and their molecular structures can be described as Ni-centered, icosahedral $Ni_{11}Sb_2(CO)_{18}$ moiety with two Sb atoms lying on opposite vertices of the icosahedron, formally behaving as exotic distibine bridging ligands, each binding to a $Ni(CO)_3$ fragment (Figure 18) [90]. Recent examples of high-nuclearity metal carbonyl clusters showing a metallic icosahedral arrangement are the heptadecanuclear $Pt_{17}(CO)_{12}(PEt_3)_8$ **60** and the hexadecanuclear $Pd_{16}(CO)_{13}(PEt_3)_9$ **53** clusters. The metal framework of **60** consists of a distorted Pt_{13} centered-icosahedron with four additional caps of Pt atoms linked pairwise by Pt-Pt bonds in approximate D_{2h} symmetry [101]. An X-ray crystallographic study performed on **53** showed a $Pd_{13}(CO)_7(PEt_3)_6$ centered-icosahedral moiety with three additional $Pd(CO)_2(PEt_3)$ bridging groups, and an overall D_3 symmetry of the metal core [95].

In addition to the above more or less regular metal arrangements, more complicated metal frameworks have been found in molecular clusters. There are several factors which could have important influence in stabilizing unusual structures forbidden in the bulk where packing periodicity is required. Unusual metal arrays may arise from either the condensation or the connection of small distinct polyhedra which are stable per se because of a particular ligand distribution or because of the presence of interstitial heteroatoms of main group elements. In this way, the cluster growth can afford chains, layers, condensed structures, either by sharing atoms (vertex-, edge-, face-connections) or linking basic units through chemical bonds without sharing atoms. In addition, when interstitial heteroatoms of main group elements are present, the type of cavity, e.g. octahedral, trigonal prismatic, tetragonal (square) antiprismatic, cubic, icosahedral, and the type of condensation are both important factors in determining a certain polyhedron, particularly whether or not they

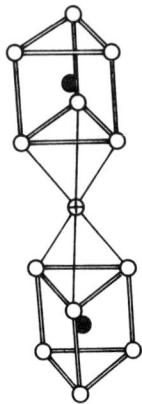

Fig. 19. The metal skeleton of [{Rh$_6$C(CO)$_{15}$}$_2$Ag]$^{3-}$ **95**. ⊕ hexacoordinated Ag atom. • interstitial C atoms.

allow a bonding interaction between the caged atoms. The molecular carbide clusters with carbon atoms fully encapsulated in a polyhedron of metal atoms are rather precise models of coordination modes for bulk transition metal carbides [177]. Furthermore, an unsystematic polyhedral growth can produce, in some cases, a final complex metallic net or polyhedron. Examples of all these situations will be discussed in the following paragraphs.

c) *Polyhedra Linked through Metal-Metal or Metal-Non-Metal Bonds*
The metal framework of the [Rh$_{12}$(CO)$_{30}$]$^{2-}$ **10** dianion is a good example of this type of linkage [58b]. The molecule is a dimeric cluster consisting of two octahedral Rh$_6$(CO)$_{15}$ units linked by a metal-metal bond supported by two μ carbonyl groups. In this sense the molecular formula could be written better as [{Rh$_6$(CO)$_{15}$}$_2$]$^{2-}$ **10**. The weakness of the linking Rh-Rh bond is the reason for the particular reactivity of this cluster. Thus, compound **10** is easily demolished upon reaction with molecular hydrogen to give the hexanuclear monohydrido derivative, [HRh$_6$(CO)$_{15}$]$^-$. The structure of the [{Rh$_6$(CO)$_{14}$(CN)$_2$}$_2${Rh(CO)$_2$}$_2$]$^{2-}$ **24** dianion [73] shows two octahedral Rh$_6$(CO)$_{14}$ fragments and two Rh(CO)$_2$ moieties linked via four dihapto CN ligands, which present the N atoms toward the Rh(CO)$_2$ fragments and the C atoms toward the cluster moieties substituting two terminal carbonyl groups on each moiety.

Another example of a molecular cluster that can be considered dimeric is the [{Rh$_6$C(CO)$_{15}$}$_2${Rh$_2$(CO)$_3$}]$^{2-}$ **25** anion. The cluster core consists of two Rh$_6$C trigonal prisms, each capped by one Rh atom onto the triangular face, giving rise to two Rh$_7$C moieties. Such units are linked by a Rh-Rh bond between the capping atoms, reinforced by three bridging carbonyl ligands [74].

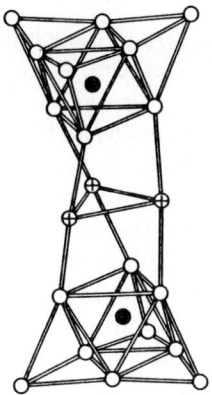

Fig. 20. The metal skeleton of $[\{Os_9C(CO)_{21}\}_2Hg_3]^{2-}$ **83**. ⊕ Hg atoms. • interstitial C atoms.

Two staggered Rh_6C prismatic units sandwiching a Ag atom are the building blocks for the structure of $[\{Rh_6C(CO)_{15}\}_2Ag]^{3-}$ **95** [130]. In this cluster the ligand arrangement is reminiscent of that of the parent $[Rh_6C(CO)_{15}]^{2-}$ dianion, with only slight deformations of the inner ligands (Figure 19).

A quite unusual situation has been found in the structure of the $[\{Re_7AgC(CO)_{21}\}_2Br]^{5-}$ **69** anion. An X-ray structure analysis showed two Re_7AgC moieties of a trans-bicapped octahedral geometry joined through a bromine interaction with the two Ag atoms [108].

The metal frame of the $[\{Ru_6C(CO)_{16}\}_2Tl]^-$ **1** cluster consists of two octahedral Ru_6C units linked together via one edge to a central Tl atom which shows an almost square-planar configuration [49b]. The tetrahedral distortion from a planar arrangement of the four Tl-Ru bonds is determined by the non-bonding interactions between the carbonyl ligands on the adjacent Ru_6C units.

A very similar metal array is present in the $[\{Ru_6C(CO)_{16}\}_2Hg]^{2-}$ **75** dianion, where the linkage between the two Ru_6C units arises via a central Hg atom [113]. An even more complicated situation is present in the molecular structure of the neutral $\{Ru_5C(CO)_{14}Cl\}_2(HgCl)_2$ **74** cluster. In this case two distorted bridged-butterfly $Ru_5C(CO)_{14}Cl$ units are linked by a Hg_2Cl_2 fragment [113].

In Ru and Os carbonyl systems, Hg atoms have been found directly bonded, as Hg, Hg_2 or Hg_3 units, to basic Ru or Os metal frames, giving rise to a broad variety of gigantic metal-core geometries. A very simple metal arrangement is present in the raft-like $\{Os_3Hg(CO)_{11}\}_3$ **78** cluster. The metal framework consists of a central Hg_3 triangle, with each of the edges symmetrically bridged by one Os atom of an Os_3 triangular unit [116]. A Hg_3 triangular unit links two tricapped-octahedral M_9 fragments (M = Ru, Os) in the $[\{Ru_9C(CO)_{21}\}_2Hg_3]^{2-}$ **76** [114] and $[\{Os_9C(CO)_{21}\}_2Hg_3]^{2-}$ **83** [121] clusters. The resulting molecules may be envisaged as being formed by two decanuclear clusters, $[M_{10}C(CO)_{24}]^{2-}$, each having lost a $M(CO)_3$ group, thus generating two tricapped octahedral fragments which are fused together by a doubly face-capping $\mu_6 - \eta^3$ Hg$_3$ unit (Figure 20).

Instead of the Hg$_3$ fragment as in **76** and **83**, a Hg$_2$ fragment, is sandwiched between two Os$_9$ clusters in the [{Os$_9$C(CO)$_{21}$}$_2$Hg$_2$]$^{2-}$ **81** dianion [119] and in the [{Os$_9$C(CO)$_{21}$}$_2$Hg$_2$]$^{4-}$ **82** tetraanion [120]. Compound **81** can be photochemically prepared starting from **83**, by reversibly losing one Hg atom, without any further degradation. Such a series of Os-Hg clusters has been recently completed with the synthesis of the [{Os$_{10}$C(CO)$_{24}$}$_2$Hg]$^{2-}$ **84** derivative. This cluster shows an unprecedented metal framework consisting of two $[Os_{10}C(CO)_{24}]^{2-}$ units linked by a single bridging Hg atom [122]. The Hg atom is four-coordinated with a geometry intermediate between tetrahedral and square planar coordination.

The glueing role of Hg is significant in stabilizing unusual situations as found in the compounds {Co$_3$Hg$_3$(CO)$_9$}$_2$Hg$_3$ **88** and Mo$_4$Hg$_4${Mo(CO)$_3$Cp}$_4$ **68**. The former has a metal skeleton which can be described as a triangular prism having Co atoms at each corner and Hg atoms at the midpoint of each edge [125]. The metal framework of **68** consists of a cuban-like cluster, having in alternate positions Hg and Mo atoms, each Hg atom bearing a Mo(CO)$_3$Cp group [107].

A linkage between two distinct polyhedra belonging to different metallic elements has been found in the Os$_{10}$C(CO)$_{24}${Au(AuPCy$_3$)$_3$} **79** cluster [117]. In this case the molecular structure presents two linked Au$_4$ and Os$_{10}$ units, with the Os frame retaining the metal geometry of the parent $[Os_{10}C(CO)_{24}]^{2-}$ ν_2 tetrahedron. The Au frame is formed by a basal [Au(PCy$_3$)]$_3$ triangle capped by one Au atom which, at the same time, bridges an outer edge of the Os$_{10}$ core. As a remarkable feature, this 'naked' Au atom is in an unusual five coordinate situation.

d) *Vertex-, Edge-, Face-Shared Polyhedra*
Another possibility of cluster growth arises from the condensation of small polyhedra via vertex-, edge-, or face-sharing, into a total polyhedron, without the necessity of a close-packing situation. The most impressive examples of this type of clusters are the inorganic oligomers of the series $[Pt_3(CO)_6]_n^{2-}$ ($n =$ 4 **57** [99b], $n =$ 5 **58** [99b], $n =$ 6 **61** [99c], and $n =$ 10). This class of compounds is based on an expanded stacking of Pt$_3$(CO)$_6$ triangles, in a slightly staggered configuration generating a sequence of twisted trigonal prismatic units. In such compounds the interlayer Pt-Pt bonds are long (3–3.1 Å) compared to the intralayer ones (2.7 Å). A non-rigidity of the metal polyhedra due to the free rotation of the triangular layers and intermolecular exchange of such triangular units, has been observed in $[Pt_{12}(CO)_{24}]^{2-}$ **57** by using ^{195}Pt- and ^{13}C-NMR spectroscopy [151]. The $[Pt_{15}(CO)_{30}]^{2-}$ **58** derivative may be considered as being formed by the fusion, via a face-sharing, of two doubly prismatic $[Pt_9(CO)_{18}]^{2-}$ fragments. The distortion from a regular prismatic

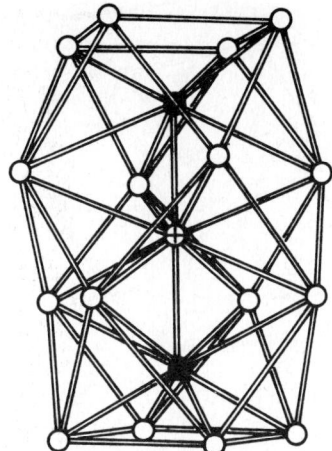

Fig. 21. The metal skeleton of $[Rh_{17}S_2(CO)_{32}]^{3-}$ **31**. ⊕ interstitial Rh atom. • interstitial S atoms.

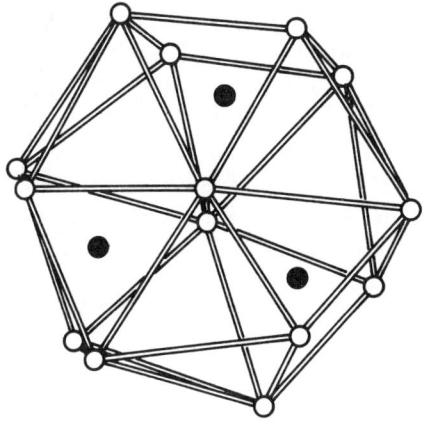

Fig. 22. The metal skeleton of $[Co_{14}N_3(CO)_{26}]^{3-}$ **6**. • interstitial N atoms.

stacking, which would afford a simple hexagonal packing, arises from a compromise between steric ligand effects and electronic effects which would favour a regular trigonal-eclipsed metal geometry. Another example of an expanded layered array of metal atoms is represented by the metal framework of the $[Rh_{17}S_2(CO)_{32}]^{3-}$ **31** trianion [79]. The metal skeleton is composed of 16 Rh atoms at the corners of four parallelly stacked squares, giving rise to three condensed square-antiprisms sharing square faces (Figure 21). The two S atoms are located at the centre of the two outer square-antiprisms, while an interstitial Rh atom lies in the centre of the inner square-antiprism.

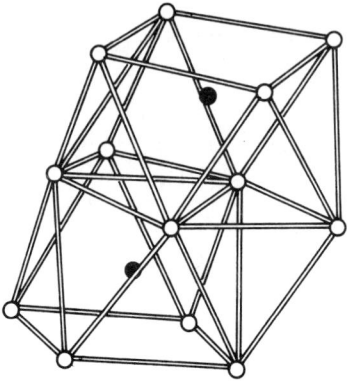

Fig. 23. The metal skeleton of $[Co_{13}C_2(CO)_{24}]^{3-}$ **5**. • interstitial C atoms.

The condensation of small polyhedra into larger polyhedra, occurs mainly when interstitial atoms stabilize a particular building-block unit. Usually in this case, no dramatic internal rearrangements occur during the condensation phase. In this sense a remarkable example is the molecular structure of the trianion $[Co_{14}N_3(CO)_{26}]^{3-}$ **6** [54]. This large cluster contains a metallic fragment composed of two superimposed centered-hexagons of Co atoms, with three interstitial N atoms located at the centre of three non-adjacent distorted trigonal-prismatic cavities (Figure 22). The metal skeleton can also be derived by a slight distortion of a fragment of simple hexagonal packing. The $Co_{14}N_3$ moiety is reminiscent of the situation present in bulk binary phases of some carbides and nitrides [177].

The $[Co_{13}C_2(CO)_{24}]^{4-}$ **4** [52] and $[Co_{13}C_2(CO)_{24}]^{3-}$ **5** [53] anions have quite similar structures, which can be described as a 3-layer (4–5–4) stack of metal atoms (Figure 23). However, the metal frames are better described as two trigonal prisms sharing one vertex, mutually rotated so that the metal atoms bonded to the shared Co atom attain a capping position over a square face of the other prism. Two further Co atoms cap a second square face on each prism, giving rise to a final metal skeleton possessing two squares in the outer layers and three fused triangles in the central one. The carbido atoms are lodged in the trigonal prismatic cavities well out of bonding distance from each other. This unusual geometry provides an example of the closest packing allowed in the presence of trigonal prismatic units, recalling the structure of cementite.

Condensed systems, formed by trigonal prismatic units stabilized by interstitial C atoms, can be recognized in the molecular structures of three isoelectronic and almost isoskeletal compounds $[Rh_{12}C_2(CO)_{24}]^{2-}$ **13** [61], $[Rh_{12}C_2(CO)_{23}]^{4-}$ **14** [62] and $[Rh_{12}C_2(CO)_{23}]^{3-}$ **15** [63]. Indeed the three metallic polyhedra can be described as a 3-layer system of two outer squares and a central rhombus, packed in such a way as to form two trigonal pris-

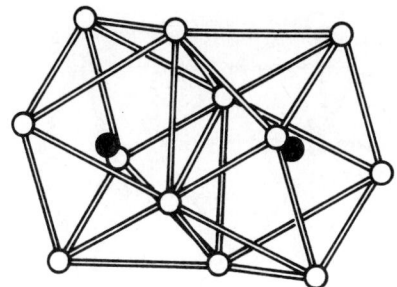

Fig. 24. The metal skeleton of $[Rh_{12}C_2(CO)_{24}]^{2-}$ **13**. • interstitial C atoms.

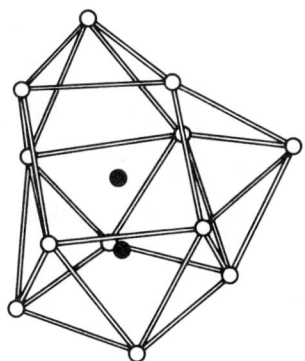

Fig. 25. The metal skeleton of $[Rh_{12}C_2(CO)_{25}]$ **12**. • interstitial C atoms.

matic cavities, rotated by 90° and crossing an edge belonging to a square face. Two interstitial carbon atoms are located within the trigonal prismatic cavities (Figure 24). The molecular structures clearly show that this series of anions originates from the condensation of two Rh_6C moieties of the parent $[Rh_6C(CO)_{15}]^{2-}$ dianion.

A different situation of polyhedra condensation, not recognizable in any close-packing array, has been found in the neutral $Rh_{12}C_2(CO)_{25}$ **12** dicarbido cluster [60]. The metal skeleton is strictly related to that of the $[Co_{11}C_2(CO)_{22}]^{3-}$ anion, showing a polyhedron which derives from a square-face sharing between a tetragonal antiprism and a trigonal prism, with two Rh atoms capping another square face of the prism and an external face of the resulting polyhedron (Figure 25). The two connected polyhedra give rise to a large well defined cavity containing a C_2 unit having a C-C distance shorter than a single C-C bond distance (1.48 Å vs. 1.54 Å). The C atoms occupy both tetragonal antiprismatic and prismatic cavities.

A similar metal array, arising from the condensation of a trigonal prism with a square antiprism by sharing a square face, has been discovered in the $[HRh_{12}N_2(CO)_{23}]^{3-}$ **16** cluster [64]. In this case, however, a marked distortion of the square antiprismatic moiety is present. The resulting polyhedron

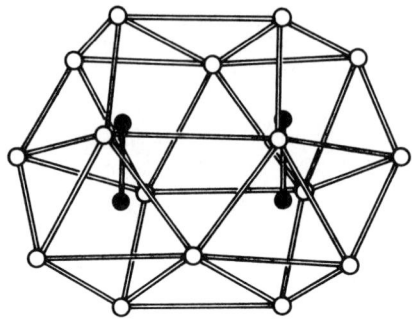

Fig. 26. The metal skeleton of $[Ni_{16}(C_2)_2(CO)_{23}]^{4-}$ **49**. • interstitial C atoms.

is further capped by two Rh atoms. Two interstitial N atoms lie in the two holes mentioned.

A progressive capping of $[Ni_{10}C_2(CO)_{16}]^{2-}$ affords the metal arrangement of the $[Ni_{12}C_2(CO)_{16}]^{4-}$ **42** [88]. The metal polyhedron of $[Ni_{10}C_2(CO)_{16}]^{2-}$ may be seen as deriving from the condensation of two μ_4-Ni capped trigonal prisms sharing a square face. The resulting polyhedron is a 3-layer (3–4–3) stack of metal atoms, which by further addition of two Ni atoms onto the two outer layers, gives rise to the observed final array. For the particular situation of the inner metal layers, the two interstitial carbide atoms are each coordinated to seven metal atoms, and, in addition, have a very short interatomic separation (1.43 Å).

A different pathway of cluster growth may be recognized in the $[Ni_{16}(C_2)_2(CO)_{23}]^{4-}$ **49** tetraanion [92]. Its metal frame consists of a unique hexadecanuclear truncated ν_2 octahedron missing the interstistial metal atom (Figure 26) but encapsulating two separated C_2 moieties having a very short interatomic C-C separation (1.38 Å). The unique large cavity of the resulting polyhedron is occupied by four interstitial carbide atoms linked in pairs.

A similar edge-sharing condensation has been found, very recently, in the heterometallic carbido species, $Rh_{10}C_2(CO)_{18}(AuPPh_3)_4$ **90** and $Rh_{10}C_2(CO)_{20}(AuPPh_3)_4$ **91** [127], which are reminiscent of the previously reported $[Ru_{10}C_2(CO)_{24}]^{2-}$ species [141]. Compounds **90** and **91** possess the same metal skeleton, consisting of two edge-sharing octahedra of Rh atoms containing an interstitial carbide atom. In both compounds, four μ_3-$[AuPPh_3]^+$ groups are coordinated to alternate faces of the bioctahedron, the main difference between the two structures being the number and the coordination mode of the carbonyl ligands.

The metal core of $Os_6Pt_7(CO)_{21}(cod)_2$ **77** [115] is also based on an edge-sharing bioctahedron not supported by interstitial atoms, with the addition of three capping metal atoms. This low-symmetry arrangement (Figure 27) is required to lodge the bulky cod ligands and the large number of CO's.

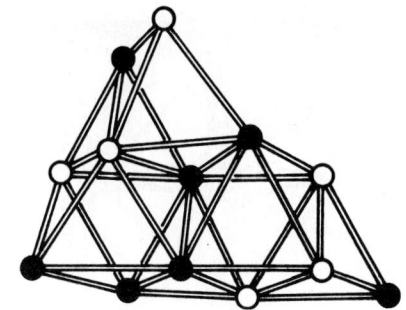

Fig. 27. The metal skeleton of $Os_6Pt_7(CO)_{21}(cod)_2$ **77**. ● Pt atoms.

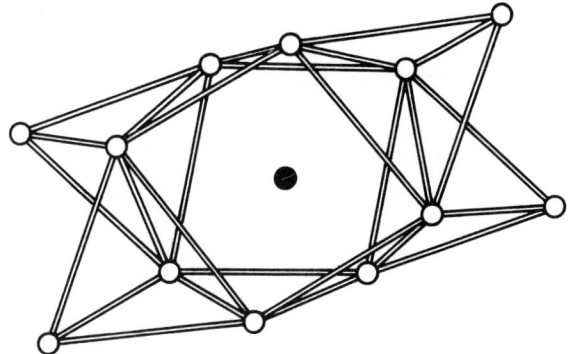

Fig. 28. The metal skeleton of $[Co_3Ni_9C(CO)_{20}]^{3-}$ **86**. ● interstitial C atom.

The $[Os_{11}C(CO)_{27}\{Cu(NCCH_3)\}]^-$ **80** monoanion is a notable example of cluster growth around a central metallic core, containing an interstitial atom of a main group element [118]. This cluster has an unusual Os_{11} core geometry which may be described as a bicapped square pyramid sharing its tetragonal face with a trigonal prism, which is also bicapped on its triangular faces and contains the carbide atom. The $[Cu(NCCH_3)]^+$ group unsymmetrically caps one of the two butterfly-like Os_4 surfaces formed by the remaining uncapped faces of the square pyramid and the capping atoms of the trigonal prism. Such a fusion between a square pyramid and a trigonal prism has also been observed in the $Rh_8C(CO)_{19}$ cluster [77b]. Usually, the condensation of the above cationic fragments into anionic carbonyl species does not affect the metal skeleton since no extra skeletal electrons are added.

The mixed Co-Ni metal carbonyl clusters, i.e., $[Co_3Ni_9C(CO)_{20}]^{3-}$ **86** [124], $[Co_3Ni_9C(CO)_{20}]^{2-}$ **87** [123] and $[Co_2Ni_{10}C(CO)_{20}]^{2-}$ **85** [123], possess the same metal frame based on a distorted square antiprismatic core, tetracapped on two alternate pairs of adjacent triangular faces (Figure 28). The two square faces are better described as parallelograms having alternate angles of about 100° and 80°, and two opposite edges elongated with respect

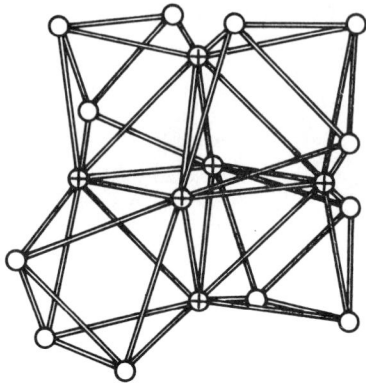

Fig. 29. The metal skeleton of $[Au_6Ni_{12}(CO)_{24}]^{2-}$ **102**. ⊕ octahedron of Au atoms.

to the other two. The four elongated edges of the prism correspond to the four capped interlayer triangular faces. The carbon atom lies at the centre of the square antiprismatic cavity. In compounds **87** and **85**, the tetragonal faces are better described as two fused non-coplanar triangular faces, the average dihedral angle being circa 160°. The Co and Ni atoms cannot be distinguished by X-ray analysis; on the basis of the metal connectivity and the number of ligands per atom, the four capping metal atoms can be labelled as Ni.

An unprecedented metal core has been found in the $[Au_6Ni_{12}(CO)_{24}]^{2-}$ **102** dianion [136]. This cluster represents the first molecular example of a Au-Ni bimetallic bonded species, The metal framework of **102** is an 18-vertex polyhedron formed by five face-fused metal octahedra (Figure 29). Alternatively, the idealized T_d geometry can be considered as arising from the condensation of four $Ni_3(CO)_6$ fragments in an antiprismatic orientation on four alternate faces of a central Au_6 octahedron. The $[Au_6Ni_{12}(CO)_{24}]^{2-}$ cluster may be considered as a segregated bimetallic phase in which the central kernel is surprisingly composed of the metal having the lower M-M bond energy. Also in this case the metal having the higher M-CO bond energy is segregated onto the surface. This situation, previously found in $[HFe_6Pd_6(CO)_{24}]^{3-}$ and in $[Ru_6Pd_6(CO)_{24}]^{2-}$, may be envisioned as a molecular analogue of the chemisorption induced surface segregation found in bimetallic crystallites [46].

Another cluster resulting from the condensation of external fragments onto an inner homometallic polyhedron, and thus mimicking a segregated bimetallic phase, is the recently characterized paramagnetic $[Fe_8Ag_{13}(CO)_{32}]^{4-}$ **71** tetraanion [110]. The anion contains a centered-cuboctahedron of Ag atoms, *i.e.*, a c.c.p. fragment of the bulk metal (Figure 30). All the triangular faces of the cuboctahedron are capped by $Fe(CO)_4$ groups, giving an overall O_h symmetry.

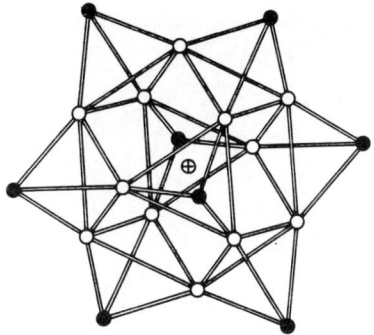

Fig. 30. The metal skeleton of $[Fe_8Ag_{13}(CO)_{32}]^{4-}$ **71**. ⊕ interstitial Ag atom. • Fe atoms.

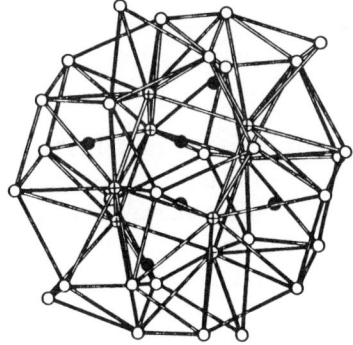

Fig. 31. The metal skeleton of $[HNi_{38}C_6(CO)_{42}]^{5-}$ **52**. • interstitial C atoms. ⊕ encapsulated cube of Ni atoms.

e) Complex Nets or Polyhedra

High-nuclearity metal carbonyl clusters having a metal framework belonging to none of the previously discussed categories, are known mainly for molecular compounds containing interstitial carbide atoms. In some cases a formal correlation between molecular carbido clusters and binary M-C phases has been found, with respect to both the carbide atom coordination and the Metal-C bonding interaction. The best example of this type of correlation is represented by the largest carbido carbonyl cluster so far structurally characterized, viz. $[HNi_{38}C_6(CO)_{42}]^{5-}$ **52** (Figure 31) [94]. The inner $Ni_{32}C_6$ core of this derivative is strictly related to a fragment of the $Cr_{23}C_6$ binary phase deriving from a three-dimensional alternation of Cr_{13} cuboctahedra and $Cr_{32}C_6$ truncated octahedra fused together along the square faces. The structure of the $Cr_{32}C_6$ moiety is identical to that of the $Ni_{32}C_6$ moiety, and consists of an inner empty cube capped on all square faces by six carbide atoms which are caged inside six square-antiprismatic cavities originating from the inner cube, in this way giving rise to a sort of truncated ν_3 octahedron. The whole

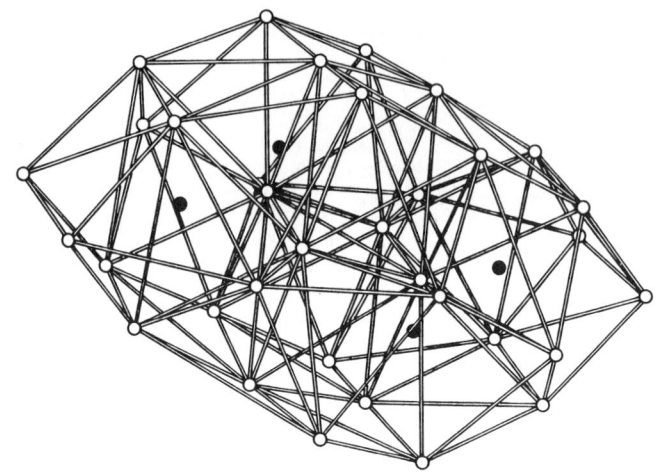

Fig. 32. The metal skeleton of $[HNi_{34}C_4(CO)_{38}]^{5-}$ **50**. ● interstitial C atoms.

polyhedron is reminiscent of the structure of $[Pt_{38}(CO)_x]^{2-}$, but, unlike the latter, the $Ni_{32}C_6$ moiety does not have a close-packed inner core. The metal polyhedron presents eight, clearly concave, hexagonal faces, six of which are μ_3-capped by Ni atoms. Since the Ni_3C hexagonal phase is the only Ni-C phase characterized so far, the preparation of new high-nuclearity carbido carbonyl clusters suggests the possibility of a molecular approach to new metal-carbon binary phases.

The closely related metal frames of $[HNi_{34}C_4(CO)_{38}]^{5-}$ **50** [93] and $[Ni_{35}C_4(CO)_{39}]^{6-}$ **51** [93] show the structural changes that may occur in small metal particles upon carbidization (Figure 32) [17]. The extremely complicated $Ni_{34}C_4$ framework consists of a Ni_{20} cubic close-packed kernel enveloping four trigonal prisms, lodging the carbide atoms, surrounded by four pentagonal bipyramids and two poly-tetrahedral moieties, which, as in the case of $[HNi_{38}C_6(CO)_{42}]^{5-}$, derives from the necessity of partially removing surfacial butterfly-like concavities. This situation probably is a compromise between a suitable accommodation for the carbide atoms in cavities larger than octahedral ones without completely losing the close-packing arrangement for the metal atoms.

The largest metal frame of Rh atoms has been found in $[Rh_{23}N_4(CO)_{38}]^{3-}$ **35** [82]. Its irregular metal skeleton is stabilized by the presence of four interstitial nitrogen atoms each connected to five Rh atoms. The two interstitial Rh atoms are connected to each other (2.571 Å) (Figure 33).

The molecular structure of the largest Pd carbonyl-phosphine cluster, $Pd_{38}(CO)_{28}(PEt_3)_{12}$ **56** [98], is another example of a very irregular metal core. This cluster has been described as a 4-layer fragment of a distorted cubic close packing with an interstitial distorted Pd_4 tetrahedron (Figure 34). However, the metal skeleton is better envisioned as a Pd_{24} atom arrangement

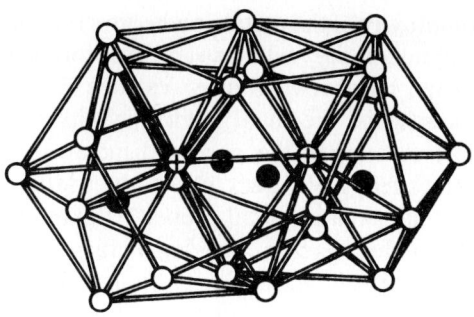

Fig. 33. The metal skeleton of $[Rh_{23}N_4(CO)_{38}]^{3-}$ **35**. ● interstitial N atoms. ⊕ interstitial Rh atoms.

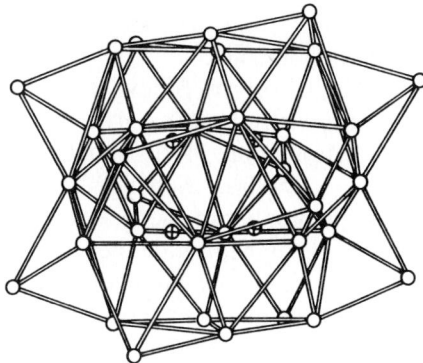

Fig. 34. The metal skeleton of $Pd_{38}(CO)_{28}(PEt_3)_{12}$ **56**. ⊕ interstitial Pd atoms.

forming approximately four planar layers with two extra interlayer Pd atoms, for an approximate overall 5-layer (4–8–2–8–4) disposition of metal atoms. The resulting Pd_{26} core is completed by 12 capping Pd atoms each bearing a PEt_3 ligand. As in previous examples, the interesting feature is that the metal skeleton tends to minimize its surfacial energy and to form a spherical metal polyhedron. This tendency is is also found in polynuclear metal aggregates occurring in amorphous metals after fast cooling of liquid droplets [175].

2.4. Electron Counting for Clusters

Many different electron counting rules, based either on empirical observations or on theoretical bases, have been successfully developed to relate the geometry, the reactivity and the spectroscopic properties of clusters. However, due to the wide variety of metal frameworks, adopted by large metal clusters, a unifying theory which accounts for all the different variations observed is still far from present realization.

An increasing complexity of the metal polyhedron is a natural consequence of an increasing metal nuclearity, with less and less correspondence between observed and calculated number of Cluster Valence Electrons (C.V.E.). As a useful guide for synthesis, the ideal theory should be easily applicable to all metals, be able to explain the observed metal geometries and then to predict the metallic arrays of new clusters. In addition, a good theory should account for the unusual physical properties exhibited by large metal clusters, resulting from the spacing and degeneracy of the energy levels close to the Highest Occupied Molecular Orbital (H.O.M.O.). Magnetism, semiconductivity, superconductivity, nucleation and melting behaviour, latent heat, optical properties and light scattering are somehow dependent on the size of a metal particle and are related to its energy levels and the occupancy of its frontier orbitals. A successful model, which correctly depicts the bonding abilities of both surfacial and interstitial metal atoms might surely be extended to small metal particles and bulk metals.

In the oldest and simplest theory by Sidgwick [178], *i.e.* the Effective Atomic Number (E.A.N.) rule, each edge of a polyhedral cluster is considered as a localized 2-centre 2-electron bond. In this situation, the total number of cluster valence electrons, N_{CVE}, in a transition metal cluster, is given by $N_{CVE} = 18N - 2E$, where N is the number of metal atoms in the cluster, and E is the number of edges of the polyhedron defined by the metal atoms. An excess of electrons with respect to E.A.N. predictions usually results in bond lengthening; viceversa, a defect of electrons results in an unsaturated cluster. However, large, polyconnected metal frames always contain more electrons than required by the E.A.N. rule, since the metal-metal bonds cannot be simply described by such a localized 2-centre 2-electrons approach.

To rationalize the disagreement between observed and calculated valence electron numbers using the E.A.N. rule, Wade [179] and Mingos [180] developed a new approach, i.e. the Polyhedral Skeletal Electron Pair Theory (P.S.E.P.T.), which describes the bonding in clusters as being delocalized over the whole molecule. The P.S.E.P.T. was originally applied to borane and carborane clusters [181] and later extended to transition metal clusters. In many cases, this approach allows prediction of the shape and the number of filled molecular orbitals for a given cluster as well as their mode of transformation during oxidation and reduction reactions. Any convex cage of N transition metal vertices, limited by triangular faces (*closo*-polyhedron), requires $7N + 1$ filled molecular orbitals. The metal skeleton is held together by $N + 1$ electron pairs, while $6N$ electron pairs are involved in metal-ligand bonding. *Nido-* or *arachno-*polyhedra can be formally derived by removal of one or two vertices from the *closo*-polyhedron having $N + 1$ or $N + 2$ vertices, respectively. A *nido*-polyhedron based cluster has $7N + 2$ filled molecular orbitals, while an *arachno*-polyhedron based cluster has $7N + 3$ filled molecular orbitals. As a consequence, it can be shown that the number of Cluster Valence Molecular Orbitals (C.V.M.O.) decreases by six for each

TABLE 3

Electronic contribution for a shared fragment

Shared unit	Subtract
Vertex	18e
Edge	34e
Triangular face	48e
Triangular face	50e[a]
Square face	62e[b]
Square face	64e[c]
'Butterfly face'	62e

[a] When both parent polyhedra are deltahedra and have nuclearity of six or greater.
[b] When either of the parent polyhedra is a deltahedron.
[c] When both parent polyhedra are three connected polyhedra.

vertex removed, consistent with the removal of the metal-ligand bonding orbitals and not of the metal-metal bonding orbitals of the vertex unit. Therefore, both the *nido-* and *arachno-*clusters have the same number of skeletal bonding electron pairs as the parent *closo-*polyhedron. The formation of *nido-* and *arachno-*clusters from *closo-*clusters is related to the formulation of the capping principle. Indeed several cluster compounds have structures which can be described in terms of polyhedra obtained by successive capping of a parent polyhedron. The capping principle, extended to a general capping situation, states that a capped polyhedron has the same number of bonding skeletal molecular orbitals as the uncapped parent polyhedron. Consequently, the electron counting for this capped polyhedron is generally $12x$ greater than for the parent polyhedron, where x is the number of capping metal atoms. The addition of 12 electrons per each metal cap does not alter the basic metal skeleton, since the extra electrons are localized in the capping metal-ligand interactions and not in the metal-metal interactions [180e]. However, for symmetry reasons, multiple capped polyhedra can have extra occupied molecular orbitals and two or four extra electrons may be expected. This situation has been found in clusters having, as metal frameworks, a tricapped trigonal prism or a hexacapped octahedron. The increase of 14 electrons, in the electron counting, upon addition of an edge-bridging group leads to the conclusion that the predicted C.V.E. counting for edge-bridged polyhedra are $A + 14x$, where A is the C.V.E. counting for the unbridged parent polyhedron and x is the number of bridging metal atoms.

TABLE 4

Summary of electron counts within P.S.E.P.T.

Polyhedral type	Electron count
Deltahedron *closo*	$14N + 2$
Deltahedron *nido*	$14N + 4$
Deltahedron *arachno*	$14N + 6$
Face-capped polyhedron	$A + 12x$
Edge-bridged polyhedron	$A + 14x$

N is the metal cluster nuclearity.
A is the electron count of the uncapped or unbridged parent polyhedron.
x is the number of capping or bridging groups.

In recent years several large metal clusters have shown metal frames which result from condensation, through vertex-, edge- or face-sharing, of small polyhedra such as tetrahedra, octahedra or trigonal prismatic units, etc. In this case, the number of C.V.E. of the resulting polyhedron can be obtained by adding the number of C.V.E. of the two fused units, minus the electronic contribution of the shared unit [180f] (see Table 3). In this sense the condensation principle is a more general case of the capping principle in that addition of a capping atom onto a triangular face can be also seen as a condensation of a tetrahedron with a *closo*-polyhedron.

For example, the number of C.V.E. for a vertex sharing bioctahedron may be calculated by adding the contribution of two separated octahedra and subtracting the contribution of a single atom (shared atom). The agreement between observed and predicted number of C.V.E. is normally fairly good, with maximum discrepancies of ±2 or ±4 electrons. Most failures are found to involve polycondensed moieties; in these clusters, it may be that an empty, antibonding molecular orbital of one of the units is stabilized by overlapping with the molecular orbitals of the condensed fragments, becoming accessible to bonding electrons; in other words, some orbitals of a simple, isolated, polyhedron may have different bonding ability when the same unit is only a component of a larger molecule of different symmetry. The electron count within P.S.E.P.T. is summarized in Table 4.

The number of bonding and antibonding orbitals and therefore the number of valence electrons for a given cluster was developed using semi-empirical Extended Hückel Molecular Orbital (E.H.M.O.) calculations. Taking into account the ligand arrangement and the resulting symmetry, this method should provide the correct number of C.V.M.O. The main disadvantage over the other approaches is the absence of predictive power for unusual geometries, such as found unexpectedly for the clusters containing interstitial elements. This method was originally introduced by Lauher [182], who considered about 20 different geometries of naked rhodium clusters and stated that if the ligands, the metal atoms and the free charges can satisfy those electronic requirements, then a stable cluster can be formed, irrespective of the actual metallic element or ligand disposition. The same approach was extended by Ciani and Sironi [147] to more than 100 geometries including either small or large carbonyl clusters, having mainly close-packed arrays. The results of the calculations showed that high-nuclearity metal clusters need $6N + X$ filled C.V.M.O., where N is the number of metal atoms and X is an adjustment factor ranging from 5 to 11 depending on the geometry, the vast majority of frames requiring $X = 7$. Also, the E.H.M.O. method can lead to wrong predictions, if the frontier orbitals are not well separated by a large gap between the Highest Occupied Molecular Orbital (H.O.M.O.) and Lowest Unoccupied Molecular Orbital (L.U.M.O.). The above theory does not take into consideration that:

(i) It is impossible to account exactly for small deformations of the metal frames observed in the solid state structures, such as variations in bond lengths or angles; these can be indicative of different metal-metal bond orders, localized electron density, steric hindrance of the surrounding ligands or small crystal packing effects.

(ii) Different metallic elements have different electronic requirements: Pt clusters are usually short of electrons since, as in the case of mononuclear square planar complexes, as much as one lone pair per metal atom can remain unavailable for bonding. This feature is not well documented for the other noble metals, which also form unsaturated 16 electron complexes. Electron counting can also be altered by ligands with anomalous behaviour, such as trisubstituted phosphine, electronegative elements, cyclopentadienyl; when these ligands prevail, exceptions to the rules frequently occur.

(iii) Close spacing of the frontier electronic levels, which may have only weakly bonding (or antibonding) character, can afford different numbers of C.V.E. with only minor deformations in the metal frames, because the variation in electron density can be distributed over the whole molecule.

The number of C.V.M.O. has been calculated by Teo [183] using the Topological Electron Counting (T.E.C.) approach, which combines Euler's Theorem with the E.A.N. rule. The Euler Theorem states that the number of edges (E) of a polyhedron possessing V vertices and F faces is given by the formula:

$$E = V + F - 2.$$

The number of C.V.M.O., according to T.E.C., can be calculated using the following expression:

$$N_{\text{CVMO}} = 8V - F + 2 + X$$

where X is the number of electron pairs in excess of the E.A.N. rule or, in other words, the number of missing antibonding orbitals.

Within the T.E.C. rule, Teo has proposed the general formula:

$$T_n = 6S_n + B_n$$

for calculating the topological electron pairs of a close-packed high-nuclearity metal-cluster, where T_n is the number of topological electron pairs, S_n is the number of surface metal atoms and B_n is the number of shell electron pairs which is coincident with the electronic contribution of the interstitial cluster [163]; in the same work the nuclearities of gigantic clusters have been calculated for many close-packed simple polyhedra.

The same approach has been proposed independently by Mingos [146], considering separately the contributions of bulk and surface metal atoms to the total number of C.V.E..

In Mingos' hypothesis high-nuclearity metal clusters are divided into three different classes on the basis of the ligand to surface metal ratio (L : M), which reflects the relative importance of the radial and tangential contributions to the skeletal bonding.

Class a: clusters with L : M $\approx$ 1.3, where radial bonding predominates, are characterized by an electron count of $12n_s + \Delta_i$, where n_s is the number of surface metal atoms and Δ_i is the electron count characteristic of the interstitial group of metal atoms.

Class b: clusters with L : M $\approx$ 1.3 - 2.0, where tangential bonding partially contributes, have an electron count of $12n_s + 24$; extra tangential orbitals of the cluster are of sufficiently low energy to be occupied.

Class c: clusters with L : M $\geq$ 2, where radial and tangential bonding are both important, are characterized by an electron count of $14n_s + 2$; this situation is reminiscent of the previously discussed P.S.E.P.T. for *closo* polyhedra, considering only the surface metal atoms.

Hume-Rothery and Westgren first observed that in many alloy systems, phases of similar crystal structures are formed with the same ratio of valency electrons to atoms, viz. electron concentration (e.c.).

Taking in consideration this idea and stressing the connection between high-nuclearity molecular metal-clusters and bulk metals, Teo [184] proposed a related method for counting electrons in large clusters, suggesting a formula that provides a way of partitioning the total number of electrons among bulk and surface metal atoms: calculated values are usually fractional and in fair agreement with the observed value.

At the present stage of knowledge, all approaches can be used to attempt a first estimation of the 'correct' electron number of new large clusters, together with the Ciani-Sironi rule, which, for such large molecules seems to be very pertinent. A summary of the above rules is shown in Table 5.

A general treatment for electron counting in cluster compounds which could encompass all situations including metal clusters containing interstitial

TABLE 5
Summary of electron counting rules

Rule	Cluster size
E.A.N. (Sidgwick)	small ($N \leq 4$)
P.S.E.P.T. (Wade, Mingos)	medium ($N \leq 5$–12)
T.E.C. (Teo)	medium ($N \leq 5$–12)
E.H.M.O. (Lauher, Ciani-Sironi)	large ($N \geq 12$)
P.S.E.P.T. (Mingos)	large ($N \geq 12$)
Hume-Rothery (Teo)	large ($N \geq 12$)

atoms of main group elements is not available at the moment. In this last case, the number of C.V.M.O. when interstitial carbon or nitrogen atoms are encapsulated in cavities larger than octahedral can be predicted by adding for each interstitial atom, 2 electron pairs to the $6N+7$ number of C.V.M.O. These extra electrons can be rationalized by considering that trigonal prismatic (90 electrons) or square antiprismatic (118 electrons) cages need two C.V.M.O. more than the corresponding more connected octahedral (86 electrons) or D_{2d} dodecahedral (114 electrons) cavities [185].

The comparison between observed and calculated C.V.E. for selected compounds, showing compact or condensed metal frameworks, is reported in Table 6. The C.V.E., using both the Mingos and Teo approaches, have been calculated differentiating the situation without interstitial metal atoms [180, 183] from that with interstitial metal atoms [146, 184], and the situation of close-packed structures from that of complex structures. The C.V.E. as from Ciani-Sironi's approach have been calculated using C.V.M.O. = $(6N+7)$ for metal frameworks not reported in [147] and C.V.M.O. = $(6N+7+2N_i+N_{i-i})$ for clusters containing interstitial atoms, where N_i is the number of interstitial atoms and N_{i-i} is the number of interactions between interstitial atoms [39].

Note added in proof

Several articles have appeared after the submission of this manuscript. Mixed ruthenium-copper carbonyl clusters have been synthesized, i.e. [$H_2Ru_8Cu_7$ $(CO)_{24}Cl_3$]$^{2-}$, [$H_2Ru_{12}Cu_6(CO)_{34}Cl_2$]$^{2-}$, [$H_4Ru_{20}Cu_6(CO)_{48}Cl_2$]$^{2-}$, [186]; interstitial boron atoms are present in [{$HRu_4(CO)_{12}B$}$_2${Cu_4Cl}], [187]. The [{$Ni_6(CO)_{11}$}$_2In(In_2Br_4OH)$]$^{4-}$ has been prepared in low yields reacting [$Ni_6(CO)_{12}$]$^{2-}$ and $InBr_3$, [188]. The first example of high-nuclearity cationic carbonyl cluster, i.e. [$Pd_{14}Au_2(PMe_3)_{11}(CO)_9$]$^{2+}$, has been prepared adding [$AuPPh_3$]$^+$ to $Pd_8(CO)_8(PMe_3)_7$ [189].

TABLE 6

Electron count for high nuclearity carbonyl metal cluster

Index No.	Cluster	Obs. C.V.E.	Calc. C.V.E. Mingos	Teo$^{(\nabla)}$	Ciani–Sironi
2	$[Os_{17}(CO)_{36}]^{2-}$	210	214	220	218
3	$[Os_{20}(CO)_{40}]^{2-}$	242	254	252	254
4	$[Co_{13}C_2(CO)_{24}]^{4-}$	177	–	–	178
6	$[Co_{14}N_3(CO)_{26}]^{3-}$	196	–	–	194
11	$H_2Rh_{12}(CO)_{25}$	160	158	162	162
12	$Rh_{12}C_2(CO)_{25}$	166	168	166	166
13	$[Rh_{12}C_2(CO)_{24}]^{2-}$	166	172	164	166
17	$[Rh_{12}Sb(CO)_{27}]^{3-}$	170	170	170	170
18	$[HRh_{13}(CO)_{24}]^{4-}$	170	170	170 (170)	170
21	$[Rh_{14}(CO)_{25}]^{4-}$	180	180	180 (180)	180
26	$[Rh_{14}N_2(CO)_{25}]^{2-}$	188	184	(180)	190
27	$[Rh_{15}(CO)_{27}]^{3-}$	192	192	(193)	192
28	$[Rh_{15}(CO)_{30}]^{3-}$	198	198	> 192 (193)	198
29	$[Rh_{15}C_2(CO)_{28}]^{-}$	200	198	198	200
30	$[Rh_{17}(CO)_{30}]^{3-}$	216	216	(223)	218
31	$[Rh_{17}S_2(CO)_{32}]^{3-}$	232	230	–	232
32	$[Rh_{22}(CO)_{37}]^{4-}$	276	276	(277)	280
33	$[H_xRh_{22}(CO)_{35}]^{5-(\lozenge)}$	276	274	(275)	278
36	$[Ir_{12}(CO)_{24}]^{2-}$	158	156	156	160
37	$[Ir_{12}(CO)_{26}]^{2-}$	162	158	162	162
38	$[Ir_{14}(CO)_{27}]^{-}$	181	178	184	188
39	$[Ni_{12}(CO)_{21}]^{4-}$	166	164	166	160
42	$[Ni_{12}C_2(CO)_{16}]^{4-}$	164	164	–	166
43	$[Ni_{12}Ge(CO)_{22}]^{2-}$	170	170	170	170
44	$[Ni_{12}Sn(CO)_{22}]^{2-}$	170	170	170	170
49	$[Ni_{16}(C_2)_2(CO)_{23}]^{4-}$	226	–	–	222
50	$[HNi_{34}C_4(CO)_{38}]^{5-}$	438	–	–	438
51	$[Ni_{35}C_4(CO)_{39}]^{6-}$	450	–	–	450
52	$[HNi_{38}C_6(CO)_{42}]^{5-}$	494	492	–	494
53	$Pd_{16}(CO)_{13}(PEt_3)_9$	204	212	–	–
54	$Pd_{23}(CO)_{22}(PEt_3)_{10}$	294	298	–	–
55	$Pd_{23}(CO)_{20}(PEt_3)_8$	286	–	–	290
56	$Pd_{38}(CO)_{28}(PEt_3)_{12}$	460	–	–	470
57	$[Pt_{12}(CO)_{24}]^{2-}$	170	170	174	–
58	$[Pt_{15}(CO)_{30}]^{2-}$	212	212	216	–
60	$Pt_{17}(CO)_{12}(PEt_3)_8$	210	210	–	–
61	$[Pt_{18}(CO)_{36}]^{2-}$	254	254	258	–
62	$[Pt_{19}(CO)_{22}]^{4-}$	238	238	244 (236)	–

TABLE 6
(Continued)

Index No.	Cluster	Obs. C.V.E.	Calc. C.V.E. Mingos	Teo$^{(\nabla)}$	Ciani–Sironi
63	$[Pt_{24}(CO)_{30}]^{2-}$	302	300	(302)	302
64	$[Pt_{26}(CO)_{32}]^{2-}$	326	324	(339)	326
65	$[Pt_{38}(CO)_x]^{2-(\Diamond)}$	470	470	(472)	470
66	$[Pt_{52}(CO)_x]^{2-(\Diamond)}$	642	642	(643)	638
70	$[HFe_6Pd_6(CO)_{24}]^{3-}$	160	158	158	160
71	$[Fe_8Ag_{13}(CO)_{32}]^{4-}$	275	266	–	–
72	$[Ru_6Pd_6(CO)_{24}]^{2-}$	158	158	158	160
73	$[Ru_9Pt_6(CO)_{28}]^{4-}$	192	192	192	198
85	$[Co_2Ni_{10}C(CO)_{20}]^{2-}$	164	162	–	162
86	$[Co_3Ni_9C(CO)_{20}]^{3-}$	164	162	–	162
90	$Rh_{10}C_2(CO)_{18}(AuPPh_3)_4$	186	186	186	182
91	$Rh_{10}C_2(CO)_{20}(AuPPh_3)_4$	190	186	186	182
93	$[Rh_{12}Pt(CO)_{24}]^{4-}$	170	170	170 (170)	170
94	$[Rh_{12}Pt_2(CO)_{26}]^{2-}$	182	180	180 (180)	180
98	$[Rh_{18}Pt_4(CO)_{35}]^{4-}$	276	274	(275)	278
100	$[Ni_9Pt_3(CO)_{21}]^{4-}$	166	164	166	160
102	$[Au_6Ni_{12}(CO)_{24}]^{2-}$	236	236	–	–
103	$[Ni_{36}Pt_4(CO)_x]^{6-(\Diamond)}$	494	492	494 (500)	494
104	$[HNi_{38}Pt_6(CO)_{48}]^{5-}$	542	542	542 (548)	542

$^{(\nabla)}$ values in brackets are calculated with Hume–Rothery rule; $^{(\Diamond)}$ for **33** ($x \approx 3$); for **65** ($x \approx 44$); for **66** ($x \approx 60$); for **103** ($x \approx 44$).

References

1. A. Simon, *Angew. Chem., Int. Ed. Engl* **27** (1988) 160.
2. A. J. Thomson, *Metalloproteins, Part 1*, Verlag Chemie, Basel (1989) p. 79.
3. M. Moskovits, *Metal Clusters*, Wiley, New York (1986) p. 185.
4. H. W. Roesky (ed.), *Rings, Clusters and Polymers of Main Group and Transition Elements*, Elsevier, New York, (1989).
5. C. E. Housecroft, *Boranes and Metalloboranes*, Ellis Horwood Ltd., Chichester (1990).
6. B. C. Guo, K. P. Kerns and A. W. Castleman, *Science* **255** (1992) 1411.
7. F. Stoddart, *Angew. Chem., Int. Ed. Engl*, **30** (1991) 70.
8. F. A. Cotton, *Quart. Rev.* **20** (1966) 389.
9. P. Chini, *Inorg. Chim. Acta Rev.* **2** (1968) 31.
10. M. Moskovits, *Acc. Chem. Res.* **12** (1979) 229.
11. (a) B. Gates, *Metal Clusters*, Wiley, New York (1986) p. 283; (b) A. Brenner, *Metal Clusters*, Wiley, New York, (1986) p. 249; (c) P. Gallezot *Metal Clusters*, Wiley, New York (1986) p. 219; (d) P. A. Jacobs *Metal Cluster in Catalysis*, Elsevier, New York (1986) p. 357.
12. E. L. Muetterties, T. N. Rhodin, E. Band, C. F. Brucker and W. R. Pretzer, *Chem. Rev.* **79** (1979) 91.

13. E. Band and E. L. Muetterties, *Chem. Rev.* **78** (1978) 639.
14. J. M. Basset and R. Ugo, *Aspects Homog. Catal.*, Vol. 3, Kluwer, Dordrecht (1978) p. 138.
15. R. Ugo and R. Psaro, *J. Mol. Catal.* **20** (1983) 53.
16. P. Chini, G. Longoni, S. Martinengo and A. Ceriotti, *Adv. Chem. Ser.* **167** (1978) 1.
17. E. M. Dry, *Catalysis Science and Technology*, Vol. 1 Springer-Verlag, New York (1981) p. 159.
18. S. C. Davis and K. J. Klabunde, *Chem. Rev.* **82** (1982) 153.
19. J. J. Burton, *Catal. Rev. Sci. Eng.* **9** (1974) 209.
20. J. C. Phillips, *Chem. Rev.* **86** (1986) 619.
21. R. D. Adams and I. T. Horvath, *Prog. Inorg. Chem.* **33** (1985) 127.
22. E. L. Muetterties, *Pure Appl. Chem.* **54** (1982) 83.
23. (a) G. Schmid, *Aspects Homog. Catal.*, Vol. 7, Kluwer, Dordrecht (1990) p. 1; (b) G. Schmid, *Chem. Rev.* **92** (1992) 1709.
24. E. L. Muetterties and R. M. Wexler, *Surv. Prog. Chem.* **10** (1983) 61.
25. K. C. Kharas and L. F. Dahl, *Adv. Chem. Phys.* **70** (1988) 1.
26. B. K. Teo and H. Zhang *J. Cluster Sci.* **1** (1990) 155.
27. D. Levine and P. J. Steinhardt, *Phys. Rev. Lett.* **53** (1984) 2477.
28. R. Kubo, *J. Phys. Soc. Japan* **17** (1962) 975.
29. R. F. Marzke, W. S. Glaunsinger and M. Bayard, *Solid State Commun.* **18** (1976) 1025.
30. B. C. Gates, L. Guczi, H. Knozinger (eds.), *Metal Clusters in Catalysis*, Elsevier, New York (1986).
31. P. Braunstein and J. Rose, *Stereochem. Organomet. Inorg. Compounds*, Vol. 3, Elsevier, New York (1989) p. 3.
32. E. R. Corey, L. F. Dahl and W. Beck *J. Am. Chem. Soc.* **85** (1963) 1202.
33. P. Chini, G. Longoni and V. G. Albano, *Adv. Organomet. Chem.* **14** (1976) 285.
34. P. Chini, *J. Organomet. Chem.* **200** (1980) 37.
35. M. D. Vargas and J. N. Nicholls, *Adv. Inorg. Chem. Radiochem.* **30** (1986) 123.
36. B. F. G. Johnson and J. Lewis *Adv. Inorg. Chem. Radiochem.* **24** (1981) 225.
37. V. G. Albano and S. Martinengo, *Nachr. Chem. Tech. Lab.* **28** (1980) 654.
38. L. Farrugia, *Adv. Organomet. Chem.* **31** (1990) 301.
39. G. Longoni, A. Ceriotti, M. Marchionna and G. Piro, *Surface Organometallic Chemistry: Molecular Approach to Surface Catalysis*, Kluwer, Dordrecht (1988), p. 157.
40. B. F. G. Johnson (ed.), *Transition Metal Clusters*, Wiley, New York (1980).
41. D. M. Mingos and D. J. Wales, *Introduction to Cluster Chemistry*, Prentice-Hall, Englewood Cliffs (1990).
42. D. F. Shriver, H. D. Kaesz, R. D. Adams (eds.), *The Chemistry of Metal Cluster Complexes*, VCH, New York (1990).
43. P. Chini, *Gazz. Chim. It.* **109** (1979) 225.
44. J. H. Sinfelt, *Acc. Chem. Res.* **10** (1977) 15.
45. W. M. H. Sachtler and R. A. Van Santen, *Adv. Catal.* **26** (1977) 169.
46. W. M. H. Sachtler, *Vide* **164** (1973) 67.
47. M. P. J. van Staveren, H. B. Brom, L. J. de Jongh and G. Schmid, *Solid State Commun.* **60** (1986) 319.
48. W. P. Halperin, *Rev. Modern Phys.* **58** (1986) 533.
49. (a) J. S. Bradley and E. W. Hill, U.S. Patent 4, 301, 086 (1981); (b) G. B. Ansell, M. A. Modrick and J. S. Bradley, *Acta Cryst*, **C40** (1984) 1315.
50. E. Charalambous, L. H. Gade, B. F. G. Johnson, J. Lewis, M. McPartlin and H. R. Powell, *J. Chem. Soc., Chem. Commun.* (1990) 688.
51. A. J. Amoroso, L. H. Gade, B. F. G. Johnson, J. Lewis, P. R. Raithby and W. T. Wong, *Angew. Chem., Int. Ed. Engl.* **30** (1991) 107.
52. V. G. Albano, D. Braga, P. Chini, G. Ciani and S. Martinengo, *J. Chem. Soc. Dalton Trans.* (1982) 645.
53. V. G. Albano, D. Braga, A. Fumagalli and S. Martinengo, *J. Chem. Soc. Dalton Trans.* (1985) 1137.
54. S. Martinengo, G. Ciani and A. Sironi, *J. Organomet. Chem.* **358** (1988) C23.
55. L. Garlaschelli, S. Martinengo and G. Ciani, personal communication.
56. A. L. Rheingold and P. J. Sullivan, *J. Chem. Soc. Dalton Trans.* (1983) 39.
57. R. L. Sturgeon, M. M. Olmstead and N. E. Schore, *Organometallics* **10** (1991) 1649.

58. (a) P. Chini and S. Martinengo, *Inorg. Chim. Acta* **3** (1969) 299; (b) V. G. Albano and P. L. Bellon, *J. Organomet. Chem.* **19** (1969) 405; (c) S. Martinengo and P. Chini, *Inorg. Synth.* **20** (1980) 215.
59. G. Ciani, A. Sironi and S. Martinengo, *J. Chem. Soc., Chem. Commun.* (1985) 1757.
60. V. G. Albano, P. Chini, S. Martinengo, M. Sansoni and D. Strumolo, *J. Chem. Soc. Dalton Trans.* (1978) 459.
61. V. G. Albano, D. Braga, P. Chini, D. Strumolo and S. Martinengo, *J. Chem. Soc., Dalton Trans.* (1983) 249.
62. V. G. Albano, D. Braga, D. Strumolo, C. Seregni and S. Martinengo, *J. Chem. Soc., Dalton Trans.* (1985) 1309.
63. D. Strumolo, C. Seregni, S. Martinengo, V. G. Albano and D. Braga, *J. Organomet. Chem.* **252** (1983) C93.
64. S. Martinengo, G. Ciani and A. Sironi, *J. Chem. Soc., Chem. Commun.* (1986) 1742.
65. J. L. Vidal and J. M. Troup, *J. Organomet. Chem.* **213** (1981) 351.
66. G. Ciani, A. Sironi and S. Martinengo, *J. Chem. Soc., Dalton Trans.* (1981) 519.
67. V. G. Albano, G. Ciani, S. Martinengo and A. Sironi, *J. Chem. Soc., Dalton Trans.* (1979) 978.
68. V. G. Albano, A. Ceriotti, P. Chini, G. Ciani, S. Martinengo and W. M. Anker, *J. Chem. Soc., Chem. Commun.* (1975) 859.
69. (a) J. L. Vidal and R. C. Schoening, *Inorg. Chem.* **20** (1981) 265; (b) G. Ciani, A. Sironi and S. Martinengo, *J. Chem. Soc., Dalton Trans.* (1982) 1099.
70. (a) S. Martinengo, G. Ciani, A. Sironi and P. Chini, *J. Am. Chem. Soc.* **100** (1978) 7096; (b) C. Allevi, B. T. Heaton, C. Seregni, L. Strona, R. J. Goodfellow, P. Chini and S. Martinengo, *J. Chem. Soc., Dalton Trans.* (1986) 1375.
71. (a) G. Ciani, A. Sironi and S. Martinengo, *J. Organomet. Chem.* **192** (1980) C42; (b) G. Ciani, A. Moret, A. Sironi and S. Martinengo, *J. Organomet. Chem.* **363** (1989) 181.
72. (a) S. Martinengo, G. Ciani and A. Sironi, *J. Chem. Soc., Chem. Commun.* (1980) 1140; (b) J. L. Vidal and R. C. Schoening, *J. Organomet. Chem.* **218** (1981) 217.
73. R. Della Pergola, L. Garlaschelli, S. Martinengo, G. Ciani, M. Moret and A. Sironi, *XVI Congresso Nazionale di Chimica*, Bologna (Italy) 9–14 October (1988), p. 337.
74. S. Martinengo, D. Strumolo, P. Chini, V. G. Albano and D. Braga, *J. Chem. Soc., Dalton Trans.* (1984) 1837.
75. S. Martinengo, G. Ciani and A. Sironi, *J. Chem. Soc., Chem. Commun.* (1991) 26.
76. J. L. Vidal, L. A. Kapicak and J. M. Troup, *J. Organomet. Chem.* **215** (1981) C11.
77. (a) V. G. Albano, M. Sansoni, P. Chini, S. Martinengo and D. Strumolo, *J. Chem. Soc., Dalton Trans.* (1976) 970, (b) V. G. Albano, P. Chini, S. Martinengo, M. Sansoni and D. Strumolo, *J. Chem. Soc., Chem. Commun.* (1974) 299.
78. G. Ciani, A. Magni, A. Sironi and S. Martinengo, *J. Chem. Soc., Dalton Trans.* (1981) 1280.
79. J. L. Vidal, R. A. Fiato, L. A. Cosby and R. L. Pruett, *Inorg. Chem.* **17** (1978) 2574.
80. S. Martinengo, G. Ciani and A. Sironi, *J. Am. Chem. Soc.* **102** (1980) 7564.
81. J. L. Vidal, R. C. Schoening and J. M. Troup, *Inorg. Chem.* **20** (1981) 227.
82. S. Martinengo, G. Ciani and A. Sironi, *J. Chem. Soc. Chem. Commun.* (1992) 1405.
83. R. Della Pergola, F. Demartin, L. Garlaschelli, M. Manassero, S. Martinengo, N. Masciocchi and P. Zanello, *Inorg. Chem.* **32** (1993) 3670.
84. R. Della Pergola, F. Demartin, L. Garlaschelli, M. Manassero, S. Martinengo and M. Sansoni, *Inorg. Chem.* **26** (1987) 3487.
85. R. Della Pergola, L. Garlaschelli, M. Manassero, N. Masciocchi and P. Zanello, *Angew. Chem., Int. Ed. Engl.* **32** (1993) 1347.
86. A. Ceriotti, P. Chini, R. Della Pergola and G. Longoni, *Inorg. Chem.* **22** (1983) 1595.
87. R. W. Broach, L. F. Dahl, G. Longoni, P. Chini, A. J. Schultz and J. M. Williams, *Adv. Chem. Ser.* **167** (1978) 93.
88. A. Ceriotti, G. Piro, G. Longoni, M. Manassero, N. Masciocchi and M. Sansoni, *New. J. Chem.* **12** (1988) 501.
89. A. Ceriotti, F. Demartin, B. T. Heaton, P. Ingallina, G. Longoni, M. Manassero, M. Marchionna and N. Masciocchi, *J. Chem. Soc., Chem. Commun.* (1989) 786.
90. (a) V. G. Albano, F. Demartin, M. C. Iapalucci, G. Longoni, A. Sironi and V. Zanotti, *J. Chem. Soc., Chem. Commun.* (1990) 547; (b) V. G. Albano, F. Demartin, M. C.

Iapalucci, F. Laschi, G. Longoni, A. Sironi and P. Zanello, *J. Chem. Soc., Dalton Trans.* (1991) 739.
91. D. Fenske and A. Hollnagel, *Angew. Chem., Int. Ed. Engl.* **28** (1989) 1390.
92. A. Ceriotti, G. Longoni, M. Manassero, N. Masciocchi, G. Piro, L. Resconi and M. Sansoni, *J. Chem. Soc., Chem. Commun.* (1985) 1402.
93. A. Ceriotti, A. Fait, G. Longoni, G. Piro, L. Resconi, F. Demartin, M. Manassero, N. Masciocchi and M. Sansoni, *J. Am. Chem. Soc.* **108** (1986) 5370.
94. A. Ceriotti, A. Fait, G. Longoni, G. Piro, F. Demartin, M. Manassero, N. Masciocchi and M. Sansoni, *J. Am. Chem. Soc.* **108** (1986) 8091.
95. E. G. Mednikov, Y. L. Slovokhotov and Y. T. Struchkov, *Metalloorg Khim* **4** (1991) 123.
96. E. G. Mednikov, N. K. Eremenko, Y. L. Slovokhotov and Y. T. Struchkov, *J. Organomet. Chem.* **301** (1986) C35.
97. E. G. Mednikov, *Metalloorg Khim* **4** (1991) 885.
98. E. G. Mednikov, N. K. Eremenko, Y. L. Slovokhotov and Y. T. Struchkov, *J. Chem. Soc., Chem. Commun.* (1987) 218.
99. (a) J. C. Calabrese, L. F. Dahl, P. Chini, G. Longoni and S. Martinengo, *J. Am. Chem. Soc.* **96** (1974) 2614; (b) G. Longoni and P. Chini, *J. Am. Chem. Soc.* **98** (1976) 7225; (c) L. F. Dahl, personal communication.
100. J. A. K. Howard, J. L. Spencer and D. G. Turner, *J. Chem. Soc., Dalton Trans.* (1987) 259.
101. S. S. Kurasov, N. K. Eremenko, Y. L. Slovokhotov and Y. L. Struchkov, *J. Organomet. Chem.* **361** (1989) 405.
102. D. M. Washecheck, E. J. Wucherer, L. F. Dahl, A. Ceriotti, G. Longoni, M. Manassero, M. Sansoni and P. Chini, *J. Am. Chem. Soc.* **101** (1979) 6110.
103. A. Ceriotti, P. Chini, G. Longoni, M. Marchionna, L. F. Dahl, R. Montag and D. M. Washecheck, *XV Congresso Nazionale di Chimica Inorganica*, Bari (Italy), 27 September–1 October (1982), p. A28.
104. A. Ceriotti, P. Chini, G. Longoni, D. M. Washecheck, E. J Wucherer and L. F. Dahl, *XIII Congresso Nazionale di Chimica Inorganica*, Camerino (Italy), 23–26 September (1980), p. A2.
105. D. A. Nagaki, Ph.D. dissertation, University of Wisconsin-Madison, 1986.
106. W. Cen, P. Lindenfeld and T. P. Fehlner, *J. Am. Chem. Soc.* **114** (1992) 5451.
107. J. Deutscher, S. Fadel and M. Ziegler, *Chem. Ber.* **112** (1979) 2413.
108. T. Beringhelli, G. D'Alfonso, M. Freni, G. Ciani and A. Sironi, *J. Organomet. Chem.* **295** (1985) C7.
109. G. Longoni, M. Manassero and M. Sansoni, *J. Am. Chem. Soc.* **102** (1980) 3242.
110. V. G. Albano, L. Grossi, G. Longoni, M. Monari, S. Mulley and A. Sironi, *J. Am. Chem. Soc.* **114** (1992) 5708.
111. E. Brivio, A. Ceriotti, R. Della Pergola, L. Garlaschelli, F. Demartin, M. Manassero and M. Sansoni, *CISCI 91*, Chianciano Terme (Italy), 6–11 October (1991), p. 682.
112. E. Brivio, A. Ceriotti, F. Demartin and M. Manassero, to be submitted for publication.
113. B. F. G. Johnson, W. Kwik, J. Lewis, P. R. Raithby and V. P. Sharan, *J. Chem. Soc., Dalton Trans.* (1991) 1037.
114. P. J. Bailey, B. F. G. Johnson, J. Lewis, M. McPartlin and H. R. Powell, *J. Chem. Soc., Chem. Commun.* (1989) 1513.
115. R. D. Adams, J. C. Lii and W. Wu, *Inorg. Chem.* **31** (1992) 2556.
116. M. Fajardo, H. D. Holden, B. F. G. Johnson, J. Lewis and P. R. Raithby, *J. Chem. Soc., Chem. Commun.* (1984) 24.
117. V. Dearing, S. R. Drake, B. F. G. Johnson, J. Lewis, M. McPartlin and H. R. Powell, *J. Chem. Soc., Chem. Commun.* (1988) 1331.
118. (a) D. Braga, K. Henrick, B. F. G. Johnson, J. Lewis, M. McPartlin, W. J. H. Nelson, A. Sironi and M. D. Vargas, *J. Chem. Soc., Chem. Commun.* (1983) 1131; (b) S. R. Drake, B. F. G. Johnson, J. Lewis, W. J. H. Nelson, M. D. Vargas, T. A. Adatia, D. Braga, K. Henrick, M. McPartlin and A. Sironi, *J. Chem. Soc., Dalton Trans.* (1989) 1455.
119. (a) L. H. Gade, B. F. G. Johson, J. Lewis, M. McPartlin, T. Kotch and A. J. Lees, *J. Am. Chem. Soc.* **113** (1991) 8698; (b) E. Charalambous, L. H. Gade, B. F. G. Johnson, T. Kotch, A. J. Lees and J. Lewis, *Angew. Chem., Int. Ed. Engl.* **29** (1990) 1137.

120. L. H. Gade, B. F. G. Johnson, J. Lewis, G. Conole and M. McPartlin, *J. Chem. Soc. Dalton Trans.* (1992) 3249.
121. L. H. Gade B. F. G. Johnson, J. Lewis, M. McPartlin and H. R. Powell, *J. Chem. Soc., Chem. Commun.* (1990) 110.
122. L. H. Gade, B. F. G. Johnson, J. Lewis, M. McPartlin and H. R. Powell, *J. Chem. Soc., Dalton Trans.* (1992) 921.
123. A. Ceriotti, R. Della Pergola, G. Longoni, M. Manassero, N. Masciocchi and M. Sansoni, *J. Organomet. Chem.* **330** (1987) 237.
124. A. Ceriotti, R. Della Pergola, G. Longoni, M. Manassero and M. Sansoni, *J. Chem. Soc., Dalton Trans.* (1984) 1181.
125. (a) J. M. Ragosta and J. M. Burlitch, *J. Chem. Soc., Chem. Commun.* (1985) 1187; (b) J. M. Ragosta and J. M. Burlitch, *Organometallics* **7** (1988) 1469.
126. W. Cen, K. J. Haller and T. P. Fehlner, *Inorg. Chem.* **30** (1991) 3121.
127. A. Fumagalli, S. Martinengo, V. G. Albano, D. Braga and F. Grepioni, *J. Chem. Soc., Dalton Trans.* (1993) 2047.
128. A. Fumagalli, S. Martinengo and G. Ciani, *J. Chem. Soc., Chem. Commun.* (1983) 1381.
129. A. Fumagalli, S. Martinengo, G. Ciani and A. Sironi, *XVI Congresso Nazionale di Chimica Inorganica*, Ferrara (Italy), 12–16 September (1983), p. 402.
130. B. T. Heaton, L. Strona, S. Martinengo, D. Strumolo, V. G. Albano and D. Braga, *J. Chem. Soc., Dalton Trans.* (1983) 2175.
131. V. G. Albano, A. Fumagalli, F. Grepioni, S. Martinengo and M. Monari, *J. Chem. Soc., Dalton Trans.* (1994) 1777.
132. (a) A. Fumagalli, S. Martinengo, G. Ciani, N. Masciocchi and A. Sironi, *XVIII Congresso Nazionale di Chimica Inorganica*, Como (Italy), 6–10 October (1986), p. 185; (b) A. Fumagalli, S. Martinengo, G. Ciani, N. Masciocchi and A. Sironi, *Inorg. Chem.* **31** (1992) 336.
133. A. Ceriotti, F. Demartin G. Longoni, M. Manassero, M. Marchionna, N. Masciocchi, G. Piva, M. Sansoni, B. T. Heaton and G. Piro, *XVII Congresso Nazionale di Chimica Inorganica*, Cefalù (Italy), 15–19 October (1984), p. 85.
134. A. Ceriotti, F. Demartin, G. Longoni, M. Manassero and M. Marchionna, to be submitted for publication.
135. A. Ceriotti, F. Demartin, G. Longoni, M. Manassero, G. Piva, G. Piro, M. Sansoni and B. T. Heaton, *J. Organomet. Chem.* **301** (1986) C5.
136. A. J. Whoolery and L. F. Dahl, *J. Am. Chem. Soc.* **113** (1991) 6683.
137. A. Ceriotti, F. Demartin, P. Ingallina, G. Longoni, M. Manassero, M. Marchionna, N. Masciocchi and M. Sansoni, *The Chemistry of the Platinum Group Metals*, Sheffield (England) 12–17 July (1987), p. O30.
138. A. Ceriotti, F. Demartin, G. Longoni, M. Manassero, M. Marchionna, G. Piva and M. Sansoni, *Angew. Chem., Int. Ed. Engl.* **24** (1985) 697.
139. J. A. Connor, *Topics Current Chem.* **71** (1977) 71.
140. G. Ciani, G. D'Alfonso, M. Freni, P. Romiti and A. Sironi, *J. Chem. Soc., Chem. Commun.* (1982) 339.
141. C. T. Hayward, J. R. Shapley, M. R. Churchill, C. Bueno and A. L. Rheingold, *J. Am. Chem. Soc.* **104** (1982) 7347.
142. T. Chihara, R. Komoto, K. Kobayashi, H. Yamazaki and Y. Matsuura, *Inorg. Chem.* **28** (1989) 964.
143. P. F. Jackson, B. F. G. Johnson, J. Lewis, W. J. Nelson and M. McPartlin, *J. Chem. Soc., Dalton Trans.* (1982) 2099.
144. P. Chini, S. Martinengo and G. Giordano, *Gazz. Chim. It.* **102** (1972) 330.
145. B. T. Heaton, L. Strona, S. Martinengo, D. Strumolo, R. J. Goodfellow and I. H. Sadler, *J. Chem. Soc., Dalton Trans.* (1982) 1499.
146. D. M. P. Mingos, *J. Chem. Soc., Chem. Commun.* (1985) 1352.
147. G. Ciani and A. Sironi, *J. Organomet. Chem.* **197** (1980) 233.
148. A. Ceriotti, G. Longoni, M. Manassero, M. Perego and M. Sansoni, *Inorg. Chem.* **24** (1985) 117.
149. A. Ceriotti, G. Longoni, M. Manassero, N. Masciocchi, L. Resconi and M. Sansoni, *J. Chem. Soc., Chem. Commun.* (1985) 181.
150. A. Fumagalli, S. Martinengo and G. Ciani, *J. Organomet. Chem.* **273** (1984) C46.

151. C. Brown, B. T. Heaton, A. D. Towl, P. Chini, A. Fumagalli and G. Longoni, *J. Organomet. Chem.* **181** (1979) 233.
152. V. G. Albano, D. Braga and S. Martinengo, *J. Chem. Soc., Dalton Trans.* (1981) 717.
153. P. Chini, G. Longoni and S. Martinengo, *Chim. Ind.* **60** (1978) 989.
154. G. J. Lewis, J. O. Roth, R. Montag, L. K. Safford, X. Gao, S. Chang, L. F. Dahl and M. J. Weaver, *J. Am. Chem. Soc.* **112** (1990) 2831.
155. S. Drake, M. Barley, B. J. Johnson and J. Lewis, *Organometallics* **7** (1988) 806.
156. P. Zanello, *Stereochem. Organomet. Inorg. Compounds*, Vol. 5, Elsevier, New York (1994) p. 103.
157. J. D. Roth, G. J. Lewis, L. K. Safford, X. Jiang, L. F. Dahl and M. J. Weaver, *J. Am. Chem. Soc.* **114** (1992) 6159.
158. P. Lemoine, *Coord. Chem. Rev.*, **83** (1983) 169.
159. (a) P. Zanello, *Coord. Chem. Rev.* **83** (1983) 199; (b) P. Zanello, *Coord. Chem. Rev.* **87** (1988) 1.
160. S. Drake, *Polyhedron* **9** (1990) 455.
161. V. G. Albano, F. Demartin, M. C. Iapalucci, G. Longoni, M. Monari and P. Zanello, *J. Chem. Soc., Dalton Trans.* (1992) 497.
162. K. Machida, A. Fukuoka, M. Ichikawa and M. Enyo, *J. Chem. Soc., Chem. Commun.* (1987) 1486.
163. B. K. Teo and N. J. A. Sloane, *Inorg. Chem.* **24** (1985) 4545.
164. A. F. Wells, *Structural Inorganic Chemistry*, Clarendon Press, Oxford (1975).
165. S. Ino, *J. Phys. Soc. Japan* **21** (1966) 346.
166. (a) B. Bagley, *Nature (London)* **225** (1970) 1040; (b) B. Bagley, *Nature (London)* **208** (1965) 674.
167. A. L. Mackay, *Acta Cryst.* **15** (1962) 916.
168. J. A. R. Clarke, *Nature (London)* **211** (1966) 280.
169. G. Schmid, R. Pfeil, R. Boese, F. Bandermann, S. Meyer, G. H. M. Calis and J. W. A. van der Velden, *Chem. Ber.* **114** (1981) 3634.
170. G. Schmid, B. Morun and J. O. Malm, *Angew. Chem., Int. Ed. Engl.* **28** (1989) 778.
171. (a) G. Schmid, *Polyhedron*, **7** (1988) 2321; (b) G. Schmid, *Mater. Chem. Phys.* **29** (1991) 133.
172. A. J. Melmed and D. O. Hayward, *J. Chem. Phys.* **31** (1959) 545.
173. A. Renou and M. Gillet, *Surf. Sci.* **106** (1981) 27.
174. D. Schechtman, I. Blech, D. Gratias and J. W. Cahn, *Phys. Rev. Lett.* **53** (1984) 1951.
175. R. Fredberg and D. I. Paul, *Phys. Rev. Lett.* **34** (1975) 1234.
176. R. B. King, *Inorg. Chim. Acta* **198–200** (1992) 841.
177. L. E. Toth, *Transition Metal Carbides and Nitrides*, Academic Press, New York (1971).
178. N. V. Sidgwick and R. W. Bailey, *Proc. Roy. Soc., Ser. A* **144** (1934) 521.
179. K. Wade, *Adv. Inorg. Chem. Radiochem.* **18** (1976) 1.
180. (a) D. M. P. Mingos and M. I. Forsyth, *J. Chem. Soc., Dalton Trans.* (1977) 610; (b) D. M. P. Mingos, *J. Chem. Soc., Chem. Commun.* (1983) 706; (c) R. L. Johnston and D. M. P. Mingos, *J. Organomet. Chem.* **280** (1985) 407; (d) R. L. Johnston and D. M. P. Mingos, *J. Organomet. Chem.* **280** (1985) 419; (e) D. M. P. Mingos, *Nature Phys. Sci.* **236** (1972) 99; (f) D. M. P. Mingos, *Acc. Chem. Res.* **17** (1984) 311.
181. R. E. Williams, *Inorg. Chem.* **10** (1971) 210.
182. J. W. Lauher, *J. Am. Chem. Soc.* **100** (1978) 5305.
183. (a) B. K. Teo, *Inorg. Chem.* **23** (1984) 1251; (b) B. K. Teo, G. Longoni and F. R. Chung, *Inorg. Chem.* **23** (1984) 1257; (c) B. K. Teo, *Inorg. Chem.* **23** (1984) 1627; (d) B. K. Teo, *Inorg. Chem.* **24** (1985) 4209.
184. B. K. Teo, *J. Chem. Soc., Chem. Commun.* (1983) 1362.
185. (a) J. F. Halet and D. M. P. Mingos, *Organometallics* **7** (1988) 51; (b) J. F. Halet, D. G. Evans and D. M. P. Mingos, *J. Am. Chem. Soc.* **110** (1988) 87.
186. M. A. Beswick, M. C. Ramirez de Arellano, P. Raithby and J. Lewis, *European Research Conference on Metal Clusters in Chemistry*, Lagonissi (Greece) 28 August – 2 September 1993.
187. A. D. Hattersley and C. E. Housecroft, *The Chemistry of the Platinum Group Metals*, St. Andrews (Scotland) 11–16 July 1993.
188. F. Demartin, M. C. Iapalucci and G. Longoni, *Inorg. Chem.* **32** (1993) 5536.
189. D. M. P. Mingos, *J. Cluster Science* **3** (1992) 397.

3. LIGAND-STABILIZED GIANT METAL CLUSTERS AND COLLOIDS

G. SCHMID
Institut für Anorganische Chemie, University of Essen, Germany

3.1. Strategy for Making Giant Metal Clusters

The formation of metal clusters and colloids is associated with the genesis of a metal. If we were able to observe the development of such particles step by step, we would see the birth of a metal with its typical properties, based on the existence of an electronic band structure. To realize such ideas, we need methods to build up polyatomic species from atoms. In principal, atoms can be generated in the gas phase or in solution. In any case the single atoms must be allowed to aggregate and reactions with other molecules must be avoided.

Gas phase syntheses of metal clusters are used to generate so-called naked clusters, which are then investigated by physical methods. As they are highly reactive and tend to form larger aggregates if they touch each other, they can only be studied either in the gas phase itself or in a matrix or on a substrate like graphite. Matrices may consist of organic polymers, inorganic materials like zeolithes, or noble gases at 4 to 20 K [1]. In a matrix, the metal particles are in principle kept separated and can be observed individually.

However, there is one big disadvantage for all naked clusters: they cannot be isolated and handled on a preparative scale like 'normal' chemical compounds. Mass-spectroscopic investigations of metal clusters, prepared in a supersonic beam, prove in any case a size distribution [2]. Numerous physical methods have been used to investigate matrix isolated clusters, e.g. IR, UV, Raman, fluorescence, and Mössbauer spectroscopy as well as EXAFS, and other methods [1]. As the physical properties of small particles mainly depend on the number of atoms, unequal particles will give only averaged values, whatever the method of examination may be.

To enable the investigation of uniformly sized clusters, they have to be protected by a ligand shell. The nature of the ligand shell controls one very important property: the solubility. Hydrophobic ligands of a more or less organic character render the particles soluble in organic solvents, whereas hydrophilic ligands cause solubility in water or water-like solvents. These facts are well-known from complex chemistry and can be transferred to clusters and even to colloids, as shall be shown later.

If metal cations in solution are reduced to metal atoms, usually polycrystalline metallic precipitates are formed immediately. As a consequence, monocrystalline domains as parts of the polycrystalline material must have been developed in the course of seconds. To stop this rapid crystal growth at a particular time, the particles must be trapped by an offer of appropriate ligands. The problem is, to avoid untimely reactions of the ligands with single atoms or very small clusters. It seems obvious that the ligand concentration may indeed determine the rate of crystal growth: without any ligands bulk metal is formed, excess of ligands leads to mononuclear complexes (the so-called metal vapor synthesis of many organometallic compounds from metal atoms and free ligands uses this effect [3]). Somewhere in between these two extremes, there should be a range of conditions to generate ligand-stabilized clusters and colloids. As we know since ancient years, the formation of gold colloids in aqueous solution is based on this concept: if Au^{3+} ions are reduced in water by sodium citrate, the metal growth continues until a certain particle size is reached. These particles are then stabilized and protected by different organic materials, mainly consisting of the oxidation products of the reducing agents. What is valid for colloids which, of course, have a relatively broad size distribution, should in principle also be applicable to smaller particles, the clusters. However, we cannot expect that this method works in every case, and furthermore it has to be expected that mixtures of clusters of different size will be formed – like in the gas phase technique. But, particles with ligand shells can be separated by usual methods, due to their solubility, and can then be treated chemically into definite compounds.

To prepare well-defined ligand-stabilized clusters in solution, another condition is presumed, namely the preferred formation of so-called full-shell clusters. Full-shell clusters consist of cubic or hexagonal close-packed structures with a complete outer geometry: a central metal atom is coordinated by a first shell of 12 other atoms. The second shell, enveloping these 12 atoms, consists of 42 atoms, etc. This 'onion-like' description of a metallic structure leads to distinct numbers of atoms (magic numbers): 13, 55, 147, 309, 561 etc., or $10n^2 + 2$ atoms for the nth shell. If full-shell clusters are formed on the way from atoms to the bulk, they should be long-lived compared with other arrangements, due to their dense-packed structure and the complete outer geometry with a minimum of energy. As a consequence, they should preferably be trapped by the ligands. Of course, many other atomic arrangements may also be realized. Ligand concentration, ligand size and geometry, ligand strength, kinetic, and thermodynamic parameters, all must harmonize to succeed in cluster isolation. Too many conditions to provide a general method of synthesis? Surprisingly, in some cases the experimental conditions obviously harmonize and full-shell clusters can be ligated, isolated, and investigated. Some typical examples shall be discussed in the following section.

As transition metal colloids can be regarded as very huge clusters, we will include them into our discussion, especially as we have been successful in stabilizing colloids with the same ligands as used for the clusters. For the first time, various colloids can now be isolated on a preparative scale in the solid state. They can be redispersed in any concentration. This enables a new quality of physical and chemical investigations. Up to date, metal colloids could only be handled in very dilute solutions or in a matrix, like naked clusters.

3.2. Synthetic and Structural Examples

3.2.1. LIGAND-STABILIZED COLLOIDS

Metal particles > 10 nm shall be defined as colloids, smaller ones as clusters; however, we know that there is no distinct transition between both.

Gold colloids in aqueous solution or in glasses are well known, are easily prepared and remarkably stable under certain conditions. M. Faraday already described the reason for the bright red and purple colors of gold colloids in 1858. Nowadays, gold colloids are applied to label biological materials like lectins, antibodies, enzymes, and lipoproteins for electron microscopic investigations.

In contrast to well-defined clusters, colloids show a size distribution. In some cases the deviations from a standard size is only about 10%. Such colloids are of interest, as they can roughly be compared with huge clusters.

Metal colloids can be prepared on two different routes: by dispersion of larger metal particles or by aggregation of atoms or smaller units in solution. The stabilization of metal colloids in solution normally happens by adsorption of ionic species on their surface. Electric repulsion forces hinder the metal particles to coagulate and to precipitate. Attempts to isolate colloidal particles from solution end up with the formation of metal powders. So-called protecting colloids like polyvinyl alcohol [4–6], polyvinyl pyrrolidone [5], cyclodextrine [7], and colloidal silicic acid [8] stabilize colloids. The protective effect of such materials may be traced back to the fact that the metal colloids are fixed on the larger protecting colloids or instead, the polymers cover the colloids so that they are unable to come into contact with each other. Colloids may also be embedded in a matrix of polymers like polystyrene. However, this kind of stabilization cannot be compared with a complexation in a chemical sense, as most of the mass consists of the protecting material and not of the metal colloids. So, physical investigations and chemical reactions are only possible in a very limited manner.

If $HAuCl_4$ in diluted aqueous solution is reduced by citric acid [9] or trisodium citrate [9–15], 18 nm gold colloids are formed as a blood-red colored solution. The size distribution is only about ± 10%. Larger Au colloids are available, if such 18 nm colloids are used as nuclei and additional $HAuCl_4$ is

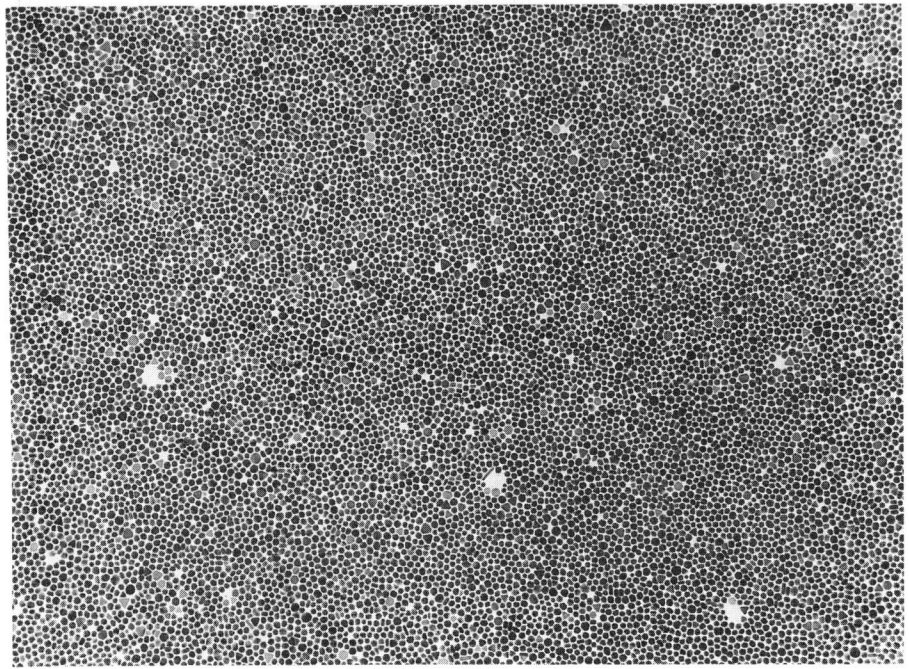

Fig. 1. A monolayer of 14 nm gold colloids, stabilized by $P(m\text{-}C_6H_4SO_3Na)_3$.

reduced by hydroxylammoniumchloride [9, 16]. The diameter of gold colloids can be determined by the formula

$$d = d_0 \sqrt[3]{\frac{n_i + n_m}{n_m}}$$

where d_0 is the particle diameter in the solution of the starting colloids, and n_i, n_m are the quantity of the ionic (n_i) and metallic (n_m) gold, respectively.

Independent of their size, gold colloids can now be stabilized by water-soluble phosphines like $(C_6H_5)_2P(m\text{-}C_6H_4SO_3Na)$ and $P(m\text{-}C_6H_4SO_3Na)_3$, to such an extent that they can be isolated as solid materials [17, 18]. Figure 1 shows a kind of dense-packed two-dimensional layer of 14 nm gold colloids. Such mono-layers and also thicker layers of stabilized colloids can be prepared on smooth surfaces like glass, paper etc.

The red color of the solution turns to metallic gold, if the water is evaporated. As can be seen from Figure 1, there is no aggregation to be observed, due to the repulsive forces of the negatively charged phosphine ligands. So, it is not surprising that the threefold sulfonated $P(m\text{-}C_6H_4SO_3Na)_3$ stabilizes much better than the mono-sulfonated $(C_6H_5)_2P(m\text{-}C_6H_4SO_3Na)$, since in solution the Na^+ cations dissociate completely, generating highly negatively

Fig. 2. 36 nm Gold colloids with characteristic patterns at low coverages.

charged colloids. If gold colloids are deposited from very dilute solutions, characteristic patterns are formed. Figure 2 shows such a pattern of 36 nm Au colloids.

An explanation for the formation of such ornaments is given in Figure 3 [19]. Due to the surface tension of the very small droplets, which are formed during the drying process, the colloidal particles move to the peripheries and so concentrate around the centers of the droplets.

Figure 13 shows a high resolution transmission electron microscopic (HRTEM) image of a single Au colloid, stabilized by $P(m-C_6H_4SO_3Na)_3$ [20] (see below on page 123).

The dimension is about 13.2×10.5 nm. It has a polycrystalline structure, as is typical for colloids. However, the more interesting detail is the ligand shell, which can be observed as an amorphous layer around the particle. The thickness of this layer corresponds well with a double-layer of $P(m-C_6H_4SO_3Na)_3$ molecules, the dimension of which can be deduced from an X-ray structure of the related $(C_6H_5)_2P(m-C_6H_4SO_3Na)$ [21]. The distances between the colloidal particles in Figures 1 and 2 also agree well with the presence of such double-layers. Free lying colloids sometimes show thicker ligand shells, if there was an excess of ligand molecules in the solution.

In principle, palladium and platinum colloids can be isolated from aqueous solution by corresponding stabilization processes as described for gold. However, with these metals the nitrogen ligand $p-H_2N-C_6H_4-SO_3Na$ has proved to be better than the phosphines. Platinum colloids in general show a less

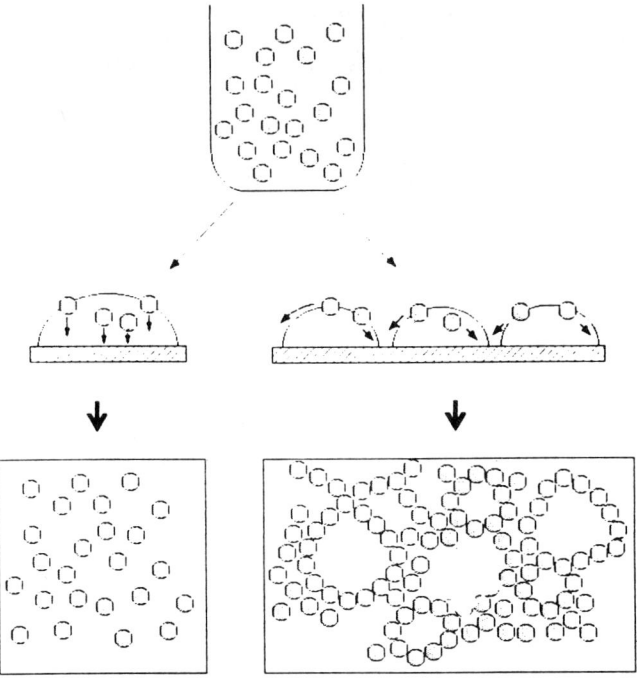

Fig. 3. Model for the formation of experimentally observed colloidal patterns on the basis of Monte Carlo simulations. The colloids in very small droplets move to the peripheries of the droplets at the surface of the support, as shown in the right part of the image.

good size distribution, compared with gold or palladium. Figure 4 shows a typical pattern of stabilized Pt colloids. Whereas the pattern agrees well with literature data, we have never observed the monodispersive character, described by some authors [22, 23].

From a chemical and physical point of view, bimetallic colloids are much more interesting than mono-metallic ones, since they can be regarded as models for the formation of alloys. As mentioned, unstabilized gold colloids can be used as crystal germs to generate larger particles in a second reduction step. If this following reduction is carried out with another metal like palladium or platinum, bimetallic colloids can be prepared very easily. Such bimetallic colloids with layered structures are stabilized by the same procedure as for the mono-metallic species. If the outer metal consists of Au, $P(m-C_6H_4SO_3Na)_3$ is used, $p-H_2N-C_6H_4-SO_3Na$ is best suited if Pd and Pt cover gold [24], similar as for the mono-metallic particles. The thickness of the covering metal layer again only depends on the concentration of H_2PtCl_6 and H_2PdCl_4, which are used as educts. If the Pd layer on gold becomes very thin, e.g. only 2–3 layers, then $p-H_2N-C_6H_4-SO_3Na$ is no longer suited to stabilize the colloids. On the contrary, we need $P(m-C_6H_4SO_3Na)_3$! If the Pd

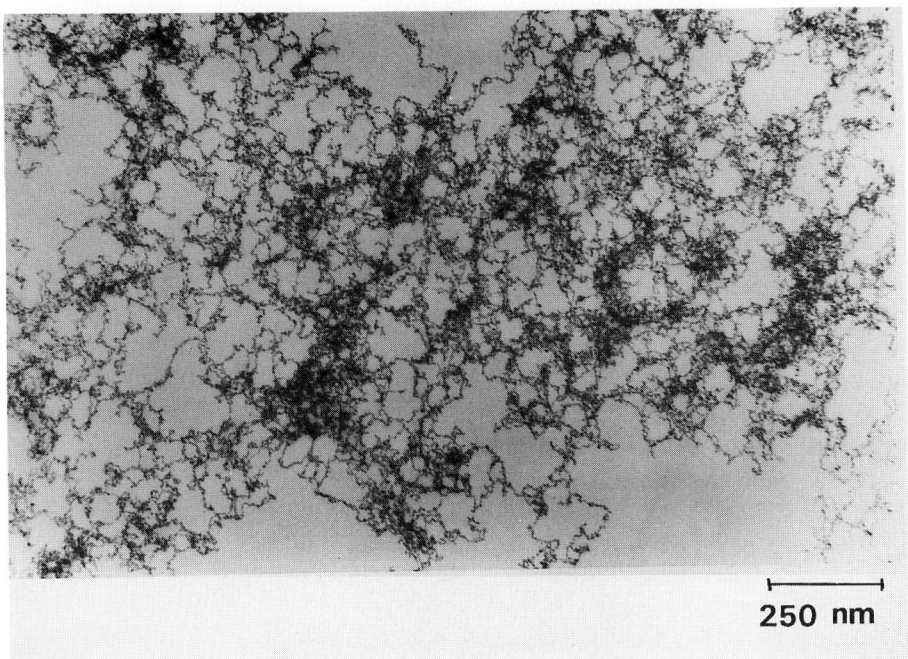

Fig. 4. Platinum colloids, stabilized by p-H_2N-$C_6H_4SO_3Na$.

layers become thicker, a mixture of both ligand types is preferred [25]. These preliminary results cannot yet be understood in detail, but we believe that migration processes are responsible for these findings. Coring and mixing processes have already been observed [26].

If gold colloids are covered with platinum and palladium, two different behaviors are observed: platinum on gold looks like an agglomerate, palladium on gold looks homogeneous [24]. Figure 5 shows two Au/Pt particles. The heterogeneous character of both is obvious.

EDX (Energy Dispersive X-Ray) analyses indeed prove that the outer parts of the colloids only consist of about 5 nm Pt particles, whereas the inner core is formed by the starting 18 nm Au nucleus. As the inserted optical diffractograms show, the Pt particles in colloid A all have the same crystallographic orientation, in B they are randomly oriented. The reason for this must be that the inner gold nucleus in A is monocrystalline, in B it is polycristalline. The fact that platinum is added in the form of more or less individual particles to the gold surface may be due to the broad miscibility gap, which is observed for the combination Au/Pt between 2% and 85% of Au. By contrast, gold and palladium are miscible in any ratio.

Figure 6 shows two Au/Pd colloids. The optical diffractogram in 6a shows the polycrystalline character of the covering Pd layer, however, it is not built

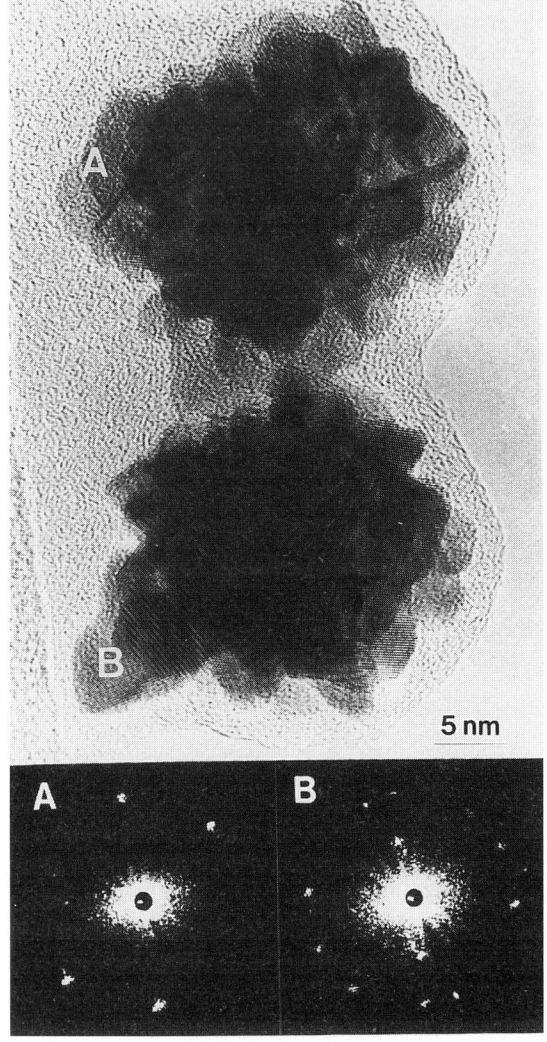

Fig. 5. Two different types of Au/Pt particles with intergranular structures. 5 nm Pt colloids cover a 18 nm gold nucleus. The optical diffractogram of A indicates the same crystallographic orientation of all Pt particles, whereas in B there is a random orientation.

up by single particles like Au/Pt, but by one compact layer with a smooth surface. Probably the Au nucleus is polycrystalline too. In Figure 6b a completely homogeneous Au/Pd colloid is shown, with a perfect monocrystalline structure. One would conclude that the factor determining the structure is the monocrystalline character of the Au nucleus.

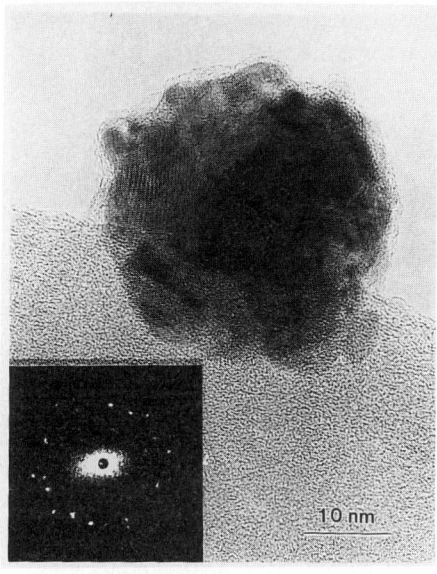

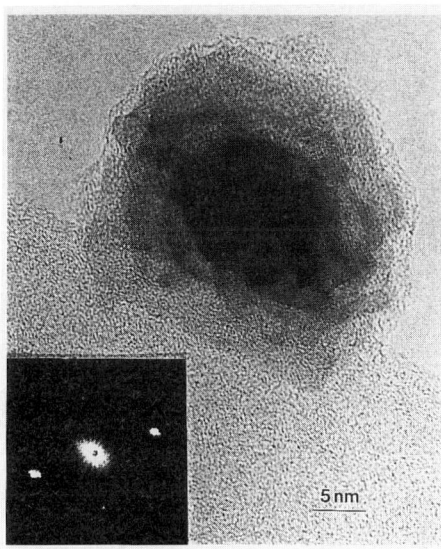

Fig. 6. Two types of Au/Pd colloids. (a) shows a polycrystalline Pd layer on the gold nucleus; in (b) a monocrystalline Au/Pd particle is shown. Both particles are built up by compact Pd layers in contrast to Au/Pt particles as shown in Figure 5.

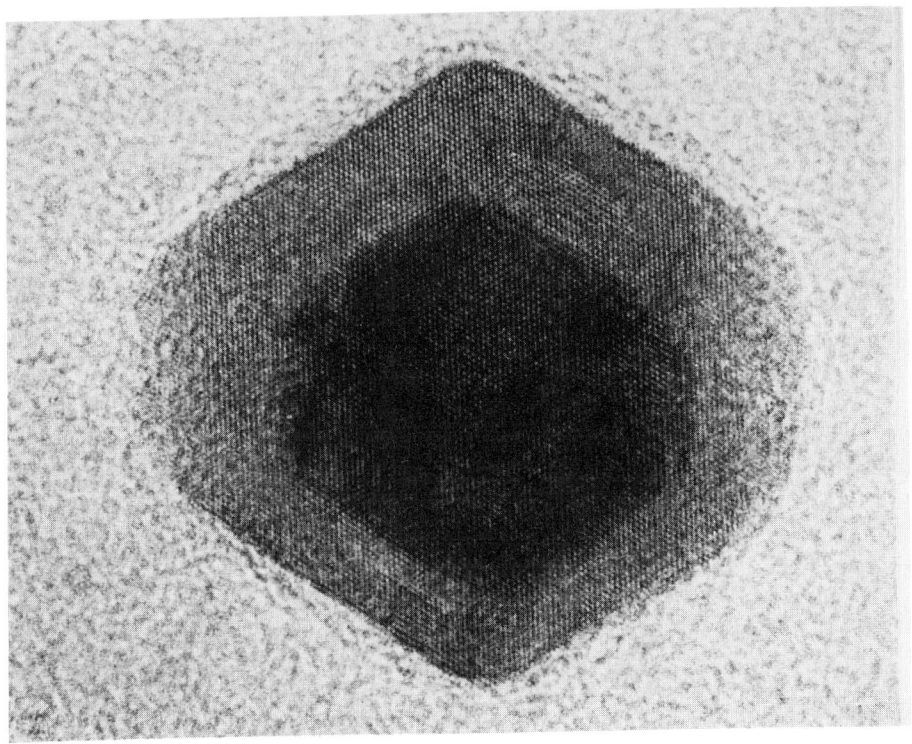

Fig. 7. A perfectly formed Au/Pd colloid.

Another perfectly formed Au/Pd colloid is shown in Figure 7. The dark gold core is covered by Pd atoms of exactly the same crystallographic orientation [27].

3.2.2. LIGAND-STABILIZED CLUSTERS

The formation of colloids by reduction of metal cations in solution, as well as the subsequent stabilization by appropriate ligands, is comparatively simple, since the 'natural' size distribution, the predominant polycrystalline character, and the not precisely defined ligand shell are 'accepted' and so facilitate the experimental conditions. On the way from the atom to the bulk, the preparation of colloids is therefore the least problematic step.

Particles, which may be considered as relatively well defined clusters, have recently been observed by us [28]. If Pd(II) acetate in acetic acid solution is reduced by hydrogen in the presence of small amounts of phenanthroline, followed by addition of oxygen, various giant palladium clusters can be isolated. One fraction mainly consists of a mixture of 31.5 and 36.0 Å clusters. HRTEM investigations show that most of these particles are monocrystalline

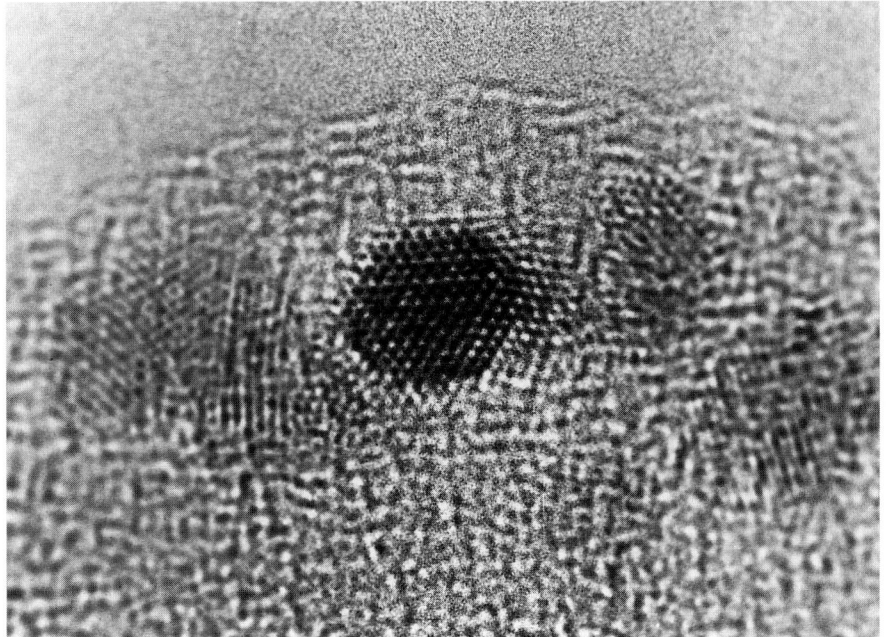

Fig. 8. An eight-shell Pd cluster with fcc structure.

and show 15 or 17 atomic fringes, respectively, corresponding with seven- and eight-shell clusters. As yet it is not possible to give a precise formula. The analytical data of a fraction mainly consisting of 36.0 Å clusters indicate the idealized formula, $Pd_{2057}phen_{84}O_{\sim 1600}$. Due to the addition of oxygen, uncoordinated surface atoms are stabilized in such a way that they become air stable. The oxidation state of this surface oxygen is not yet known. The clusters can best be dissolved in pyridine-water mixtures.

Figure 8 shows the metal core of a seven-shell cluster with fcc structure like in the bulk palladium metal.

Another fraction of this synthetic procedure contains a much better characterized cluster of the formula $Pd_{561}phen_{36}O_{190-200}$ [29, 30]. This five-shell cluster is the largest cluster up to now, which could be isolated in an acceptable purity. It should be mentioned that the number of palladium and oxygen atoms may still vary to a certain (small) extent. However, HRTEM investigations show cluster molecules very uniform in size and with a symmetry and size agreeing well with the assumption of a 5 shell cuboctahedral metal core. The first five-shell palladium cluster has been described by Moiseev et al. in 1985 [31] as a product of a very similar reaction, but with a different ligand shell.

The reason for the solubility of these palladium clusters in pyridine-water mixtures has been studied in detail at $Pd_{561}phen_{36}O_{190-200}$. The com-

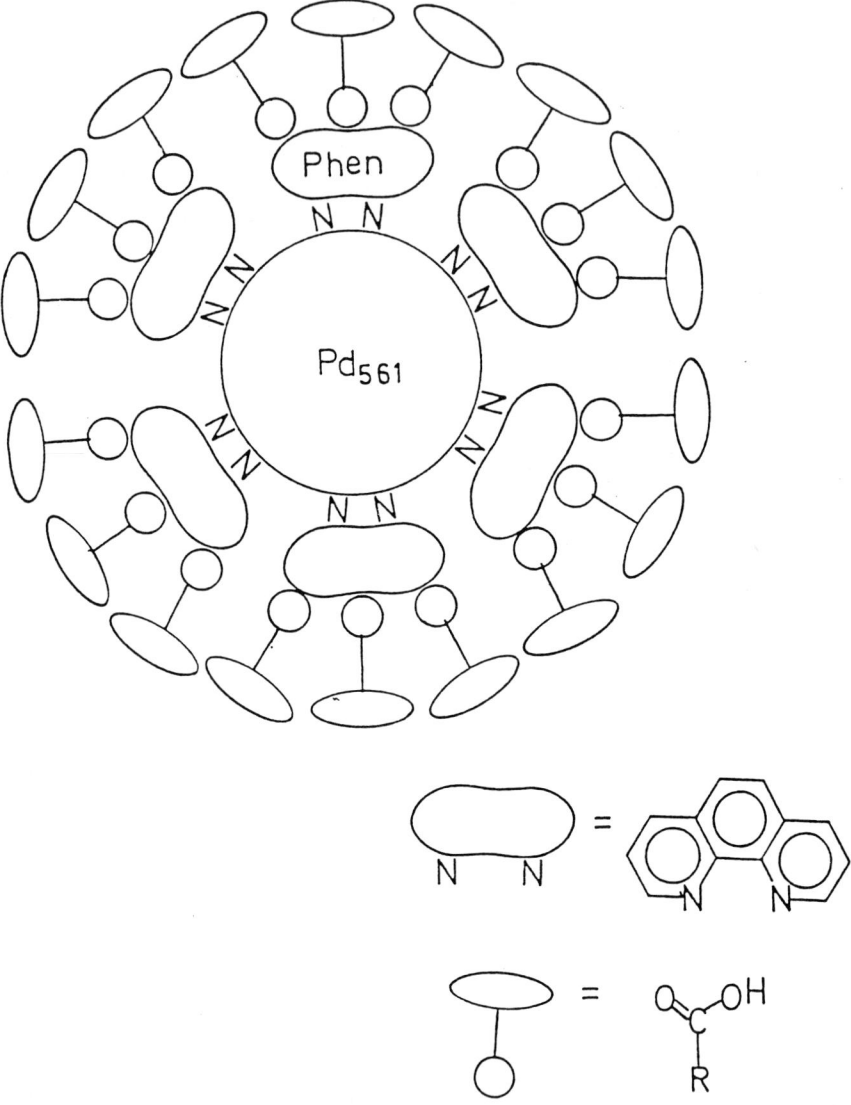

Fig. 9. Illustration of the two ligand shells around a Pd_{561} cluster formed by hydrophobic-hydrophobic interaction.

pound itself is almost insoluble in water. By addition of pyridine, acetic acid or other water-soluble carbonic acids, its solubility increases spontaneously. The reason is the formation of a second ligand shell due to the hydrophobic-hydrophobic interactions between the first ligand shell and the added molecules. Their hydrophilic groups mediate the attraction with water. Consequently, trichloroacetic acid cannot act as a solubilizer, due to the lack

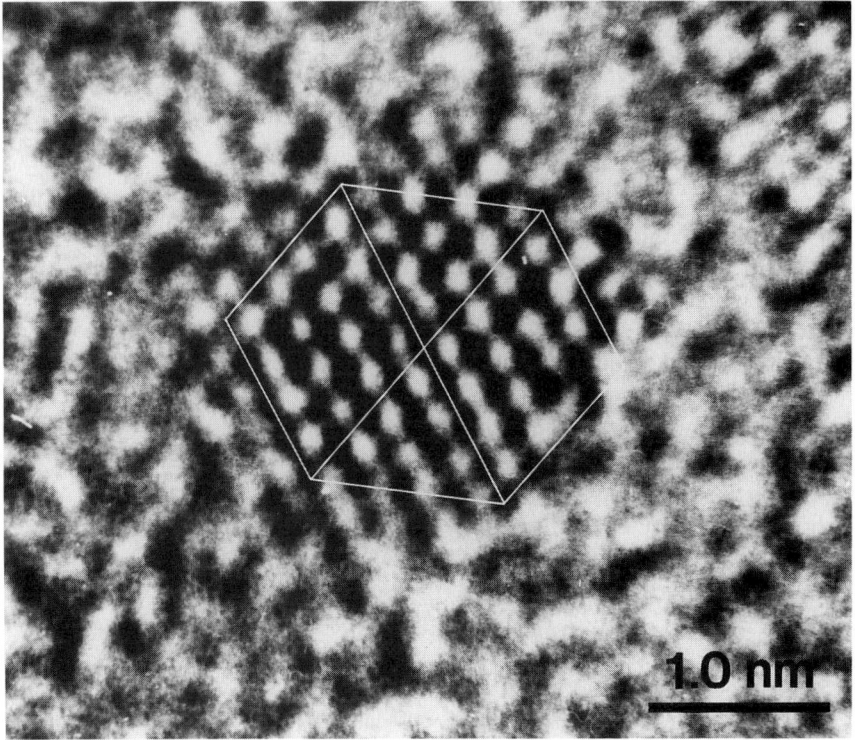

Fig. 10. View on a single Pt_{309} cluster with atomic resolution in (110) direction.

of a hydrophobic group [32]. In Figure 9 the conditions on the Pd_{561} surface are illustrated.

This behavior reminds of the characteristics of colloids, which also tend to adsorb double- or multi-layers of ligands on their surface. The weaker interaction of the second ligand shell with the cluster can be demonstrated by the centrifugation of solutions. During the centrifugation process, the molecules loose their second ligand shell and again become insoluble.

Initial attempts to image cluster molecules by means of the scanning tunneling microscopy (STM) have been carried out with this five-shell cluster. Of course, it cannot be expected to achieve atomic resolution, as it is known for smooth surfaces. However, the imaging of $Pd_{561}phen_{36}O_{190-200}$ cluster molecules succeeded in a kind of 'molecular resolution' [33, 34]. Figure 14 (below, page 123) shows an assembly of spheric looking molecules. Their diameter of about 30 Å agrees well with the expected data.

The use of ligands with their own hydrophilic groups makes large clusters water-soluble without additional solubilizers. Such a ligand is the phenanthroline derivative phen*:

If Pt(II) acetate is reduced by hydrogen in the presence of phen*, followed by cautious oxidation, the four-shell cluster $Pt_{309}phen^*_{36}O_{30\pm10}$ is formed as a water-soluble, black powder [35, 36]. Figure 10 shows a HRTEM image of a single cluster nucleus with a perfect fcc structure and with the outer geometry of a four-shell cuboctahedron.

The fcc structure of the five- and four-shell clusters can also be proved by means of X-ray powder diffractions. The single particles are already large enough to diffract X-rays and so produce broad Bragg spots with peak centers identical with those of the bulk metal.

The smallest type of giant clusters which should be discussed here, is represented by the series of two-shell clusters of the formula $M_{55}L_{12}Cl_x$. (M = Au, Pt, Rh, Ru; L = PR_3, AsR_3; $x = 6, 20$). Again, metal salts are reduced in solution in the presence of appropriate ligands L. However, as these ligands have an organic nature, the reduction is carried out in organic solvents and with B_2H_6 as a reducer instead of H_2. B_2H_6 also binds excess of ligands as H_3B-L adducts, so that the concentration of L is kept low [36–38].

For instance, if $(C_6H_5)_3PAuCl$ is reduced in benzene or toluene solution by diborane, $Au_{55}[(C_6H_5)_3P]_{12}Cl_6$ can be isolated as air-stable brown-colored solid in about 20% yield. It is soluble in dichloromethane, THF etc. Other two-shell clusters with M = Pt, Rh or Ru are extremely air sensitive and are available only in very low yields.

Imaging of these smaller particles by HRTEM is very difficult, as under the influence of the 400 keV electron beam the clusters move and tend to coalesce to larger particles. In some cases HRTEM pictures could be achieved. Figure 11 shows a Pt_{55} cluster core in the (110) direction with atomic resolution.

Using $Au_{55}[(C_6H_5)_3P]_{12}Cl_6$, initial experiments to achieve more than a 'molecular resolution' by means of STM have been started [39, 40]. First results look very promising, though not all details can be understood.

Figure 12 shows four single $Au_{55}[(C_6H_5)_3P]_{12}Cl_6$ molecules on a tungsten selenide surface with low resolution. It may be of interest to mention how the individual (single) cluster molecules have been anchored on the WSe_2 surface. Normally, a droplet of a very dilute solution is put on the support and then the solvent is evaporated. Here, the surface of the support has been covered with the solid cluster material. After shaking down all visible powder, some molecules are found to remain still fixed on the surface and so can be

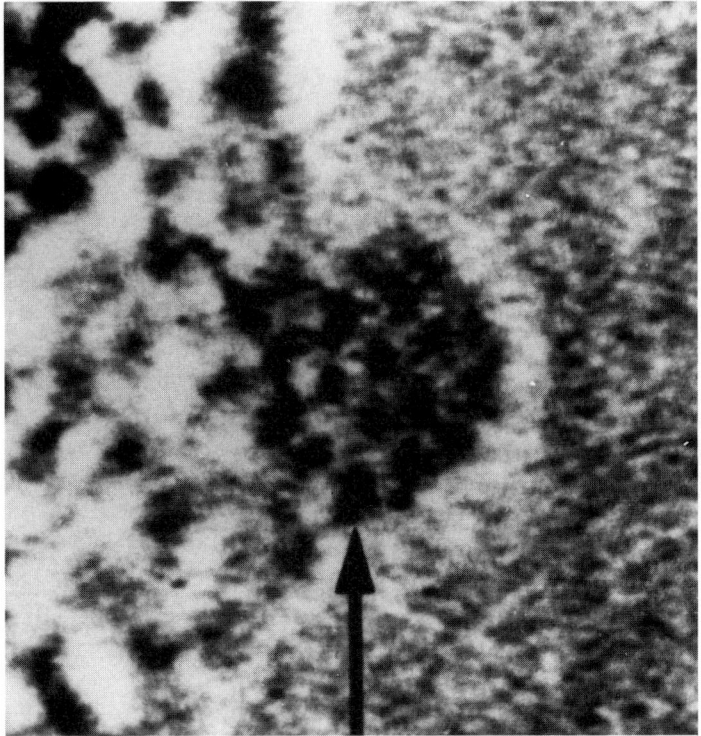

Fig. 11. A Pt$_{55}$ cluster with atomic resolution in (110) direction.

used for STM imaging. This is a very remarkable procedure, which indicates very weak intermolecular forces between the cluster molecules, and which seems to be in line with the fact that attempts at crystal growth have never been succesful for this kind of clusters.

What can we expect to observe, if the resolution becomes better than in Figures 14 or 12? If the tip of the microscope scans a single molecule, it mainly should get into contact with the ligand electrons, but not to the deeper lying metal atoms. So, one may expect a kind of a two-dimensional electron density profile of the ligands.

Figure 15a shows a high-resolution image of a $Au_{55}[(C_6H_5)_3P]_{12}Cl_6$ molecule on graphite. The different colors symbolize various electron densities on the cluster surface. In Figure 15b a computer simulated image, viewed in probably the same direction is shown. The agreement cannot be perfect, as the experimental information is much more complicated. However, the comparison of Figure 15b with 15a shows that the interpretation of the STM images should roughly be correct. Of course, each of the numerous STM images which have been registered will be different, since an infinite number of orientations of the ball-like molecules on the support is possible.

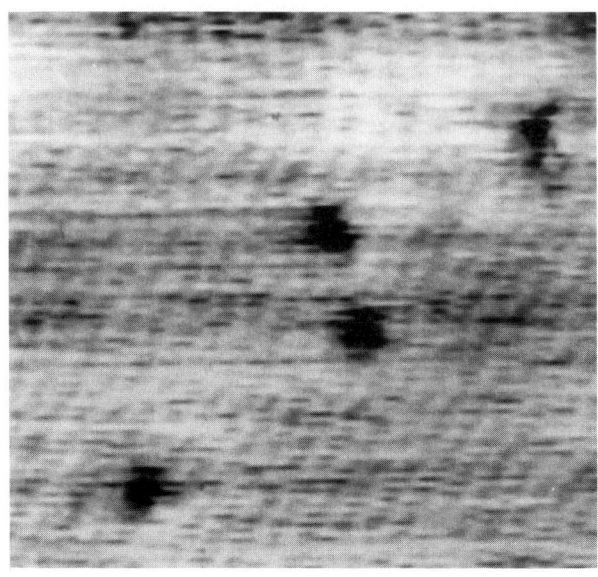

Fig. 12. STM images of single $Au_{55}[P(C_6H_5)_3]_{12}Cl_6$ clusters.

3.3. Chemical Properties

The tendency of the clusters discussed above, including the colloids, to decompose in solution limits the investigation of their chemical properties. The sensibility of ligand-stabilized colloids is mainly directed to changes in the electrolyte concentration. The clusters easily loose ligands in solution and so enable aggregation processes, finally resulting in bulk-metal formation. This is especially valid for the two-shell clusters. The phen and phen* protected Pd and Pt clusters are more stable.

From extensive ^{31}P NMR investigations of phosphine stabilized M_{55} clusters in solution we know that the phosphine ligands are highly fluxional, i.e. the ligand molecules rapidly change positions on the metal cluster surface. Equilibrium processes between coordinated and dissociated ligand molecules have been studied in the case of $Au_{55}[(C_6H_5)_3P]_{12}Cl_6$. The contact time between the $P(C_6H_5)_3$ molecules and the surface gold atoms has been determined to be 3 μsec [41].

The mobility of these ligands allows substitutional reactions. If a CH_2Cl_2 solution of $Au_{55}[(C_6H_5)_3P]_{12}Cl_6$ is stirred with an aqueous solution of $(C_6H_5)_2P(m-C_6H_4SO_3Na)$, the cluster molecules move quantitatively into the water phase, due to the complete exchange of the $(C_6H_5)_3P$ ligands by $(C_6H_5)_2P(m-C_6H_4SO_3Na)$. The stability of $Au_{55}[(C_6H_5)_2P(m-C_6H_4SO_3Na)]_{12}Cl_6$ in solution is much better, compared with the triphenylphosphine substituted cluster, probably because of the quantitative dis-

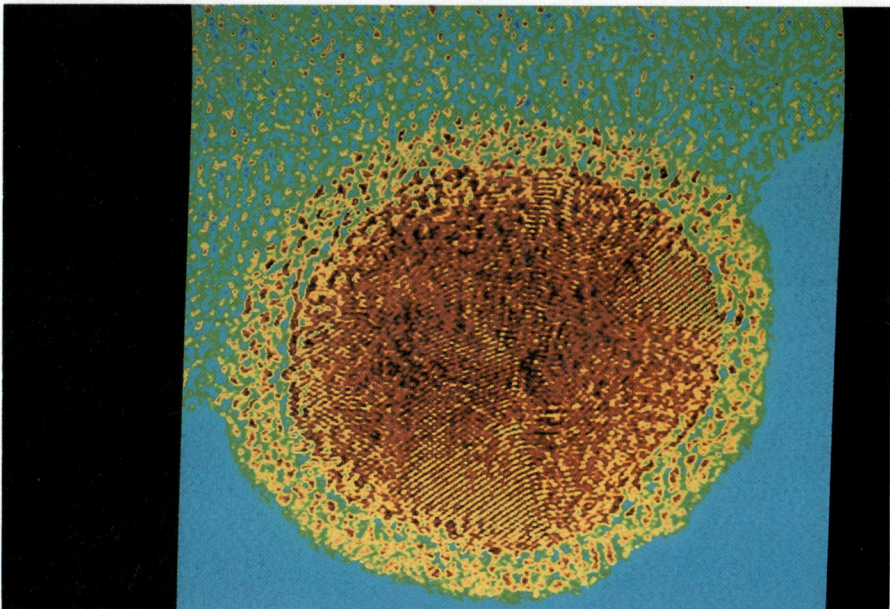

Fig. 13. HRTEM image of a single gold colloid with 13.2 × 10.5 nm. The polycrystalline structure as well as the ligand shell, consisting of P(m-$C_6H_4SO_3Na$)$_3$ molecules, can be observed.

Fig. 14. STM image of $Pd_{561}phen_{36}O_{190-200}$ cluster molecules in low resolution.

sociation of the Na$^+$ cations, generating the twelvefold negatively charged anion $\{Au_{55}[(C_6H_5)_2P(m\text{-}C_6H_4SO_3)]_{12}Cl_6\}^{12-}$ [42].

From preparative thermal decomposition reactions and from DSC measurements of $Au_{55}[(C_6H_5)_3P]_{12}Cl_6$, we know that it decomposes quantitatively to metallic gold and $[(C_6H_5)_3P]_2AuCl$ [43], following the equation

$$Au_{55}[(C_6H_5)_3P]_{12}Cl_6 \xrightarrow{156°C} 6[(C_6H_5)_3P]_2AuCl + 49Au.$$

In solution this reaction occurs at much lower temperatures. The heat of decomposition has been determined to 112 J/g or 1590 kJ/Mol cluster. These data allow the calculation of the nearest neighbor Au-Au interactions. The value of 76.1 kJ is considerably higher than in bulk gold with 61.3 kJ and agrees well with EXAFS measurements [43].

Another kind of cluster degradation can be observed on M_{55} clusters in CH_2Cl_2 solution. If they get into contact with Pt electrodes, to which a DC voltage of 20 V is applied, they decompose into a metallic looking powder, which consists mainly of superclusters of the type $[(M_{13})_{13}]_n$. This very surprising behavior can be understood, if we assume that by the contact with the electrodes the cluster molecules are polarized to such an extent that the surface metal atoms are stripped off together with the ligands. Consequently, free M_{13} particles are generated, which then aggregate to multiples and so become stabilized in a novel kind of metal structure. From X-ray powder diffraction investigations we know that the M_{13} building blocks are linked via the triangle faces to form a kind of dense-packed structure of M_{13} building blocks [29, 44]. However, these structures are thermodynamically unstable and change in the course of weeks into more or less amorphous metal powders.

The M_{55} degradation process and the construction of the superclusters is demonstrated in Figure 16.

The individual character of the M_{13} nucleus inside an M_{55} cluster is not only suggested by this amazing behavior, but has also been demonstrated by TOF-SIMS and PDMS investigations of $Au_{55}[(C_6H_5)_3P]_{12}Cl_6$ [45, 46]. The Secondary Ion Mass Spectroscopy (SIMS) works with primary ions in the keV region, the Plasma Desorption Mass Spectrometry (PDMS) uses ^{252}Cf fission fragments in the MeV region to bombard solid materials with formation of secondary ions, which can be determined by a Time of Flight (TOF) mass spectrometer. Both methods, applied on $Au_{55}[(C_6H_5)_3P]_{12}Cl_6$, give results, which not only agree with the observations of the electrochemical degradation, but which give a unique series of superclusters of the form $(Au_{13})_n^{+/-}$. Again, the outer metal shell, together with the ligands, is destroyed to give various metal ligand fragments, which can be registered in the low-mass range. In the process of forming the actually investigated sample, the generated Au_{13} clusters aggregate: the thicker the sample layer is, the higher is the value for n. Table 1 summarizes the measured and calculated values for the peak centers of $(Au_{13})_n^+$ in the positive SIMS spectra. PDMS gives only the first three series, but with identical fragments.

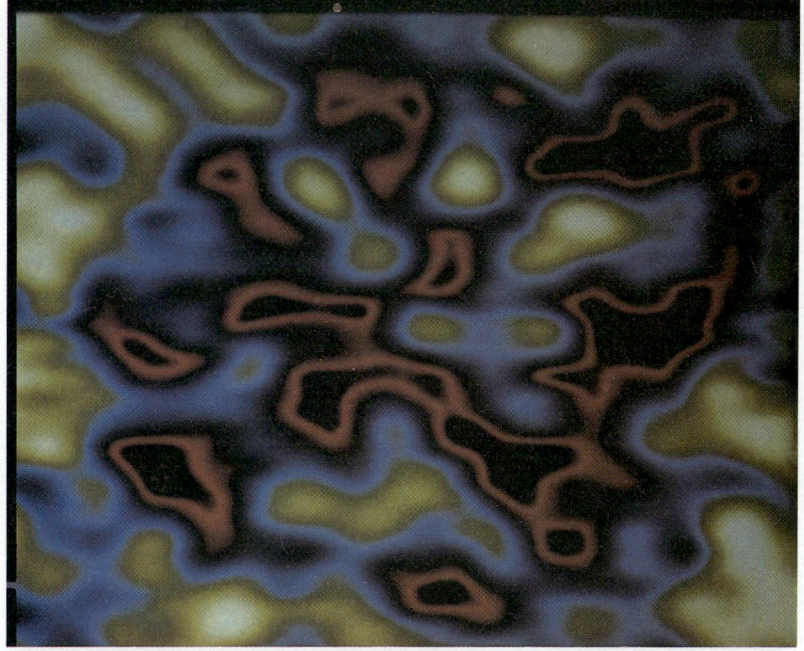

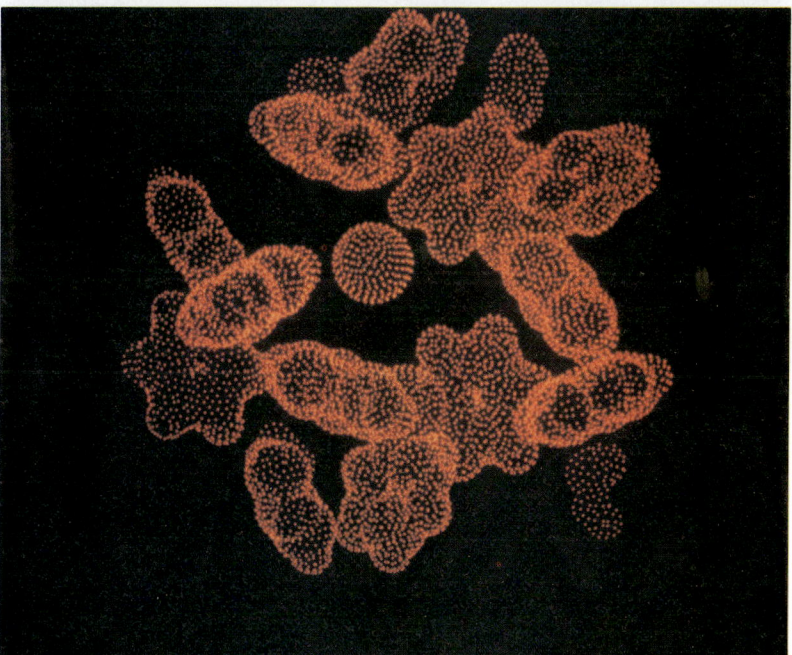

Fig. 15. (a) High-resolution STM image of a $Au_{55}[P(C_6H_5)_3]_{12}Cl_6$ molecule, mainly showing an electron density profile of the $P(C_6H_5)_3$ ligands and of chlorine. (b) A computer simulated electron density profile of the same molecule and probably viewed in the same direction.

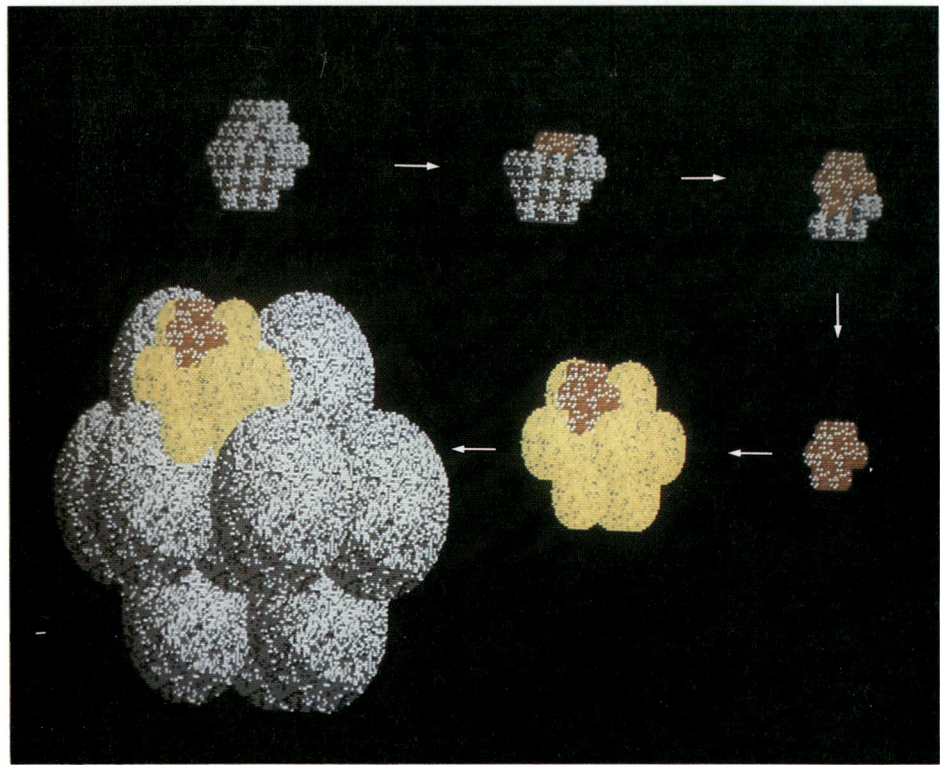

Fig. 16. Formalized degradation of M_{55} clusters. The ligand shell, together with the outer metal shell, is lost. The generated M_{13} clusters (red) aggregate to form novel super clusters $(M_{13})_{13}$ (yellow) and $[(M_{13})_{13}]_{13}$ (blue).

TABLE 1

Measured and calculated values[a] for the peak centers of $(Au_{13})_n^+$ in the positive SIMS spectra of $Au_{55}(PPh_3)_{12}Cl_6$

First series			Second series			Third series		
measd	calcd	n	measd	calcd	n	measd	calcd	n
7700	7682	3	12800	12803	5	17900	17924	7
18000	17924	7	22500	23045	9	38400	38408	15
33300	33287	13	43500	43530	17	58900	58893	23
			84500	84499	33	79400	79378	31
						99500	99862	39
						119000	120347	47
						139000	140831	55

[a] In unified atomic mass units.

Perhaps, it is not only by chance that the largest fragment at 139.000 u corresponds with $(Au_{13})_{55}$, a two-shell cluster of Au_{13} clusters!

The concept of a 'cluster in the cluster', where here it is Au_{13} in Au_{55}, can be transferred to the larger clusters Pt_{309} and Pd_{561}. ^{195}Pt NMR investigations and susceptibility measurements (see Chapters 8 and 10) indicate that also for these full-shell clusters the properties of the inner-core metal atoms differ appreciably from those at the surface of the metal clusters. These differences are attributed to the effects of the ligand coordination on the electronic structure of the outermost metal atoms.

3.4. Catalysis

The marked tendency of large clusters to decompose in solution prevents their application in homogeneous catalysis. On the other hand, such huge metal clusters are potentially unique catalysts, since they combine the properties of a usual organometallic compound with structural characteristics of heterogeneous metal catalysts, as they are frequently applied in practice. However, the appliction of metal cluster compounds to homogeneous catalysis has so far led to disappointing results, even for small and medium sized clusters. Only a few examples have been described in the literature where cluster molecules ascertainably have acted as catalysts [47, 48]. The sucessful use of large clusters as catalysts requires their heterogenization by anchoring them on supports. If the clusters are strongly fixed on a surface, they cannot aggregate, even if they loose some of their ligands during the anchoring process. But since the number of successful applications of heterogeneous metal catalysts is already extremely numerous [48–51], the question may arise why metal cluster compounds should be of interest as heterogenous catalysts. In an important paper concerning the function of metal particles as catalysts, van Hardeveld and van Montfoort pointed out that the adsorption of molecular nitrogen on Ni, Pd and Pt crystallites only succeeds with particles between $\sim$ 15 and 70 Å in diameter [52]. The maximum of adsorptive activity was observed for 18–25 Å clusters. The reason for this behavior is explained by a maximum of so-called B_5-sites on these particles. The B_5-site is a position where an adsorbed atom comes into contact with five metal atoms of the cluster. This finding is only valid under the presumption that the metal clusters are crystallographically ordered and that they are shaped in such a manner that their free energy is minimized. This means that the number of metal-metal bonds is a maximum. To fulfill these conditions, clusters with optimized adsorption forces should have a spherical shape.

The clusters under discussion are in the size region of 10 to 35 Å. Following the concept of van Montfoort and van Hardeveld, this is exactly the region where the best results could be expected. Established metal catalysts show a broad size-distribution, with average size usually outside the mentioned range of 10–35 Å. One can assume that only parts of the metal particles really act

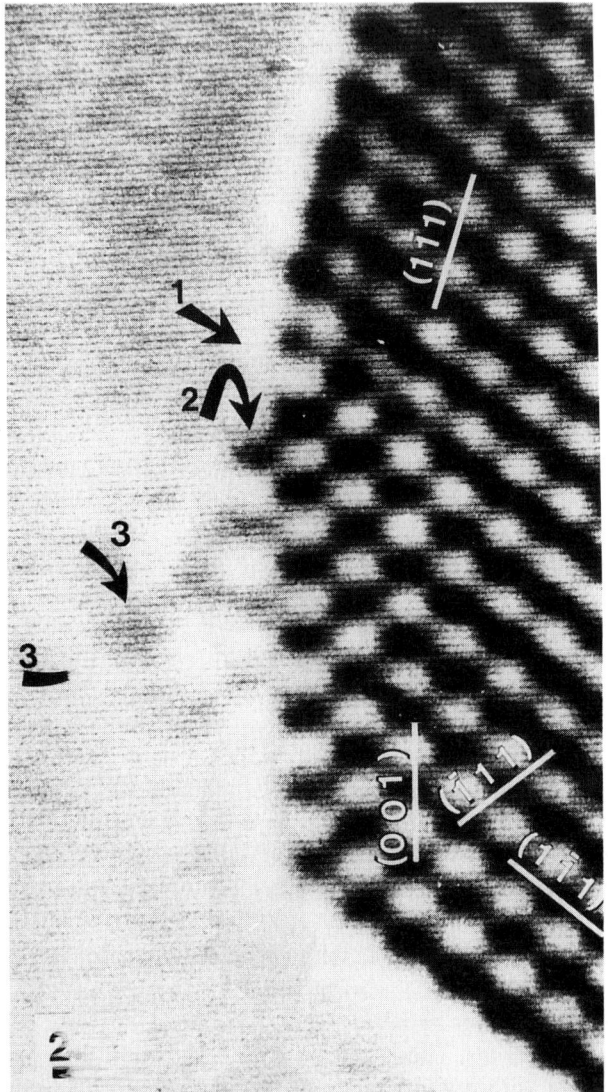

Fig. 17. Edge of a 75 Å gold colloidal particle with clouds of atoms outside a (001) face.

catalytically. From numerous electron microscopic investigations we know that, at least under the conditions in the electron microscope, the surface atoms of small metal particles are highly fluxional. This causes a continuous renewal of the surface and for this reason of the catalytic activity.

Special activities have been observed above (001), (100), (110), and (331) faces, respectively, but only rarely above (111) [53]. Figure 17 shows a 75 Å gold crystallite with clouds of atoms outside a (001) face.

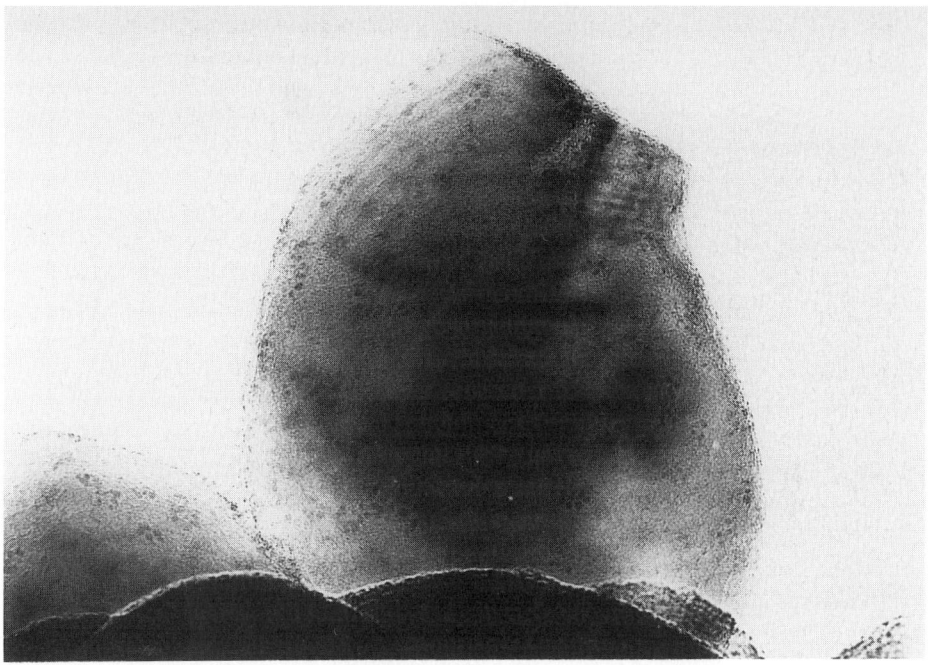

Fig. 18. Rh_{55} clusters on TiO_2. Groups of 6–10 cluster molecules are arranged around the entrances of micropores.

Atomic rearragenments are also observed above vertices. Such faces and vertices are in a permanent exchange with atoms in the gas phase. Processes in heterogeneous metal catalysis at higher temperatures might be compared with these events.

Rh_{55} clusters as well as 30–35 Å Pd clusters can be strongly fixed on supports like TiO_2 or Na-Y-zeolite. Detailed investigations of Rh_{55} on these supports indicate that the clusters are anchored on the openings of the micropores, and due to this interaction with the support they are, as expected, considerable stabilized [28, 54]. Na-Y-zeolite can be covered for 5% by Rh_{55} clusters, TiO_2 to a maximum of 1.5%. BET measurements of the undoted TiO_2 showed a surface of 21 m^2 g^{-1} and pore radii between 8.5 and 10.5 Å. This material looses its adsorption capacity quantitatively, if it is doted with 1.5% by weight of Rh_{55} clusters. This finding can only be interpreted by a closing of the micropores by the cluster molecules, like a bottle by a stopper, as the clusters have diameters of 10–11 Å, including the ligand shell. If such materials are heated to 200°C, the original adsorption capacity is quantitatively regained. This procedure can be followed by means of HRTEM. Figure 18 shows a TiO_2 particle, doted with 1.5% of $Rh_{55}[P[tert-Bu)_3]_{12}Cl_{20}$. Groups of 6–10 molecules are arranged together, probably around micropore entrances.

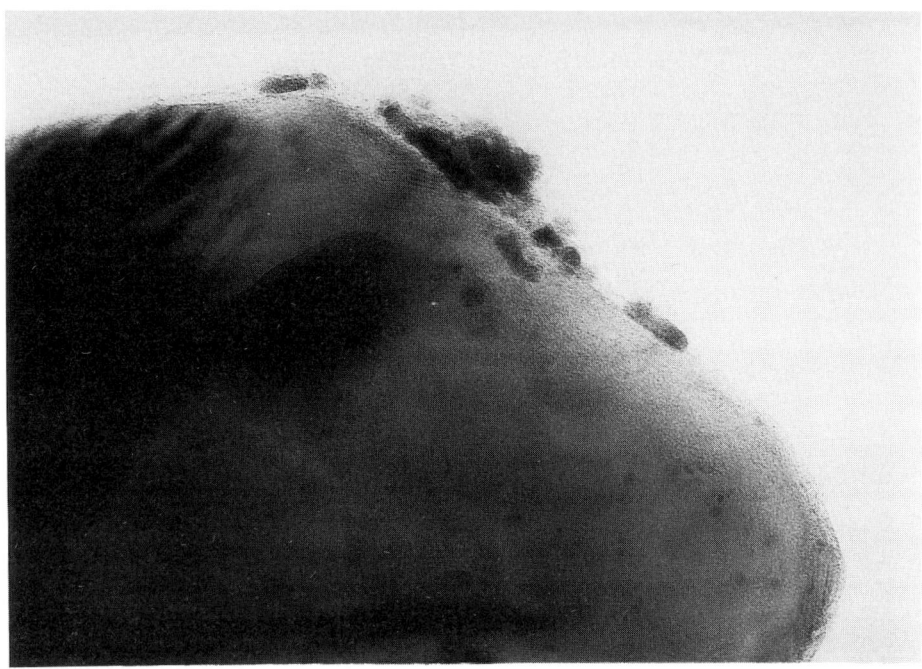

Fig. 19. A TiO$_2$ particle, originally doted with Rh$_{55}$ clusters, after heating to 200°C. Large aggregates are formed on the surface.

We suggest that after the anchoring of the first cluster molecule on a micropore, some others are attracted by van der Waals forces (see Section 3.2.1). Figure 19 shows such a particle after heating to 200°C. All cluster molecules have left their original positions to aggregate to larger particles with less or no catalytic activities.

Supported Rh$_{55}$ clusters have been used for hydroformylation reactions. For instance, propene is converted to butyric aldehyde in a heterogenous reaction. The results are impressive: 1 g of the catalyst (1% weight of cluster) has been used to prepare 3 kg of butyric aldehyde ($n : i = 1:1$) at 300–100 bar and 100–120°C, without any decrease of activity! This corresponds with a turnover of about 500 Mol$_{propene}$ × Mol$_{Rh}^{-1}$ × min^{-1}.

The mixture of 31.5 and 36.0 Å Pd clusters (see Section 3.2.2) can also be anchored on TiO$_2$. Their solubility in water enables also another catalyst preparation than the usual one by adsorption from solution: an aqueous solution of the clusters is used to hydrolyze Ti(O-i-C$_3$H$_7$)$_4$. TiO$_2$·aq is precipitated, embedding the cluster molecules. After drying, a TiO$_2$ powder is isolated, which is completely interspersed with cluster molecules. Such a catalyst may be of advantage, if it is stressed mechanically, as abrasion continuously gives new active surfaces. Table 2 compares three catalysts in the

TABLE 2

Comparison of different Pd catalysts for the hydrogenation of 1-hexene. A: 31.5 and 36.0 Å Pd clusters *on* TiO_2, B: 31.5 and 36.0 Å Pd clusters *in* TiO_2 (prepared by coprecipitation), and C: commercially available Pd catalyst (5% Pd on γ-Al_2O_3, FLUKA)

Catalyst	Pd/Educt Ratio (%)	p (bar)	T (°C)	t (min)	products (%)
A	0.012	1	20	360	hexane (100)
B	0.012	1	20	360	hexane (97.2) isomeres (2.8)
C	0.012	1	20	360	hexane (29.7) isomers (21.6) 1-hexene (48.7)

simple hydrogenation reaction of 1-hexene: TiO_2 with adsorbed Pd clusters (A), TiO_2 with embedded clusters, prepared by coprecipitation (B), and a commercially available Pd catalyst (C).

As can be seen, the catalysts A and B quantitatively form hexane at 1 bar of hydrogen pressure in the course of 360 min, whereas the usual catalyst (C) only forms ca 30% hexane and large amounts of isomers. The fact that a given amount of catalyst B acts as good as the same amount of A means in fact that B must be more active, because most of the metal particles are incorporated in the TiO_2 matrix and cannot act catalytically. Table 3 summarizes some further hydrogenation reactions of alkenes and of hexyne. Besides the fact that different olefins can be hydrogenated quantitatively, there are some selective hydrogenations of diolefins. 1,3-cyclooctadiene is hydrogenated at 30°C to cyclooctene, at 100°C the reaction is continued to give cyclooctane. The hydrogenation of dicyclopentadiene to dicyclopentene is finished after 280 min, after 420 min it is completely hydrogenated to dicyclopentane. The hydrogenation of 1-hexyne to 1-hexene is completed before the second reaction step, the formation of hexane, begins.

3.5. Outlook

It is difficult to make any prediction on the synthetic development of giant clusters by chemical methods. Up to now, successful preparative work is still characterized by a remarkable degree of accident. Though some experience has been acquired during the last decade, there is still no reliable procedure, how to synthesize clusters of a certain size. However, the preparation of a series of full-shell clusters has helped to gain valuable knowledge on metal

TABLE 3

Hydrogenation reactions of alkenes and 1-hexyne with 31.5 and 36.0 Å Pd clusters on TiO$_2$ (Catalyst A)

Olefin	Pd/Olefin	p (bar)	T (°C)	t (min)	Products	(%)
1-hexene	0.012	1	20	360	hexane	(100)
1,3-COD	0.010	1	30	300	cyclooctene	(100)
cyclooctene	0.010	1	100	240	cyclooctane	(100)
dicyclopentadiene	0.009	1	105	280	dicyclopentene	(96.2)
		1	105	420	dicyclopentane	(98.9)
acroleine	0.008	100	55	11	propanal	(100)
crotonaldehyde	0.008	220	85	60	butanal	(95.1)
1,4-cyclohexadiene	0.010	1	80	450	cyclohexane	(99.8)
		100	50	23	cyclohexane	(98.0)
1,3-cyclohexadiene	0.010	100	50	16	cyclohexene	(85.9)
					cyclohexane	(13.7)
1-hexyne	0.035	1	20	180	1-hexane	(97.5)
					hexane	(2)
					isomers	(0.5)

formation. The undisturbed aggregation of atoms in an inert medium, followed by trapping by appropriate ligands, seems the best suited route to go. To generate metal atoms by the reduction of ions in solution is the most simple method. Other routes, for instance the vaporization of metals and their subsequent aggregation in solution in the presence of ligands, could be a promising alternative.

There is no doubt that clusters in the size range of 10–40 Å are exciting species for physicists and chemists. The proof of the onset of metallic behaviour could be produced by the cluster compounds $Pt_{309}phen^*_{36}O_{30}$ and $Pd_{561}phen_{36}O_{190-200}$ (see Chapters 6, 8 and 10). From a chemical point of view, the application of large clusters as heterogenized catalysts is the most promising concept. As could be shown in initial attempts, the uniform size of full-shell clusters appears to speed up catalytic reactions immensely. The embedding of cluster molecules in supports by coprecipitation opens routes to novel applications.

$Au_{55}[(C_6H_5)_3P]_{12}Cl_6$ has been used to produce extremely thin conducting paths by the bombardment of thin cluster layers by laser or ion beams [55]. The access to quantum dots via clusters is another interesting perspective and numereous groups around the world are engaged in achieving this pretentious goal.

References

1. G. A. Ozin and S. A. Mitchell, *Angew. Chem. Int. Ed. Engl.* **22** (1983) 674.
2. E. Choi and R. P. Andres, *Physics and Chemistry of Small Clusters*, NATO ASI Series B: Physics, Vol. 158, Plenum Press, New Yoek (1987).
3. M. J. McGlinchey and P. S. Skell, *Cryochemistry*, J. Wiley (1976).
4. H. Hirau, Y. Nakao and N. Toshima, *J. Macromol. Sci. Chem.* **A13** (1979) 633.
5. H. Hirau, Y. Nakao and N. Toshima, *ibid* **A13** (1979) 27.
6. H. Hirau, Y. Nakao and N. Toshima, *ibid* **A13** (1978) 1117.
7. M. Komiyama and H. Hirau, *Bull. Chem. Soc. Jpn.* **56** (1983) 2833.
8. Y. Yonezawa, T. Sato, M. Ohno and H. Hada, *J. Chem. Soc. Faraday Trans* **1** (1987) 1559.
9. J. Turkevich, P. C. Stevenson and J. Hillier, *Disc. Faraday Soc.* **11** (1951) 55.
10. M. Mabuchi, T. Takenaka, Y. Fujiyoshi and N. Uyeda, *Surf. Sci.* **119** (1982) 150.
11. G. Frens, *Kolloid-Z. u. Z. Polymere* **250** (1972) 736.
12. D. G. Duff, A. C. Curtis, P. P. Edwards, D. A. Jefferson, B. F. G. Johnson and D. E. Logan, *Angew. Chem. Int. Ed. Engl.* **26** (1987) 676.
13. D. G. Duff, A. C. Curtis, P. P. Edwards, D. A. Jefferson, B. F. G. Johnson and D. E. Logan, *J. Chem. Soc. Chem. Comm.* (1987) 1264.
14. P. L. Freund and M. Spiro, *J. Phys. Chem.* **89** (1985) 1074.
15. J. Westerhausen, A. Henglein and J. Lilie, *Ber. Bunseng. Phys. Chem.* **85** (1981) 182.
16. R. Zzigmondy and P. A. Thiessen, *Das kolloidale Gold*, Akademische Verlagsgesellschaft, Leipzig (1925).
17. G. Schmid and A. Lehnert, *Angew. Chem. Int. Ed. Engl.* **28** (1989) 780.
18. G. Schmid, A. Lehnert, U. Kreibig, Z. Adamczyk and P. Belouschek, *Z. Naturforsch.* **45b** (1990) 989.
19. P. Belouschek, private communication.
20. J.-O. Bovin and J.-O. Malm, unpublished results.
21. R. Boese and G. Schmid, unpublished results.
22. D. N. Fulong, A. Laumikonis and W. H. F. Sasse, *J. Chem. Soc. Faraday Trans. 1.* **80** (1984) 571.
23. R. F. Marzke, W. S. Glaunsinger and M. Bayard, *Solid State Chem.* **18** (1976) 1025.
24. G. Schmid, A. Lehnert, J.-O. Malm and J.-O. Bovin, *Angew. Chem. Int. Ed. Eng.* **30** (1991) 852.
25. H. West and G. Schmid, unpublished results.
26. J. Bradley, private communication.
27. D. A. Jefferson, unpublished results.
28. G. Schmid and M. Harms.
29. G. Schmid, *Polyhedron* **7** (1988) 2321.
30. G. Schmid, *Materials Chem. Phys.* **29** (1991) 133.
31. M. N. Vargaftik, V. P. Zagarodinkov, I. P. Stolyarov, I. I. Moiseev, V. I. Likhobolov, D. I, Kochubey, A. L. Chuvilin, V. I. Zaikowsky, K. I. Zamaraev and G. I. Timofeeva, *J. Chem. Soc. Chem. Comm.* **1985**, 937.
32. G. Schmid, in *Aspects of Homogeneous Catalysis*, Vol. 7, R. Ugo (ed.) (1990).
33. H. A. Wieringa, C. C. Soethout, J. W. Gerritsen, L. E. C. van de Lemput, H. van Kempen and G. Schmid, *Advanced Materials* **2** (1990) 482.
34. L. E. C. van de Leemput, J. W. Gerritsen, P. H. H. Rongen, R. T. M. Smokers, H. A. Wieringa, H. van Kempen and G. Schmid, *J. Vac. Sci. Technol.* **B9** (1991) 814
35. G. Schmid, B. Morun and J.-O. Malm, *Angew. Chem. Int. Ed. Engl.* **28** (1989) 778.
36. G. Schmid, R. Boese, R. Pfeil, F. Bandermann, S. Meyer, G. H. M . Calis and J. W. A. van der Velden, *Chem. Ber.* **114** (1981) 3634.
37. G. Schmid, U. Giebel, W. Huster and A. Schwenk, *Inorg. Chem. Acta* **85** (1984) 97.
38. G. Schmid and W. Huster, *Z. Naturforsch.* **41b** (1986) 1028.

39. C. Becker, Th. Fries, K. Wandelt, U. Kreibig and G. Schmid, *J. Vac. Sci. Technol.* **B9** (1990) 810.
40. Th. Fries, K. Wandelt, K. Fauth, G. Schmid and R. Boese, unpublished results.
41. G. Schmid, *Structure and Bonding*, **62** (1985) 51.
42. G. Schmid, N. Klein, L. Korste, U. Kreibig and D. Schönauer, *Polyhedron* **7** (1988) 605.
43. R. E. Benfield, J. A. Creighton, D. G. Eadon and G. Schmid, *Z. Phys. D* **12** (1989) 533.
44. G. Schmid and N. Klein, *Angew. Chem. Int. Ed. Engl.* **25** (1986) 922.
45. H. Feld, A. Leute, D. Rading, A. Benninghoven and G. Schmid, *J. Amer. Chem. Soc.* **112** (1990) 8166.
46. H. Feld. A. Leute, D. Rading, A. Benninghoven and G. Schmid, *Z. Physik D* **17** (1990) 73.
47. L. Markó and A. Vizi-Orosz, in *Studies in Surface Science and Catalysis*, Metal Clusters in Catalysis, Vol. 29, Elsevier, Amsterdam (1986).
48. B. Walther, *Z. Chem.* **29** (1989) 117.
49. B. C. Gates, in *Studies in Surface Science and Catalysis*, B. C. Gates, C. Guczi and H. Knözinger (eds.), Metal Clusters in Catalysis, Vol. 29, Elsevier, Amsterdam (1986).
50. R. Psaro and R. Ugo in [49].
51. H. Knötzinger and B. C. Gates in [49].
52. R. van Hardeveld and A. van Montfoort, *Surface Sci.* **4** (1966) 396.
53. J.-O. Bovin, L. R. Wallenberg and D. J. Smith, *Nature* **317** (1985) 47.
54. G. Schmid, R. Küpper, H. Hess, J.-O. Malm and J.-O. Bovin, *Chem. Ber.* **124** (1991) 1889.
55. P. Hoffmann, private communication.

4. THEORY OF ELECTRONIC PROPERTIES OF METAL CLUSTERS AND PARTICLES

D. E. ELLIS
Northwestern University, Evanston, IL 60208, U.S.A.

4.1. Why Are Metal Particles Interesting?

In this chapter, the word 'particle' is used in a general sense, referring to naked clusters, to ligated molecular units, and to structural fragments of an extended system such as a supported catalyst. One theme of theoretical interest has been the effort to understand and describe how specific properties such as ionization potentials evolve from the atomic limit, through M_n n-atom clusters, arriving at the bulk solid. On the experimental side it has been noted that important properties such as chemical reactivity are size and shape dependent, offering opportunities to study and control materials properties. Cluster compounds provide a nearly ideal testing ground for the comparison of theories and experiment, since the local coordination can be precisely determined and a single structure can be characterized. Among others, the ceramics and metallurgical industries are beginning to exploit the possibilities of novel materials formed by condensation and compaction of particle vapors. Lessons learned from study of cluster compounds can be applied to analysis of composite particulate materials.

This brief article makes no pretense of completeness nor of a unified view of the phenomena and techniques used in this vast field. Instead, a discussion is given of the current state of theories of electronic structure, using examples to illustrate some main points of interest. The reader may consult several of the fine recent reviews for more detailed treatments [1–3]. A main focus of theoretical discussion will be spectroscopic properties, and what they tell us about the local chemical environment and bonding interactions. The magnetic properties which have been of particular interest to theorists for many years are seen to be intimately related to the local chemical bonding. The dynamical and thermal properties are also briefly mentioned here, since they are fundamentally interrelated with the electronic states. Finally, an assessment is made of some emerging methodologies which offer the possibility of theoretical modelling with real predictive power.

4.2. Model Hamiltonians

Model Hamiltonians or semiempirical approaches have the advantage of simplicity and computational efficiency. Qualitative models can be constructed, starting from an enumeration of bonding, antibonding, and nonbonding molecular orbitals (MO) permitted by a given cluster symmetry. In general, maximum stability is expected when bonding orbitals are filled and antibonding orbitals are empty; this leads to well-known electron counting rules [4]. By combining interpretations of experiment with qualitative MO schemes one may be able to rationalize a large body of data, as in the polynuclear $(d^4)_N$ complexes of Re, Mo and W [5]. However, since the density of metal levels increases with nuclearity, and interactions with ligands introduce large uncertainties in level ordering, it becomes necessary to perform explicit MO calculations.

In circumstances where the symmetry of the cluster is unknown, or many geometries may lie close in energy, the qualitative schemes may remain useful as a means of cataloging possible outcomes of spectroscopic measurements. As the cluster size increases, drastic simplifications are required in order to make predictions of relative stability, and by implication relative abundance of different species. Viable methodologies are discussed below; we would like to note here that geometrical packing arguments combined with 'chemical probe' experiments provide alternative interpretations and a challenge to MO theories [6].

In the hands of Hoffmann and collaborators [7] the Extended Hückel Theory (EHT) has provided much insight into the qualitative features of metal-metal and metal-ligand bonding of cluster compounds. Making use of minimal or near-minimal basis sets, the EHT models provide an attractive interpretation of effective atomic configurations and bonding charge distribution. Since the matrix elements of the Hamiltonian can be chosen optimally to fit spectroscopic data, the model can be 'calibrated' against a reference system, and then used to extrapolate into new territory. Of course the biases and preconceptions of the investigator are built into the parameter selection, and the range of safe extrapolation is unknown *a priori*.

The familiar secular equation

$$(\mathbf{H} - \mathbf{ES})\mathbf{C} = \mathbf{0} \tag{1}$$

may be solved in the atomic orbital basis C to obtain expansion coefficients of the cluster molecular orbitals (MO) Ψ,

$$\Psi_\mathbf{i} = \Sigma \chi_\mathbf{j} \mathbf{C}_\mathbf{ji}. \tag{2}$$

A Mulliken population analysis of the occupied eigenvectors provides the effective atomic configurations, and input for possible further iterations of the potential. Populations of individual levels $f_{\nu nl}(i)$ can be identified with atomic site ν and quantum numbers nl, and in conjunction with a line-broadening function $L(E, Gk\ E_i)$ permit the calculation of Partial Densities of States (PDOS),

$$D_{\nu nl}(E) = \Sigma f_{\nu nl}(i) L(E, Gk\ E_i). \tag{3}$$

PDOS curves give a simple visualization of the distribution of component atomic states over the MO energy spectrum, replacing the tedious examination of composition of individual MOs which contribute to a 'band' of states. The DOS provide an important point of contact with experimental photoelectron, optical, X-ray, and electron energy-loss data. When the dipole transition matrix elements are sufficiently slowly varying, the DOS (or its convolution to include a range of initial and final states) can be used to interpret both position and amplitude of spectral features. The presence of spectral features not seen in the DOS may be a useful indication of two-electron or many-body excitations, thus focussing attention for further study.

When theoretical oscillator strengths f_{ij} connecting initial state i and final state j are available, the absorption cross-section

$$\sigma(\hat{e}, E) \sim \Sigma f_{ij}(\hat{e}) n_i (1 - n_j) L(E, Gk\ E_{ij}) \tag{4}$$

can provide the most detailed interpretation. Here $\hat{e}$, E are the photon polarization and energy, n_i are occupation numbers and e_{ij} is the electronic excitation energy. Coupled with experimental studies using polarized radiation in crystalline samples, detailed absorption studies can help to disentangle metal-ligand interactions of rather complicated transition metal complexes [8].

As an example of historical importance in understanding the relation of cluster properties to those of the ordered crystal, we may consider studies made on molybdenum chalcogenide clusters, chains and extended solids [9]. Here EHT calculations were made both on isolated clusters like $(Mo_6S_8)^{4-}$ and on the extended structure of an Mo_6S_8 'crystal component' found in Chevrel-phase compounds with chemical formula MMo_6X_8. Here M is a metal and X = S, Se, Te, or halogen. The crystal band structure studies were carried out by introducing the k-space Bloch-wave combinations of atomic orbitals which decompose the secular equation (1) into block-diagonal form, one block for each k value [10]. Theoretical band structures are generally validated by comparison with photoelectron spectra, and with X-ray absorption and emission cross-sections. The level of agreement may be reasonable, but still leaves many questions to be resolved by 'local probes' such as EXAFS, NMR, ESR, and Mössbauer spectroscopy. Unfortunately, few theoretical studies are carried out to the level required for such detailed comparisons with experiment.

The intimate relationship between the isolated cluster MOs and the network of interacting clusters is seen in Figure 1.

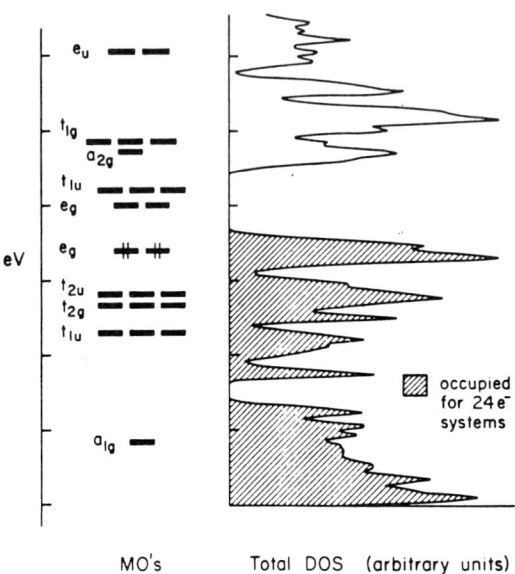

Fig. 1. Mo-based MOs for the $(Mo_6S_8)^{4-}$ cluster and total DOS for an Mo_6S_8 crystal component of a Chevrel-phase compound [7].

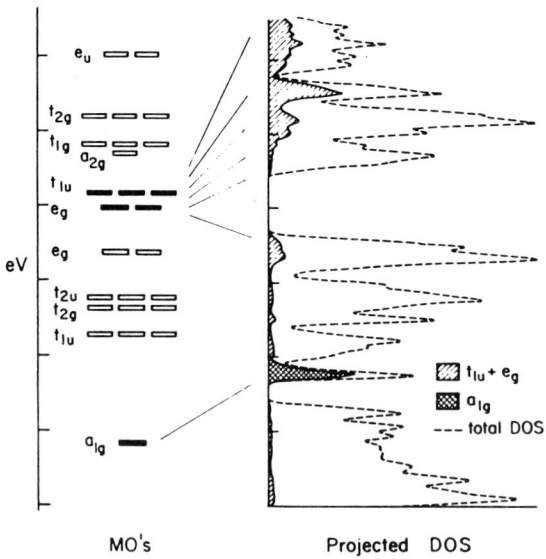

Fig. 2. PDOS curve for z^2 type frontier orbitals of the Mo_6S_8 cluster in the Mo_6S_8 crystal, in comparison with total DOS [9].

The utility of the PDOS analysis is seen in Figure 2, where *cluster* MOs are used as the projection basis to reveal effects of intercluster interactions.

In EHT and other MO theories, the size of the secular equation (1) increases as N, the number of basis functions, and the computation time increases ultimately as N^3. This approach becomes cumbersome as the cluster size increases and the computational load becomes heavy, even when the matrix elements are 'free'. The k-space band structure approach exploits periodic symmetry of a crystal, and reduces the problem to N_k uncoupled equations each of size N_c, where N_k is the number of k-vectors chosen to sample the bands and N_c is the isolated cluster size. Actually, in self-consistent treatments, the equations are coupled by the potential matrix elements and k-space integrations must be included in the iterative loop. As N_c becomes large and the internal structure of the cluster becomes complex, additional types of partitioning or fragment analysis become worthwhile.

It is therefore interesting to consider another model Hamiltonian approach, often called Recursive Tight Binding (RTB), which is capable of treating very large and infinite systems to reasonable accuracy [11]. An important concept underlying the RTB idea is the fact that a Hermitean matrix such as $\mathbf{H}$ can be reduced to tridiagonal form by a set of orthogonal transformations which amounts to a particular choice of basis. The eigenvalue spectrum and other properties can be obtained in the form of partial fraction expansion [12]. This methodology is very closely related to the Green's function approach, where site- and atomic orbital-based spectral densities are emphasized rather than individual MO contributions.

In the simplest RTB applications, it is customary to choose a nearest-neighbor interaction model. Then the topological ramifications of the particle structure can be worked out in detail through the partial fraction expansion. There is, however, no need to limit the RTB scheme to short-range interactions, and some interesting applications have been made recently to transition metal particles, chains, surfaces, and defective solids [13, 14]. For example, Jena et al. have carried out first principles Local Density (LD) calculations on transition metal dimers, and have fitted their results in a Slater–Koster tight-binding parametrization [15]. The variation of matrix elements $\langle il|\mathbf{H}|jm\rangle$ with internuclear separation R_{ij} was fitted with a simple exponential form; here l, m identify individual spin orbitals of the LCAO expansion basis. The n-th moments of the (i,l) PDOS

$$\mathbf{M}_n^{i,l} = \langle i,l|\mathbf{H}^n|i,l\rangle \tag{5}$$

are then calculated as a sums over closed loops of n steps which originate at site i. The partial fraction expansion of the PDOS in terms of the moments gives the results exemplified in Figure 3.

The RTB densities of states are in remarkably good agreement with first principles band structures for the bulk and general features of photoemission data [16], and reproduce the well-known band narrowing at surfaces and in low-dimensional geometries.

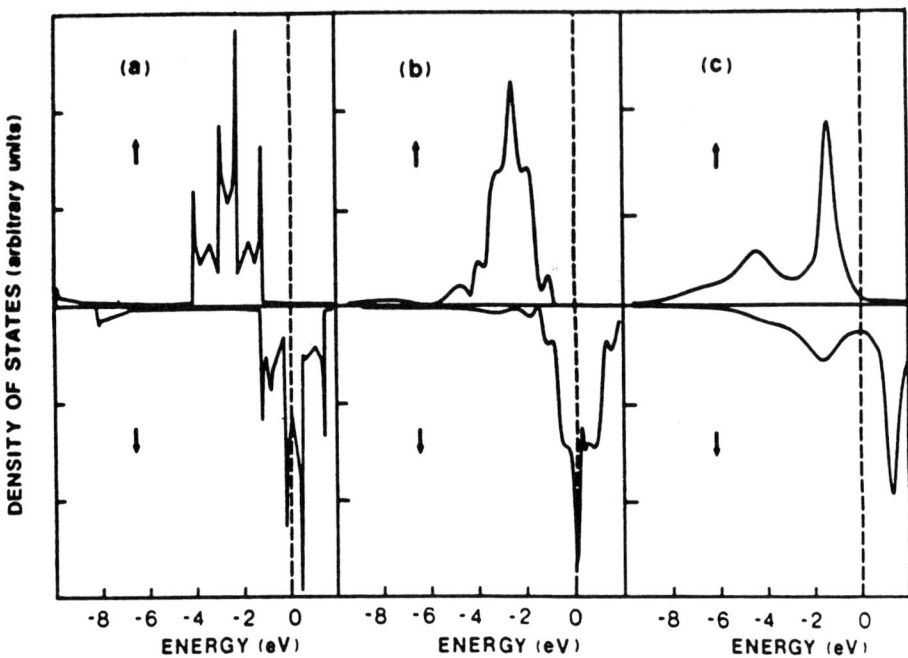

Fig. 3. Spin-resolved DOS for iron in RTB model; (a) linear chain, (b) (100) monolayer, (c) bcc bulk [13].

One of the great advantages of the RTB scheme is that it is not restricted to periodic systems; it can be equally well applied to impurity, defect-state and amorphous structures. In a study of the role of oxygen impurities on stacking-fault defects in transition metals, Wang et al. used the RTB method to probe the PDOS of Ni atoms at various distances from the impurity [14]. This pioneering work showed how to calculate contributions to the stacking-fault energy from individual sites out to fourth-nearest neighbors, bringing us closer to an atomistic theory of mechanical and structural properties. The fact that the calculated defect energy is close to that of experiment [17] is very encouraging; however, we are reminded that this and many other critical properties have a dynamical component, in this case defect-related vibrational energy [18]. The challenge of integrating models for electronic properties with dynamical and thermodynamical effects is one of the more interesting current problems of theory.

The LCAO matrix elements were obtained from the empirical methodology of Harrison [19]; it would be most interesting to carry out further RTB analyses of this type using first-principles matrix elements from cluster calculations.

4.3. Traditional Quantum Chemical Methods

In this section we briefly discuss the rigorous quantum chemical methods, their successes, and their limitations. The Hartree–Fock (HF) approach, with its single determinant wavefunction formed from optimized MOs forms the basis for much of what we know about molecular electronic structure [20]. The optimization is carried out so as to minimize the total energy, for a given nuclear geometry, thus giving a powerful tool for studying structure and cohesion. The computational limitations of HF are well known, arising principally from the N^4 increase in the number of two-electron Coulomb and exchange matrix elements with basis size N. The use of frozen cores and effective core potentials to focus effort upon the valence MOs is a strategy which has carried the HF scheme to a point where metal clusters can be effectively treated. Despite its well-known failure to treat Coulomb correlations, and imbalanced treatment of exchange and correlation energies, the HF approach continues to provide useful insights.

A good illustration is provided by the recent work of Rao and Jena on the electronic structure and geometries of heteroatomic clusters [21]. Here spin-unrestricted HF calculations were made on Li_nMg and Li_nAl clusters, with $n \leq 7$, using a Gaussian-type orbital (GTO) basis. Some correlation corrections were included in the total energy by perturbative inclusion of two-electron excitations at the end of the HF procedure. The atomic binding energies and ionization potentials (see Table 1) provide a rich data base for comparison with experiment. Since the geometries of all but a few gas-phase clusters are unknown, the prediction of structures and their modification with 'alloying' are very useful. The rather satisfactory agreement of ionization energies, where experimental data are available [22], gives some confidence in the atomization energies, defined here as the energy gained by binding an additional atom to the cluster:

$$\mathbf{E_B} = \mathbf{E(M_nX)} - \mathbf{E(M_n)} - \mathbf{E(X)}. \tag{6}$$

The relative stability of clusters of different composition and in particular the famous even-odd alternation of stability [23] can thus be successfully addressed at the HF level, and indeed by other MO theories as well. Comparison with relative abundance data in molecular cluster beams is somewhat delicate, since kinetic conditions and thermodynamic factors are known to play an important role. Here systematic synthesis and characterization of mixed metal cluster compounds is needed and will make a major contribution to our understanding of intermetallic bonding in the 'cluster state'.

The HF wavefunction has to be augmented or replaced to obtain more accurate treatment of electron correlations which are important for bonding and cohesion in metal clusters. The Multi-Configuration Self-Consistent Field (MCSCF) approach gives a minimal extension to a multi-determinant wavefunction [20]; however, computational experiments with many determinant Configuration Interaction (CI) wavefunctions in transition metal dimers

TABLE 1

Total energy and atomization and binding energies, vertical ionization potentials (IP) of Li_n, Li_nMg and Li_nAl clusters [21].

Cluster	Total energy (au)	Atom. energy (eV)	Adatom Binding (eV)	ΔE_N (eV)	IP (eV)
LiAl	-248.4102	9.58	-9.58	-9.58	5.00
Li_2Al	-255.8659	11.09	-10.11	-1.52	2.02
Li_3Al	-263.2354	13.24	-11.85	-2.17	3.60
Li_4Al	-270.8025	14.79	-11.91	-1.55	3.02
LiMg	-206.2159	3.20	-3.20	-3.20	1.70
Li_2Mg	-231.7448	6.70	-5.73	-3.50	3.67
Li_3Mg	-221.1651	7.25	-5.85	-0.55	3.14
Li_4Mg	-228.6283	8.96	-6.09	-1.71	3.60
Li_5Mg	-236.0559	9.71	-6.19	-0.74	1.99
Li_6Mg	-243.5297	11.71	-6.66	-2.00	3.72
Li_7Mg	-250.9491	12.23	-6.06	-0.52	2.05
Li	-7.4002			0.00	4.86
Li_2	-14.8371	1.00		-1.00	4.44
Li_3	-22.2534	1.43		-0.44	3.19
Li_4	-29.7085	2.93		-1.49	3.84
Li_5	-37.1374	3.70		-0.78	3.59
Li_6	-44.5870	5.05		-1.34	4.23
Li_7	-52.-324	6.28		-1.23	2.95

and trimers [24] show that there are difficulties in generating a sufficiently converged expansion. Other more 'intuitive' approaches such as Generalized Valence Bond [25] (GVB), Many-Body Perturbation Theory (MBPT) and Coupled Cluster Perturbation Theory (CCPT) [26] allow theorists to custom-build model wavefunctions (often implicit) which hopefully contain the most important components of correlation. These 'experiments' allow us to test effects of extending single-particle descriptions which contain some averaged effects of interelectronic correlation (as in effective Hamiltonian and Local Density methods) to more rigorous and detailed models. The acid test of these experiments is a comparison with experimental spectroscopic and structural features.

As an example, let us discuss an application of GVB theory to study the take-up of hydrogen by small platinum particles [27]. Cluster compounds containing the Pt_4 moiety are well known and much studied, including 'metallic bonding' cases like $Pt_4(CO)_5(PMe_2Ph)_5$ and 'covalent bonding' cases such

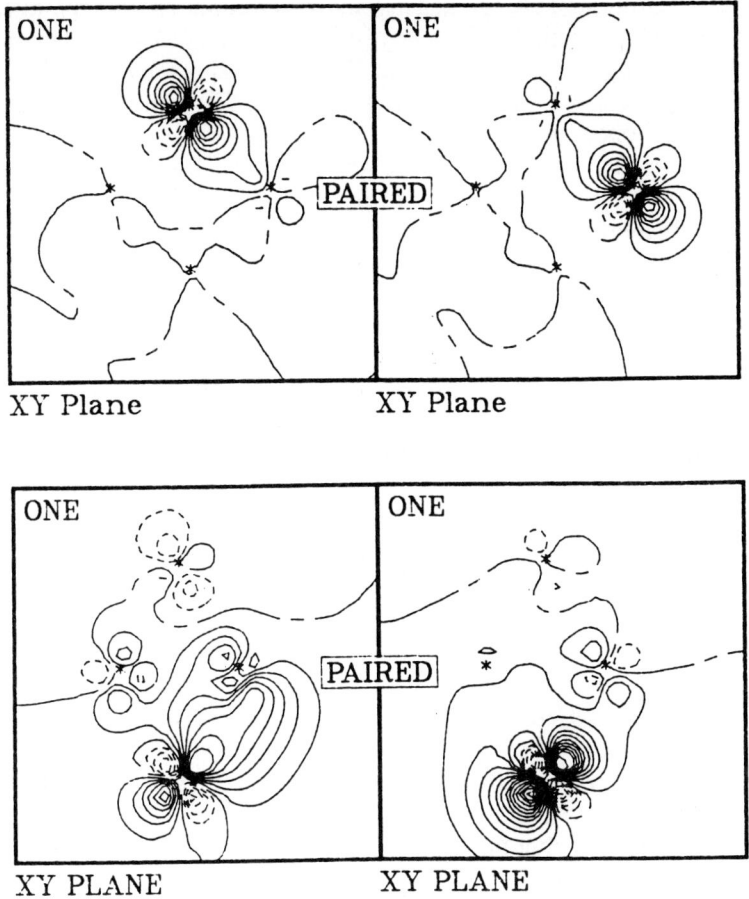

Fig. 4. GVB Pt-Pt bond orbitals in planar Pt$_4$: (a) ground state, (b) excited state [27].

as Pt$_4$(O$_2$CCH$_3$)$_8$ whose X-ray structures have been precisely determined [28]. The molecular beam studies of Riley *et al.* [29] and Cox *et al.* [30] on adsorption of hydrogen, water, and other simple molecules onto Pt$_n$ particles provide a particularly attractive ground for developing theoretical models of ligation to the metal. The modelling of naked particle + adsorbate can hopefully reveal basic features which would be obscured in the cluster compound environment. The experimental observation that hydrogen takeup tends to saturate at the Pt$_n$H$_{2n}$ composition [29] provides an immediate challenge. The observation of 'abnormally large' deuterium uptake on small transition metal clusters [30] forces us to recognize that simple explanations, while useful, seldom tell the entire story.

Local Density studies were made of Pt$_4$ and a variety of Pt$_4$H$_n$ complexes, and the high relative stability of the Pt$_4$H$_8$ case was verified. However, the

LD level structure and charge density analysis did not answer the question: why is this so? A detailed GVB study of the Pt$_4$ 'core' was made in order to develop a valency model for platinum. The essential results are summarized in Figure 4, where the 'covalent' ground state GVB orbitals are contrasted with 'metallic' excited state orbitals.

The square-planar ground state is found to form four two-electron two-center bonds which tie up all unpaired electrons of the d^9s^1 Pt atom, resulting in a relatively inert closed-shell singlet. A low-lying excited state with energy 10 kcal/mole, at the beginning of a manifold of excitations, has a planar rhombus structure. The bonds here are localized on the edges between atoms; four one-electron two-center bonds account for the s-electrons and leave the d-electrons in a nonbonding state. The low-lying states of the variously coupled d^9 manifold are thus available to interact with incoming ligands; the singly occupied $d(z^2)$ orbitals being most obvious.

The basic lesson derived from this work is that a concerted geometrical and electronic transition 'prepares' the metal particle for ligation; a simple bond-counting then rationalizes the observed hydrogen saturation. The irrelevance of the ground state geometry and wavefunction to complex formation is eye-opening and reminds us of the dangers of focusing always on ground state structures. The existence of such excited state 'metallic' bond structures has been pointed out by Goddard *et al.* in work on alkali metals, silver, and nickel [31]. The simple answer to the question "Why does LD succeed in predicting this behavior?" seems to be that the effective exchange and correlation potentials used crack open the $5d$ shell, making it an active participant in bonding, without invoking the details of multiplet structure. The gold-cluster case discussed below reinforces this notion.

It is not surprising that the most complete applications of *ab initio* theory to particles have involved alkali metal systems. The very extensive CI studies on ground and excited states of Li$_n$ and Na$_n$, $n \leq 8$, and (Na$_n$)$^-$, $n \leq 5$, by Koutecky *et al.*, have set a standard for future work on more complex metals. In connection with the photoelectron detachment spectra of the anions, and the photodepletion spectra of the neutrals, it is possible to identify major electronic transitions and most common geometries. Details of this work have been given in recent reviews [32].

4.4. Density Functional Approaches

Some aspects of the Density Functional (DF) approach to electronic structure have been mentioned above; the goal of this section is to give a sketch of the DF scheme and its current direction of evolution. DF theory is a many-body theory which seeks to find feasible ways of approximating the interelectronic exchange and correlation interactions. In its most popular and practical Local Density form, an effective potential $\mathbf{V_{xc}}(\rho(\mathbf{r}))$ is obtained which depends only on the (spin) density at the field point $\mathbf{r}$ [33]. While more elaborate nonlocal forms of effective potential are known and are beginning to be used, the great number of LD results for molecular and solid state systems testifies to the power of the scheme [34].

Within the realm of molecular and cluster modelling there exist a number of LD implementations, which differ in their approach to representing the wavefunction, approximately generating the potential, and solving the Schrödinger (or Dirac) equations. A partial list of the methods in use today would include: Scattered Wave (MS-Xα) [35], LCAO Discrete Variational (DV-Xα) [36], and GTO Variational [37]. These and related methods seek to solve the single-particle electronic equations in the field of *fixed* nuclei. The problem of determining geometries and interatomic potential surfaces is treated by distinct and sequential SCF calculations for different nuclear coordinates. Recently new methodologies have been introduced in which nuclear motions are treated simultaneously, in a process sometimes called 'simulated annealing' [38]. A time variable is introduced, which is sometimes treated as a dummy variable in a search for the minimum energy configuration, and sometimes treated as 'real' in an effort to describe dynamical effects. Although applications have been limited mostly to simple systems such as Si_n and Na_n for technical reasons, dynamical approaches are sure to take on greater importance in the future.

In the following section, some examples are given of LD results which throw light on current problems of cluster electronic structure.

4.4.1. ALKALI METAL CLUSTERS

The most primitive, and yet strikingly successful LD model of alkali metal clusters arises from the nearly free electron, or jellium, model. Here the particle is treated as an electron gas confined to a spherical or ellipsoidal potential well, thus averaging over ionic core potentials [39]. The 'magic numbers' n associated with M_n particles of high relative abundance in molecular beams can be explained to first order by closings of the jellium electronic shells, exactly as in nuclear (or atomic) shell-structure theory [40]. While the abundance data do seem to depend somewhat on the beam temperature [41], evidence for shell-closings is also found in ionization spectra versus particle size, in addition to the well-understood even-odd spin-pairing alternation.

Particle shape information can be obtained within the jellium models by introducing the ion cores as a perturbation, or by directly inserting the spherical average of the core potentials [42]. Mixed metal clusters such as $ZnNa_8$ can be treated by inserting additional spherical regions with potentials parametrized to describe the 'impurity' atom [43]. This scheme is adequate to alter the uniform jellium level structure to bring it into agreement with experiment [44] and with DV-Xα MO predictions [45].

Among the many MS-Xα studies on alkali clusters, the work of Martins *et al.* [46] is notable for showing the quantitative agreement between theoretical and experimental ionization trends, and for pointing out the importance of

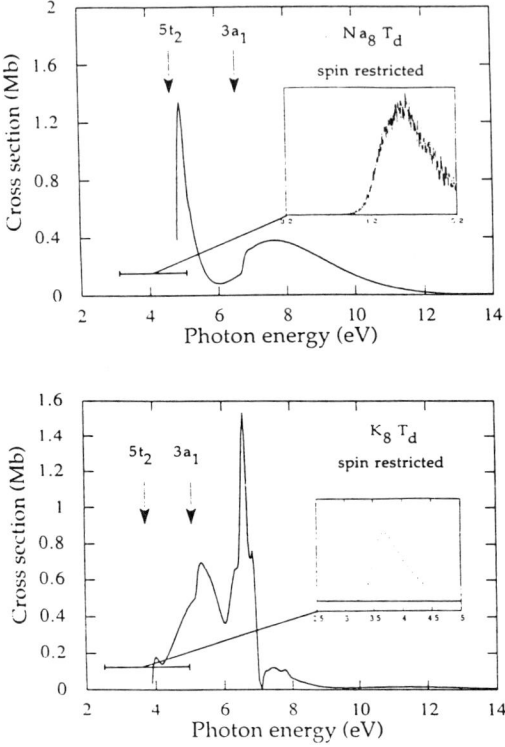

Fig. 5. Predicted total photoionization cross-section of eight atom clusters in T_d symmetry: (a) sodium, (b) potassium [47].

geometry relaxation in the excited state. Recently, Wästberg and Ros'en have presented the photoionization cross-sections in detail for Na_n and K_n, $n \leq 8$, using the MS-Xα continuum orbital approach [47]. These calculations which predict shape resonances in the UV cross-section due to quasi-bound excited states, present a challenge to experiment which may be resolved by use of tunable synchrotron radiation sources.

Finally, let us note the applications of simulated annealing, or Born–Oppenheimer dynamics, to Na_n and Na_n^+ clusters, $n \leq 4$, by Barrett et al. [48] using an LD pseudopotential. This work is particularly interesting in its discussion of photodissociation at finite temperature.

4.4.2. Cluster Compounds

To follow up on our previous discussion of $5d$-element bonding, let us consider some recent work on main-group-element centered octahedral gold clusters, carried out in the DV-Xα scheme [49]. The discovery of carbon-centered gold cluster compounds raises some fundamental questions about metal–ligand bonding and stability. It is possible to construct plausible MO

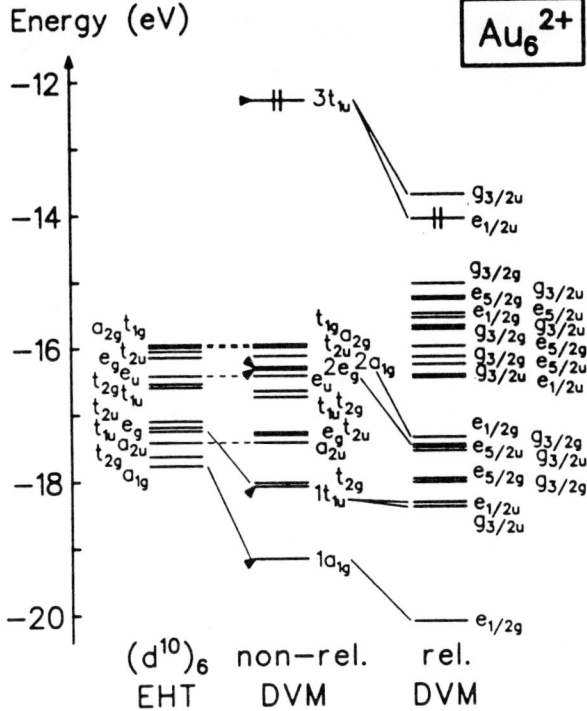

Fig. 6. Comparison of Hückel levels (5d shell excluded) and DV-Xα relativistic and nonrelativistic levels for Au_6^{2+} cluster [49].

diagrams and electron-counting schemes either with the central atom as a donor, C^{4+}, or as an acceptor, C^{4-}. The LD calculations show that carbon is an acceptor, although not to the extent suggested by formal valency arguments. Early analyses proceeded on the plausible assumption of an inert $5d^{10}$ gold configuration, with all of the 'action' taking place in the $6s$ and possibly $6p$ shells. The LD results show that, on the contrary, the bonding interactions crack the $5d$ shell, whose electrons become active participants in the bond formation, with the expected expulsion of antibonding states from the occupied portion of the spectrum. A fully relativistic Dirac–Slater treatment of the system shows that the well-known indirect shielding effects stabilize the s, p manifold at the expense of the d manifold. This further increases the reactivity of the $5d$ shell, increasing its effective bandwidth and hybridization with Au s- and p-electrons, and their mixing with ligand levels.

Comparison with photoelectron and single crystal optical absorption spectra would be most interesting, in order to validate and develop further the electronic structure model.

TABLE 2

Mulliken population analysis of selected molecular orbitals with Au $6s$ and Au $5d(z^2)$ character, and total valence populations for the cluster $\{[(H_3P)au]_6C\}^{2+}$ [49].

MO		Energy (eV)	C s	C p	Au $s+p$	Au d	PH_3
$1a_{1g}$	nr[a]	-20.13	0.70		0.05	0.21	0.05
	rel[b]	-19.54	0.65		0.12	0.17	0.05
$1t_{1u}$	nr	-16.06		0.19	0.02	0.47	0.33
	rel	16.53		0.29	0.03	0.25	0.43
$1e_g$	nr	-15.25			0.02	0.37	0.61
	rel	-15.57	0.00		0.02	0.17	0.80
$2a_{1g}$	nr	-14.41	0.11		0.05	0.16	0.69
	rel	-14.65	0.11		0.08	0.06	0.75
$3a_{1g}$	nr	-12.45	0.00		0.50	0.40	0.10
	rel	-13.07	-0.01		0.33	0.65	0.03
$3t_{1u}$	nr	-11.96		0.03	0.22	0.43	0.32
	rel	-12.07		0.07	0.23	0.59	0.10
$3e_g$	nr	-11.10			0.17	0.57	0.26
	rel	-10.67	0.00		0.16	0.76	0.08
$4t_{1u}$	nr	-9.76		0.36	0.30	0.21	0.13
	rel	-10.22		0.30	0.12	0.39	0.19
Total	nr		1.67	4.44	0.77	9.57	7.97
	rel		1.55	4.96	0.99	9.31	7.95

[a] Nonrelativistic calculation.
[b] Relativistic calculation.

4.4.3. TRANSITION METAL PARTICLES: FREE AND EMBEDDED

A strong effort to understand the level structure and related chemistry of small transition metal particles has been driven by analogies to well-studied surface chemisorption and interest in catalytic processes. An entire chapter could well be devoted to LD work in this area; instead we must select a few typical results. A notable sub-theme of theoretical work here is the relation of magnetic properties to chemical state.

The work of Dunlap on Fe_{13} provides useful insights on the relationship between spin and structural stability [50]. The relative stability of the icosahedral structure over bulk-related fcc and hcp structures is demonstrated in GTO variational calculations. The very small total energy differences which separate one optimized geometry from another would not surprise someone used to empirical potential models; however they remind theorists of the

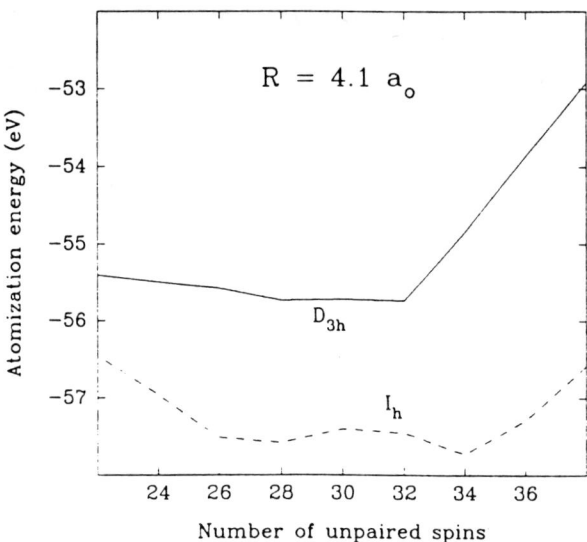

Fig. 7. Atomization energies of hcp (solid line) and icosahedral (dashed line) isomers of Fe_{13} cluster [50].

critical nature of potential approximations and basis truncations which are commonly used in LD applications.

DV-Xα work on interatomic potentials derived from small transition metal particles and consequences for bulk and surfaces has been recently presented [51]. A nice example of the effects of chemisorption and alloying on the magnetism of transition metal clusters is found in the work of Salahub and Raatz [52], who used the MS-Xα approach. They treated $Ni_{14-n}Cu_nCO$ clusters with varying structural arrangements, finding a reduction of Ni magnetic moment with both Cu and CO content, in qualitative agreement with experiment on bulk and surface environments. The moment reduction was found to be a 'local' effect (as confirmed by other workers on similar clusters), traceable to specific orbital mixing. A concerted effect was noted, whereby alloying could lead to a local increase in moment, contrary to the general trend. The importance of such studies for the design of magnetic materials is rather obvious.

Similar phenomena have been explored in high nuclearity carbonylated nickel clusters, $[Ni_{32}C_6(CO)_{32}]^{n-}$ and $[Ni_{44}(CO)_{48}]^{n-}$ with $0 \leq n \leq 6$ by Rösch et al. using the GTO approach [53], following up previous small cluster studies [54]. They used the characteristic quenching of the local magnetic moment by ligands to study the transition from molecular to bulk metallic magnetic behavior. Figures 10 and 11 demonstrate dramatically the transition from the paramagnetic bare Ni_{44} cluster with exchange-splitting of $\sim$ 1 eV

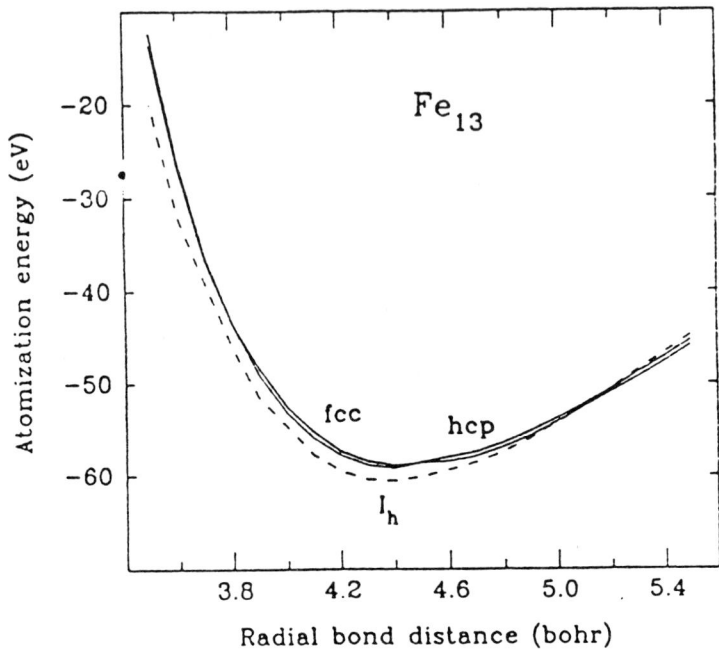

Fig. 8. Spin-optimized atomization energy curves for Fe_{13} clusters: (solid line) fcc and hcp; (dashed line) icosahedral [50].

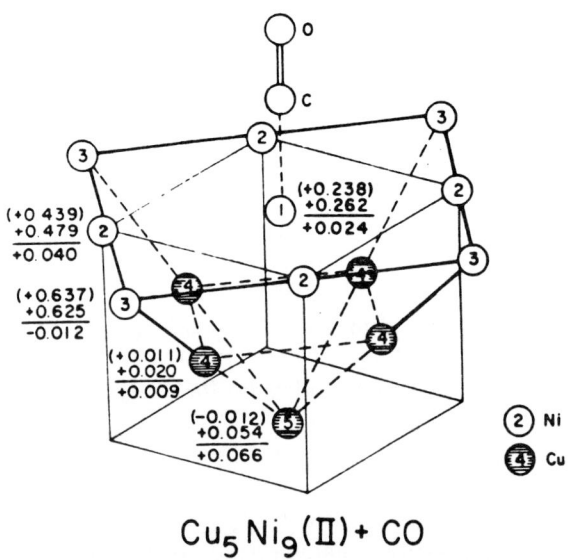

Fig. 9. Calculated magnetic moments (Bohr magneton) for Cu_5Ni_9 before (in parenthesis) and after chemisorption of CO; the difference for each atom is given below the solid line [52].

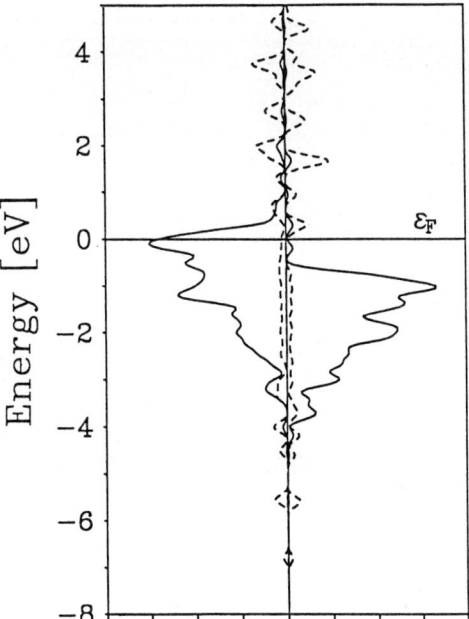

Fig. 10. Density of states (arbitrary units) of the bare Ni_{44} from spin-polarized LD calculations: (left side) minority spin; (right side) majority spin. E_f marks the cluster Fermi energy. Dashed line is $4sp$ contribution; solid line is $3d$ contribution [53].

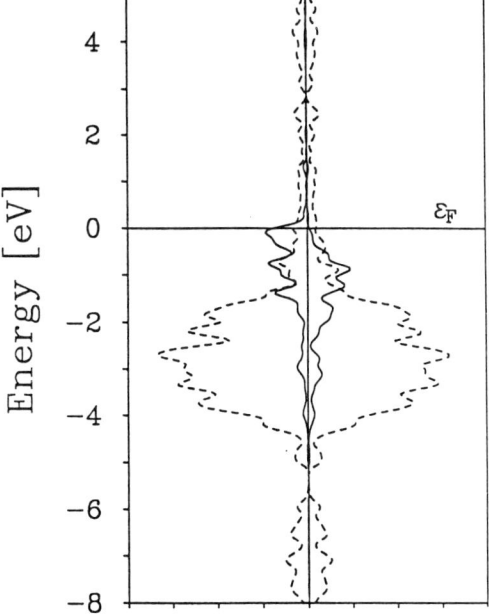

Fig. 11. Density of states (arbitrary units) for spin-polarized model of $[Ni_{44}(CO)_{48}]^{6-}$: (left side) minority spin, (right side) majority spin. Only the Ni $3d$ contribution is shown. Dashed line represents 'surface atoms'; solid line represents 'bulk' atoms [53].

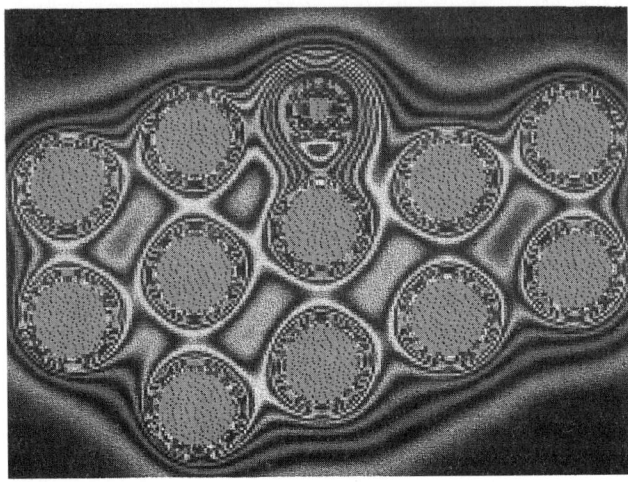

Fig. 12. Valence charge density contours (equal intervals) for P: (111)Fe chemisorption state near equilibrium distance. The surface normal is in the z-direction; a 36 atom cluster is embedded in the semi-infinite bulk crystal [56].

to the essentially diamagnetic behavior of $[Ni_{44}(CO)_{48}]^{6-}$. The diamagnetic nature of saturated carbonyls is well established.

Our final example will be drawn from some recent work on the role of sulphur, phosphorus and other atoms on embrittlement of steels [55, 56]. Here a comparison of the electronic structure around the impurity atom in different environments: free surface, interstitial, and model grain boundary is made. In a sense one is studying the structure of 'microcompounds' with different geometry and host environment. The covalent P-Fe bonds and the metallic Fe-Fe bonds are depicted in valence charge density maps for a 36 atom model of chemisorption on a (111)Fe surface in Figure 12. The highly local nature of the chemisorbed state and the essentially negligible perturbation of iron atoms beyond the second neighbor are evident. The corresponding charge density maps for a 35 atom embedded cluster model [57] of a periodic grain boundary is given in Figure 13. In this figure, the grain boundary runs along the vertical axis, the direction of strongest P-Fe interaction. Perturbation of three surrounding Fe in the axial plane (only one atom is visible in the figure) is also seen to be strong, with visible effects extending out to the second neighbor region.

Detailed PDOS analysis shows quantitatively how neighboring iron atoms are perturbed by covalent bonding to the P. In Figure 14 the phosphorus PDOS for surface and grain boundary environments are compared; one can see a marked accumulation of bonding states at the bottom of the P $3p$ band (5–7 eV below $E_f = 0$ energy) in the grain boundary case. In general one expects that formation of strong Fe-P bonds will occur at the expense of

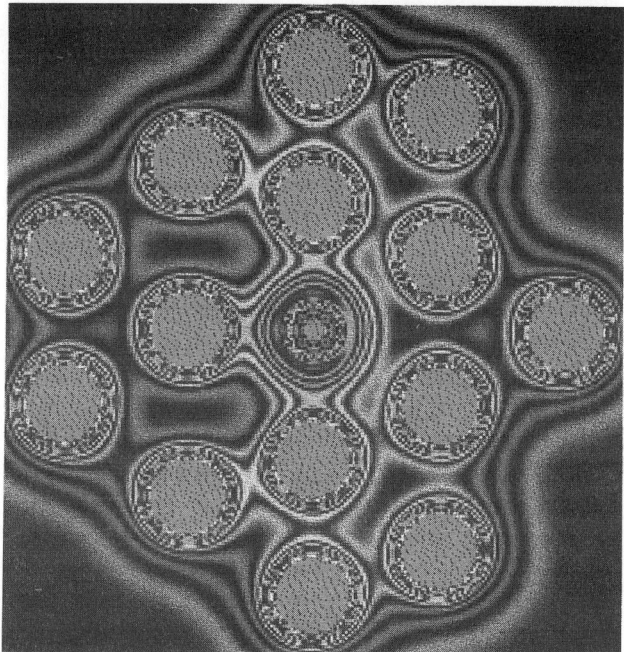

Fig. 13. Valence charge density contours (equal intervals) for a 36 atom fragment of a periodic model grain boundary in iron, containing P. Distances and geometry were chosen on the basis of molecular dynamics modelling and analogy to phosphide compounds [56].

Fe-Fe bonding; this gives a rationale for the observed weakening and fracture at grain boundaries.

In order to make a more quantitative but simple measure of changes in bond strength, one can exploit the empirical correlation between overlap charge and bond strength [58]. Equivalently, one can integrate the charge density over conveniently defined 'bond volumes' to obtain a bond charge. The P-Fe bond charges at the grain boundary are found to be almost equal to those at the surface; thus charge withdrawal from the Fe-Fe bonds is essentially additive, depending on the local coordination. Studies on interstitial and substitutional sites are needed to confirm this idea. In addition, it is possible to directly integrate the LD energy density function over bond volumes, to obtain a direct measure of the changes in cohesion induced by the impurity [59].

4.5. Summary

An overview of the status of electronic structure models for metal particles has been given, from the point of view of a participant. Rather than make the futile attempt of encyclopedic coverage, I have selected examples which seem to be both simple and interesting, and conform to the 'story line' which

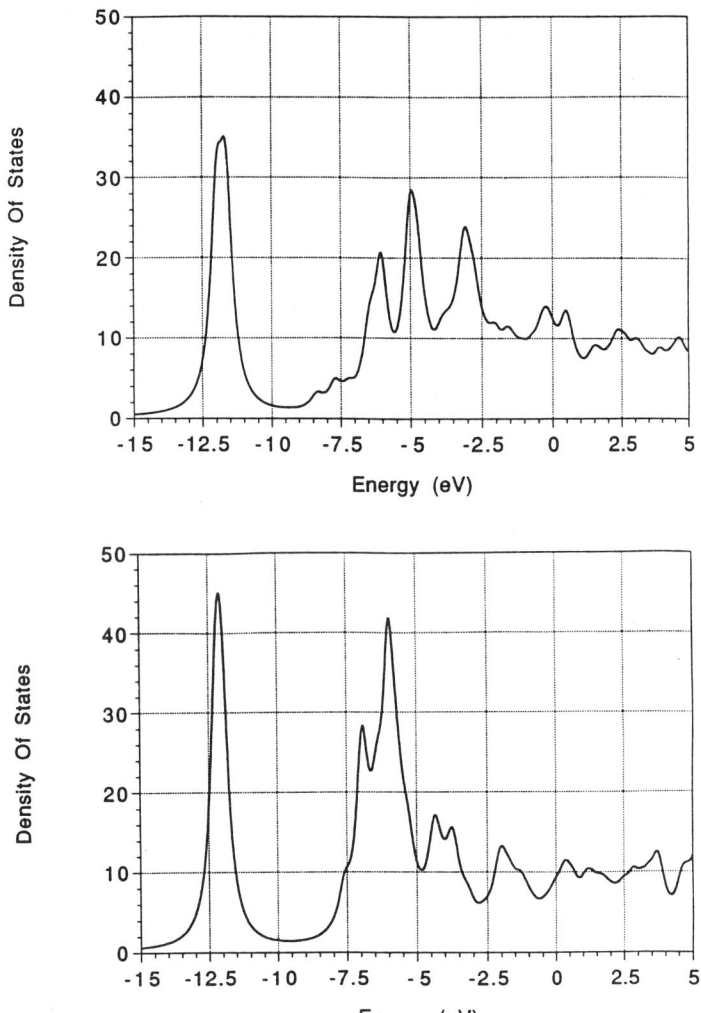

Fig. 14. Projected density of states (states/atom-eV) for phosphorus: (a) at (111)Fe surface; (b) within periodic grain boundary [56].

was chosen. Apologies are due to the many colleagues whose fine work was not mentioned here.

From the present perspective, cluster compounds form a proving ground where theory and experiment can confront each other with fewer uncertainties than in most environments. Because the basic cluster component typically has a degree of symmetry which should be manifest in experiments, different kinds of MO theory can be profitably executed and compared. Understanding the basic component, one can then proceed to examine the coupled-cluster and solid state effects. Lessons learned here can be immediately applied to the more complex world of finite-temperature cluster beams, supported particles on surfaces, and composite materials. A great deal has been learned in the past ten years about the structure of 'simple metal' particles, and about the interplay of magnetism and chemical bonding. The present-day interactions between theory and experiment show great promise for unravelling the mysteries of more complex materials.

Acknowledgements

This work was supported in part by the Office of Naval Research, Contract No. N00014-90-J-1363, and by the MRL Program of the National Science Foundation, at the Materials Research Center of Northwestern University, under Award No. DMR-9120521. Thanks are due to Manfred Kappes, Gregory Olson and Arne Ros'en for helpful discussions. Calculations were performed in part at the National Center for Supercomputing Applications, under NSF support.

References

1. D. R. Salahub, in *Comparison of Ab Initio Quantum Chemistry with Experiment for Small Molecules. The State of the Art*, W. Weltner, Jr. and R. J. Van Zee (eds.), D. Reidel Publ. Co., Dordrecht (1985).
2. J. Koutecky and P. Fantucci, *Chemical Revs.* **86** (1986) 539.
3. Moskovits, M., *Metal Clusters*, Wiley-Interscience, New York (1986).
4. B. F. G. Johnson (ed.), *Transition Metal Clusters*, Wiley, New York (1980),
5. T. C. Zietlow, M. D. Hopkins and H. B. Gray, *J. of Solid State Chem.* **57** (1985) 112.
6. For example: H_2OCu_n mass spectra give evidence of icosahedral packing: B. J. Winter, E. K. Parks and S. J. Riley, *J. Chem. Phys.* **94** (1991) 8618.
7. M. Elian and R. Hoffmann, *Inorg. Chem.* **14** (1975) 1058; D. L. Thorn and R. Hoffmann, *Inorg. Chem.* **17** (1978) 126; R. Hoffmann, *Angew. Chem. Int. Ed. Engl.* **21** (1982) 724, and references therein.
8. For example, X.-L. Liang, S. Flores, D. E. Ellis, B. M. Hoffman and R. Musselman, *J. Chem. Phys.* **95** (1991) 403.
9. T. Hughbanks and R. Hoffmann, *J. Am. Chem. Soc.* **105** (1983) 1150.
10. R. Hoffmann, *Solids and Surfaces, A Chemists's View of Bonding in Extended Structures*, VCH Publishers, New York (1988).
11. R. Haydock, in *Sol. Sta. Phys.*, Vol. 35, H. Ehrenreich, F. Seitz and D. Turnbull (eds.), Academic, New York (1980), p. 216; D. W. Bullet, *ibid*, p. 129.
12. J. P. Gaspard and F. Cyrot-Lackmann, *J. Phys.* **C6** (1973) 3077.
13. Feng Liu, M. R. Press, S. N. Khanna and P. Jena, *Phys. Rev.* **B39** (1989) 6914.
14. C.-Y. Wang, S.-Y. Liu and L.-G. Han, *Phys. Rev.* **B41** (1990) 1359.
15. J. C. Slater and G. F. Koster, *Phys. Rev.* **94** (1954) 1498.
16. However, the reader should note the unresolved discrepancy between magnetic exchange splittings found in UPS and in virtually all band calculations.
17. G. Thomas *et al.*, *Transmission Electron Microscopy of Materials*, Wiley, New York (1979).
18. Z.-B. Zheng, Z. Shao and C.-Y. Wang, *J. Phys. F. Met. Phys.* **18** (1988), 1137.

19. W. A. Harrison, *Electronic Structure and the Properties of Solids*, Freeman, San Francisco (1980).
20. H. F. Schaefer, III (ed.), *Methods of Electronic Structure Theory*, Plenum, New York (1977).
21. B. K. Rao and P. Jena, *Phys. Rev.* **B37** (1988) 2867.
22. M. M. Kappes, P. Radi, M. Schar and E. Schumacher, *Chem. Phys. Lett.* **119** (1985) 11, and references therein.
23. W. D. Knight, K. Clemenger, W. A. deHeer, W. A. Saunders, M. Y. Chou and M. L. Cohen, *Phys. Rev. Lett.* **52** (1984) 2141.
24. I. Shim and K. A. Gingerich, in *Physics and Chemistry of Small Clusters*, P. Jena, B. K. Rao and S. N. Khanna (eds.), Plenum, New York (1987), p. 523.
25. F. W. Bobrowicz and W. A. Goddard, III, in *Modern Theoretical Chemistry*, H. F. Schaefer, III (ed.), Plenum, New York (1977), Vol. 3, p. 79.
26. J. Cizek, *J. Chem. Phys.* **45** (1966) 4256; R. J. Bartlett and G. D. Purvis, *Physica Scripta* **21** (1978) 255; R. J. Bartlett, *J. Phys. Chem.* **93** (1989) 1697.
27. D. E. Ellis, J. Guo, H.-P. Cheng and J. J. Low, *Adv. Quantum Chem.* **22** (1991) 125.
28. R. G. Vranka, L. F. Dahl, P. Chin and J. Chatt, *J. Am. Chem. Soc.* **91** (1969) 1574; P. Moor, P. S. Pregosis, L. M. Venanzi and L. M. Welch, *Inorg. Chem. Acta* **85** (1984) 103; M. A. A. F. de C. T. Carrondon and A. C. Skapski, *J. Chem. Soc., Chem. Commun.* (1976) 410.
29. H/Pt, unpublished data, S. Riley, private communication; The ratio H/Nb tends to 1.3: J. L. Elking, F. D. Weiss, J. M. Alford, R. T. Laaksonen and R. E. Smalley, *J. Chem. Phys.* **88** (1988) 5215; The ratio H/M tends to be less than 1.0 for Fe, Ni, Co presumably since bulk hydrides are not stable: E. K. Parks, I. K. Liu, S. C. Richtsmeier, L. G. Pobo and S. J. Riley, *J. Chem. Phys.* **82** (1985) 5470; E. K. Parks, G. C. Nieman, L. G. Pobo and S. J. Riley, *J. Chem. Phys.* **91** (1987) 2671; T. D. Klots, B. J. Winter, E. K. Parks and S. J. Riley, *J. Chem. Phys.* **91** (1990) 2110; T. D. Klots, B. J. Winter, E. K. Parks and S. J. Riley, *J. Chem. Phys.* **95** (1991) 8919.
30. D. M. Cox, P. Fayet, R. Brickman, M. Y. Hahn and A. Kaldor, *Catal. Lett.* **4** (1990) 271.
31. M. H. McAdon and W. A. Goddard, III, *J. Chem. Phys.* **88** (1988) 277; M. M. Goodgame, Ph.D. Thesis, CalTech, Pasadena CA (1983).
32. V. Bonačić-Koutecký, P. Fantucci and J. Koutecký, *Chem. Rev.* **91** (1991) 1035; Na_n: V. Bonačić-Koutecký, P. Fantucci and J. Koutecký, *Phys. Rev.* **B37** (1988) 4369; Li_n: I. Boustani, W. Pewestorf, P. Fantucci, V. Bonačić-Koutecký and J. Koutecký, *Phys. Rev.* **B35** (1987) 9437; see also [2].
33. W. Kohn and L. J. Sham, *Phys. Rev.* **A140** (1965) 1133; N. M. March and S. Lundqvist (eds.), *Theory of the Inhomogeneous Electron Gas*, Plenum, New York (1983); O. Gunnarson and B. I. Lundqvist, *Phys. Rev.* **B13** (1976) 4274.
34. T. Ziegler, *Chem. Rev.* **91** (1991) 651.
35. K. H. Johnson, *Adv. Quant. Chem.* **7** (1973) 143; J. C. Slater, *The Self-Consistent Field for Molecules and Solids*, McGraw-Hill, New York (1974), Vol. 4.
36. E. J. Baerends, D. E. Ellis and P. Ros, *Chem. Phys.* **2** (1973) 41.
37. B. I. Dunlap, J. W. D. Connolly and J. R. Sabin, *J. Chem. Phys.* **71** (1979) 3396.
38. P. Ballone, W. Andreoni, R. Car and M. Parrinello, *Phys. Rev. Lett.* **60** (1988) 271; R. Car, M. Parrinello and W. Andreoni, in *Microclusters*, S. Sugano, Y. Nishina and S. Ohnishi (eds.), Springer-Verlag, Berlin (1987), p. 134.
39. W. A. deHeer, W. D. Knight, M. Y. Chou and M. L. Cohen, in *Solid State Phys.*, H. Ehrenreich and D. Turnbull (eds.), Academic, New York (1987), Vol. 40, p. 93.
40. M. M. Kappes, M. Schar, U. Rothlisberger, C. Yeretzian and E. Schumacher, *Chem. Phys. Lett.* **143** (1988) 251, and references therein.
41. E. C. Honea, M. L. Homer, J. L. Persson and R. L. Whetten, *Chem. Phys. Lett.* **171** (1990) 147.
42. M. J. Lopez, M. P. Iniguez and J. A. Alonso, *Phys. Rev.* **B41** (1990) 5636; M. P. Iniguez, M. J. Lopez, J. A. Alonso and J. M. Soler, *Z. Physik D* **11** (1989) 163.

43. M. J. Lopez, A. Mananes, J. A. Alonso and M. P. Iniguez, *Zeit. Physik D* **12** (1989) 237; J. M. Lopez, A. Ayuela and J. A. Alonso, in *Ohio Supercomputer Workshop on Theory and Applications of Density Functional Approaches to Chemistry*, Springer-Verlag, Berlin (1991).
44. S. Pollack, C. R. C. Wang, T. Dahlseid and M. M. Kappes, *Nuclear Physics Concepts in the Study of Atomic Cluster Physics*, Proceedings of the 88th WE-Heraeus-Seminar, Bad Honnef, F.R.G. (1991); in *Lecture Notes in Physics*, Vol. 404, R. Schmidt, H. Lutz and R. Dreizler (eds.), Springer-Verlag, Berlin (1992), and private communication.
45. D. E. Ellis (unpublished).
46. J. M. Martins, J. Buttet and R. Car, *Phys. Rev.* **B31** (1985) 1804.
47. B. Wästberg and A. Ros'en, *Z. Phys. D* **18** (1991) 267.
48. R. N. Barnett, U. Landman, A. Nitzan and G. Rajagopal (to be published).
49. A. Görling, N. Rösch, D. E. Ellis and H. Schmidbaur, *Inorg. Chem.* **30** (1991) 3986.
50. B. I. Dunlap, *Phys. Rev.* **A41** (1990) 5691.
51. H.-P. Cheng and D. E. Ellis, *J. Chem. Phys.* **94** (1991) 3735; D. E. Ellis, J. Guo, H.-P. Cheng and J. J. Low, *Adv. Quantum Chem.* **22** (1991) 125.
52. D. R. Salahub and F. Raatz, *Int. J. Quantum Chem. Symp.* **18** (1984) 173; F. Raatz and D. R. Salahub, *Surf. Sci.* **146** (1984) L609.
53. N. Rösch, L. Ackermann, G. Pacchioni and B. I. Dunlap, *J. Chem. Phys.* **95** (1991) 7004.
54. G. Pacchioni and N. Rösch, *Inorg. Chem.* **29** (1990) 2901.
55. H. Eguchi, M.Sc. Thesis, Northwestern University (1990).
56. L. Sagert, D. E. Ellis and G. B. Olson (unpublished).
57. D. E. Ellis, G. A. Benesh and E. Byrom, *Phys. Rev.* **B20** (1979) 1198.
58. M. Morinaga and N. Yukawa, in *Computer Aided Innovation of New Materials*, M. Doyama, T. Suzuki, J. Kihara and R. Yamamoto (eds.), Elsevier, Amsterdam (1991), p. 803; M. Morinaga, N. Yukawa, H. Adachi and S. Kamado, *J. Less-Common Met.* **141** (1988) 295.
59. M. Press and D. E. Ellis, *Phys. Rev.* **B38** (1988) 3102.

5. X-RAY PHOTOELECTRON SPECTROSCOPY APPLIED TO PURE AND SUPPORTED MOLECULAR METAL CLUSTERS

R. ZANONI
*Dipartimento di Chimica, Università degli Studi di Roma 'La Sapienza',
Piazzale Aldo Moro, 5, I-00185 Roma, Italy*

5.1. Introduction

The class of compounds represented by molecular metal clusters offers the highest level of size-selection of particles, with nuclearities in the range of 3 to a few hundreds atoms. Molecular metal clusters provide a way of studying metal-metal and metal-adsorbate interactions in a single species. The growing interest connected to these studies is largely related, yet not confined, to the desire for a deeper knowledge of the electronic basis for catalytic behaviour. Other common applications have been reported in adsorption, corrosion, surface treatments, microelectronics.

Fundamental condensed matter physics may take advantage from metal cluster chemistry, for an understanding of both the transition from the discrete energy levels of free atoms to the continuous energy bands of bulk metals and of the chemisorption of molecules on transition metal surfaces.

Several excellent review articles and volume series deal with different aspects of photoemission spectroscopy, and the reader is referred to those contributions for a complete introduction to the technique [1-3] and to the subject of photoemission from clusters [3-6]. Photoemission studies on clusters have been, so far, mainly focussed on the electronic properties of bare metal clusters. A general extension to molecular metal clusters of several aspects, both theoretical and experimental, which have been discussed for bare metal clusters is still missing, mainly because of the complexity of a quantum mechanical treatment of large species with attached ligands. Because of the above limitation, in Section 2 the selection of issues relevant to the study of both pure and supported molecular metal clusters in the solid state by means of X-ray photoemission spectroscopy (XPS) is discussed mainly with reference to available data for bare metal clusters. Sections 3 and 4 contain a brief account of existing XPS studies and related problems on, respectively, pure and supported molecular metal clusters. Finally, Section 5 presents an outlook for future experiments.

5.2. Generalities of Photoemission Spectroscopy Applied to Pure and Supported Molecular Metal Clusters

In a photoemission experiment, the absorption of monochromatic photons of energy $h\nu$ is used to eject photoelectrons from the investigated system. In the process, a photon is annihilated and a photoelectron is emitted *in vacuo*, with creation of an ionic species. The number of experimental observables of the process depends on the characteristics of the photon source and of the detection system. If the energy, direction and polarization of the photons are known and the kinetic energy (KE), intensity, angular distribution and spin of the ejected electrons are measured, a variety of photoemission experiments, corresponding to different photoemission techniques, results. In its simplest version, a photoemission experiment is performed at fixed photon energy, by measuring the KE and number of the emitted photoelectrons. This technique is termed 'angle-integrated' (or 'conventional') photoemission.

The total energy of the ionic system produced, E_{ion}, results from different electronic, vibrational, rotational and translational energy contributions. The energy conservation equation requires that

$$h\nu + E_{in} = KE + E_{ion}$$

where E_{in} is the initial energy of the system, supposed to be neutral, in its ground state. If the binding energy (BE) or ionization energy of the ejected electron coming from a particular state is defined as the energy difference of the corresponding excited ionic state, E_{ion}, and the ground state of the system, E_{in}:

$$BE \equiv E_{ion} - E_{in}$$

we may write:

$$h\nu = KE + E_{ion} - E_{in} = BE + KE.$$

This last equation allows to derive BEs from the experimental determination of photoelectron KEs. Photoionization requires a threshold photon energy which, in the case of gaseous samples, corresponds to the first ionization energy, or ionization potential (IP), typically amounting to few eVs. In the case of solids, the threshold energy is termed the work function, Φ (typical values: 1.5–5 eV), which has to be added to the energy balance equation:

$$h\nu = BE + KE + \Phi.$$

The intensity of a photoemission process from a particular energy level is related to the probability of ionization of a particular electron and, quantitatively, to the corresponding cross-section for photoionization, σ, which is defined as the transition probability per unit time for exciting a system with a unit incident flux of photons. Photoionization cross-section values allow to perform relative quantitative analysis of the chemical composition from intensity ratios of photoemission core peaks [1].

Measurements of the angular distribution of photoelectrons are performed in angle-resolved photoemission studies [1, 2]. From the experimental determination of the KE and direction of propagation of a photoelectron one may derive the photoelectron momentum (or wave vector) k. This allows to obtain selective information on individual points in the k-space. A widely explored application in the solid state is the mapping of the band structure of a solid along the directions of the Brillouin zone [1, 2].

Photoelectron spectroscopy falls into two energy ranges: in the UPS (Ultraviolet Photoelectron Spectroscopy) regime, monochromatic sources in the ultraviolet spectral region are employed; only shallow core lines (BEs $\leq$ 50 eV) and valence electrons can be excited *in vacuo*. In the XPS regime, photon energies in the soft-X-ray range (10^2–10^3 eV) are used and core levels are accessed. Electron storage rings are sources of radiation ('synchrotron radiation') tunable in energy and bridging the gap between the UPS and XPS energy ranges. The peculiar properties of synchrotron radiation (mainly its wide and continuous energy spectrum, high intensity, and natural polarization) allows different classes of photoemission experiments to be performed, based on energy-, angular-, polarization- and spin-dependent photoelectron detection [7].

Photoelectron spectroscopy measures the excitation spectra of a system which is ionic in its final state. A BE value should be interpreted as a result of the many electron states of the ionic system, rather than of the occupied one-electron states (the orbitals) of the initial system. Since the interest is generally focused on the properties of the initial electronic state of a system, this information has to be recovered as accurately as possible from the experiment. This step requires, in general, a good knowledge of the system from a theoretical, quantum-mechanical point of view [7].

Photoemission spectroscopy has played a central role in affording experimental evidences to several basic questions regarding the electronic structure of a metal cluster. A central issue in photoemission studies of metal clusters is the determination of the minimum cluster size required for bulklike properties. As, among others, Mason has pointed out [6], cluster electronic properties may be associated with both short-range interactions and long-range order. BEs are mostly sensitive to the former term, while the shape, photon-energy dependence and width of the VB are mainly determined by the latter and converge more slowly to the bulk values. The interpretation of electron BEs measured in supported clusters is still a matter of controversy, mainly because of the different referencing procedures adopted (see below for a discussion of this topic). On the basis of measured metal BEs, the onset of a true metallic behaviour has been associated with a cluster nuclearity in the 100–200 range [8]. However, some relevant features of the photoemission spectra of supported clusters require few thousands of atoms to become comparable with those of bulk metals. This holds, e.g., for the dependence of the intensity of valence band (VB) spectral features on the photon energy $h\nu$. The spectra of ordered solids, in fact, are significantly excitation-energy dependent at low photon energies, i.e. up to $\approx$15–25 eV, while they are almost independent from $h\nu$ at the typical photon energies employed in XPS. The normal selection rule for electronic transitions via optical excitation in a solid, i.e., the conservation through the process of the electron wave vector k, is assumed to hold in the photoemission process [1]. Only vertical (or direct) transitions between occupied and empty bands will result. At high photon energies (i.e., in the 'XPS regime') the photoelectron has high free-electron character and the final state lies in a continuum region. At low photon energies, the excitation process brings the photoelectron to a final energy very close to the Fermi level, in the case of a metal, or to the top of the conduction band, for insulators. In these cases, the character of the excited states changes very rapidly with the energy and, correspondingly, the probability of the transition is strongly dependent on the photon energy. Several factors, however, tend to reduce the importance of the k selection rule in photoemission experiments, allowing indirect (or 'non-vertical') transitions to become part of the overall VB signal. In the case of a cluster, if the periodic metal structure is lost, the electron crystal momentum, $hk/2p$, which is conserved in direct transitions, is not a good quantum number and the cluster should exhibit a photoemission spectrum almost independent of the photon energy [6]. Moreover, the actual percentage of indirect transitions in the photoemission process varies on the surface and in the bulk. This will particularly affect the VB spectra of a cluster, because of the relevant surface contribution with respect to the whole cluster. As an example, Figure 1 shows the effect in the case of Au particles on carbon [9].

At the lowest coverage of Au deposited on C, several effects are visible in the spectra: intensity modulations at low photon energies in the VB features are much smaller than in the bulk, peak positions do not shift with $h\nu$ and the $5d$ spin-orbit splitting is far from the bulk value. Variations in the relative intensity of VB features become more and more significant as the coverage is increased to 5×10^{15} atoms/cm^2, where the clusters have actually begun to coalesce, but they are still not completely bulklike at 1.3×10^{15} atoms/cm^2 [9]. Therefore, a cluster nuclearity in the range of thousand atoms is required to reproduce a true bulklike behaviour as regards this class of experiments.

A chemical shift in XPS is defined as the difference between two BE values of the same core orbital of an element, measured in different compounds or atomic environments. A core BE of an atom is related to the effective charge on it. The energy required to remove an electron from an atom is expected to be inversely related to the valence electron density on the atom. A BE should increase (decrease) with increasing positive (negative) charge on that atom. A good correlation of BEs with atomic charges requires to take into account the residual charge on all the other atoms in the compound

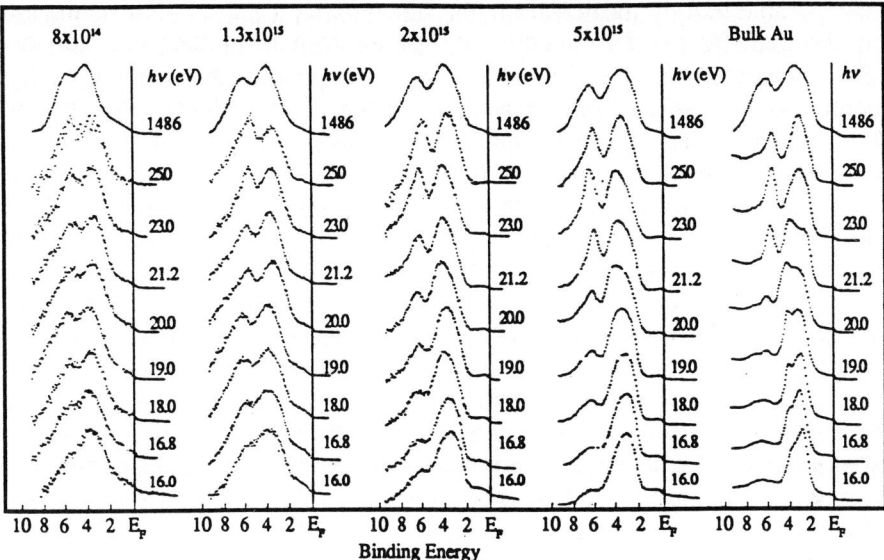

Fig. 1. Valence band photoemission spectra of gold clusters, deposited on graphite at various coverages. The spectra have been taken at different photon energies and are background subtracted. Transmission electron micrographs of clusters at 2×10^{15} atoms/cm^2 have shown an average diameter of 19 Å, which corresponds to slightly more than 100 atoms. At 5×10^{15} atoms/cm^2 the average cluster contains several hundreds atoms. (Reproduced by permission from [9]).

and to limit the correlation to systems with close structures and bonding. Several factors contribute to a core-level chemical shift. Chemical shifts of a metal atom in a cluster, which are generally measured with respect to the bulk metal, are mainly due to a modification of the electronic configuration of the emitting atom and of the electron screening of the core hole created in the photoemission process. Both effects are due to differences in electronic charge density and affect the final state. Initial-state electronic charge reorganizations following, e.g., variations in size are referred to as 'initial-state effects', while core-hole screening of the system is a 'final-state effect'. Modifications in d-electron count and d-band rehybridization with s, p bands are common effects in transition metal clusters [4]. Screening of a hole is a phenomenon which accompanies the photoemission process and acts in its same time scale. When an electron is photoemitted, the positively charged system will react through relaxation of the remaining electrons (the 'passive electrons') towards the hole, which will be screened as a result. The difference between the total energies of the primary final state and the initial state is termed the relaxation (or 'polarization' or 'screening') energy, E_r. Removal of an electron from an atom, a molecule and a solid will result in different total relaxation energies because of the different contributions to E_r. Core orbitals are strongly localized and the screening of a hole is expected to be much more effective for core than for valence orbitals. In the case of atoms, E_r may be conveniently divided into inner-shell, intra-shell and outer-shell contributions, respectively corresponding to the screening exerted by passive electrons in shells with a principal quantum number lower, equal or higher than the shell of the photoelectron [10a]. While the first term is generally negligible ($\ll 1$ eV), because the ionization of an electron in an outer shell has little influence on the inner wavefunctions, ionizations inside the same shell will result in a different average pair repulsion among the remaining electrons in that shell. The corresponding calculated values of the intrashell E_r are ≤ 5 eV [10a]. Outer-shell relaxation constitutes the larger contribution, since the removal of an electron from an inner shell will result in the increase of (almost) one unit on the nuclear charge effectively experienced by the outer electrons. As expected on the basis of the dominance of outer-shell relaxation, calculated atomic E_r's show a monotonical increase with the atomic number Z for each principal quantum number n and a monotonic decrease, for each Z, with increasing n. Values of E_r up to ≈ 50 eV have been reported for the orbitals of light atoms (helium to copper) [10b]. In the case of molecules, E_r for a given core level is expected to increase with respect to free atoms because of the added contribution to the relaxation given by extra-atomic passive electrons. Finally, in the case of a solid E_r is larger than for single molecules, because of the relaxation of the lattice in addition to the local polarization channel. The lattice contribution is obviously more efficient for a conductive solid than for an insulator [1].

For small (molecular) metal clusters there is obviously no metallic screening state, but only local relaxation contributions, examined above. This introduces the question of complete vs. incomplete screening as a function of cluster nuclearity. The screening response to a core hole may be regarded as a measure of cluster metallicity even if, as pointed out by Di Cenzo and Wertheim [5], a core hole is a localized perturbation which can distort delocalized bands. With decreasing cluster size, the mechanism of energy lowering in the photoemission final state will change from conduction electron screening to a local process. The authors in [5] set the crossover point for the change where the energy required to leave the system with a positive charge on its surface will equal the energy gained by the process of screening the core hole. As an estimate of cluster size at the crossover point, the authors propose a sphere of 10 Å diameter, since a charge of $+e$ on it will produce a potential e^2/r of 1.44 eV, which closely compares with typical values of E_r [10b]. The corresponding cluster nuclearity for the 10 Å sphere has been estimated by Mason to be ≈ 30 atoms in the case of Au, i.e. a nuclearity rather small for a true metallic behaviour [6]. At approximately the same cluster size, the spin-orbit splitting of the Au 5d VB is still closer to the atomic than to the bulk value [6].

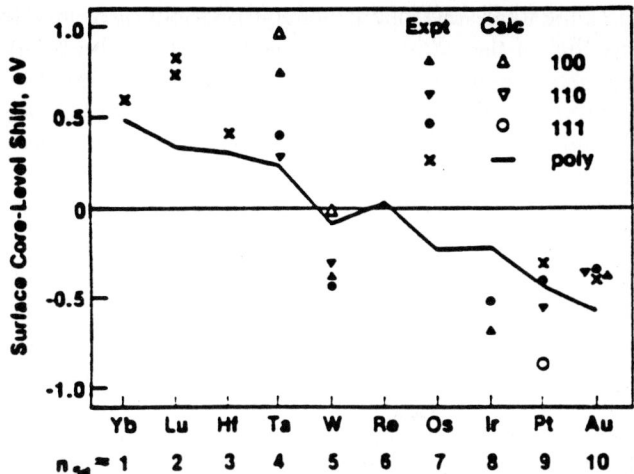

Fig. 2. Bulk-to-surface core-level shifts of the 5d elements. The solid line is an equivalent-core energy-cycle analysis. Open symbols are self-consistent field (SCF) slab-type calculations. (Reproduced by permission from [3]).

E_r values could, in principle, be directly measured from a photoemission spectrum but experimental determinations of this separation are, generally, rather complicated mainly because of the difficulty in resolving and assigning the peaks corresponding to various relaxation processes [3]. The difference in the relaxation energy between two chemical environments, ΔE_r, may be experimentally obtained from photoemission data through the Auger parameter. Wagner introduced the Auger parameter, α or ΔKE, as the sum of a core-level BE and the KE of an Auger transition (involving only core levels) for a given element [11]. An Auger transition is a radiationless decay following the primary process of photoemission. In the Auger process, a core hole, created by the primary photoelectron, is filled by an inner electron, with an energy release sufficient to ionize a second electron, the 'Auger electron'. The Auger parameter value is independent of the particular procedure followed to reference a BE. The difference between ΔBE and ΔKE is equal (within several approximations) to $2\Delta E_r$ [3].

It has been experimentally reported that, for a variety of metals on supposedly non-interacting substrates as amorphous carbon, SiO_2 and Al_2O_3, metal core lines broaden and shift with decreasing cluster size [3–5]. A parallel can be made with the case of surface core-level shifts (SCLS) in metals. A SCLS is the ΔBE for a given core level between a surface and a bulk atom. Figure 2 shows SCLS values for 5d metals [3]. It may be seen that for more than half-filled d-bands the shift is negative, i.e. a metal core level is less bound at the surface than in the bulk, while an opposite shift is operating for elements at the left of W. A possible explanation of the phenomenon has been proposed [12]: the final-state valence-charge distribution around the core hole is essentially that of the $(Z+1)$ element; for elements in the beginning of the transition series, electron screening takes place in the bonding part of the d-band and, therefore, the binding due to the conduction electrons is stronger in the final than in the initial state. The reverse will be true for the heavier transition elements, since the screening will take place in the antibonding part of the d-band. Due to the larger coordination number presented by a bulk atom with respect to a surface atom, the effect will be stronger for the former, which will present a decreased (increased) core BE for the lighter (heavier) transition metals.

The agreement between experiment and theory is roughly good, yet not exact and calls for an interplay of both initial and final state contributions to the overall effect [3]. The reason for the different behaviour of surface vs. bulk atoms in a metal has been envisaged in narrowing of the valence d band, because of reduced coordination, which causes a decreased s-d hybridization and a net flow of charge from s to d-band, to align the Fermi levels [3]. The repulsive Coulomb interaction between core and valence electrons is much larger with d electrons than with the more diffuse s or sp conduction electrons, therefore leading to a lower core BE with increasing d-band occupation. It has been reported that the experimentally obtained BE shift of metal core levels for a supported cluster is in the opposite direction to the experimentally found SCLS for the same element [3]. The discrepancy is particularly relevant, since, due to the high surface-to-bulk ratio of clusters, the expected value of the metal core BE shift for a cluster should be even larger than the corresponding SCLS for the same metal. The interpretation of the phenomenon is still controversial. Wertheim *et al.* have suggested that the most relevant effect is played by a positive unit charge left on the cluster after emission of the photoelectron *in vacuo* [5]. The positive charge would not be compensated in the time scale of the photoemission event because of poor coupling between the cluster and the substrate VB (see Section 3). An alternative explanation has been reported by Mason [4], suggesting that the dominant effect is a size-dependent change in electronic configuration in clusters. Either interpretation is able to explain several experimental findings [4, 5]. At present, a quantitative assessment of the relative importance of all possible factors affecting the experimental behaviour is still lacking.

In addition to BE shifts, a very informative observable in a photoemission experiment is the width of the metal peaks. Several papers have dealt with a theoretical analysis of the experimentally observed broadening of core peaks with decreasing cluster size [3–5, 13, 14]. The interpretation is controversial. Some attribute it to inhomogeneity in the cluster size distribution, while others refer to lifetime broadening or reduced screening [13, 14].

Cluster size induces a modification of the VB photoemission spectra. A general observation reported in the literature is a progressive narrowing of the VB width with decreasing cluster size, as a result of decreasing coordination

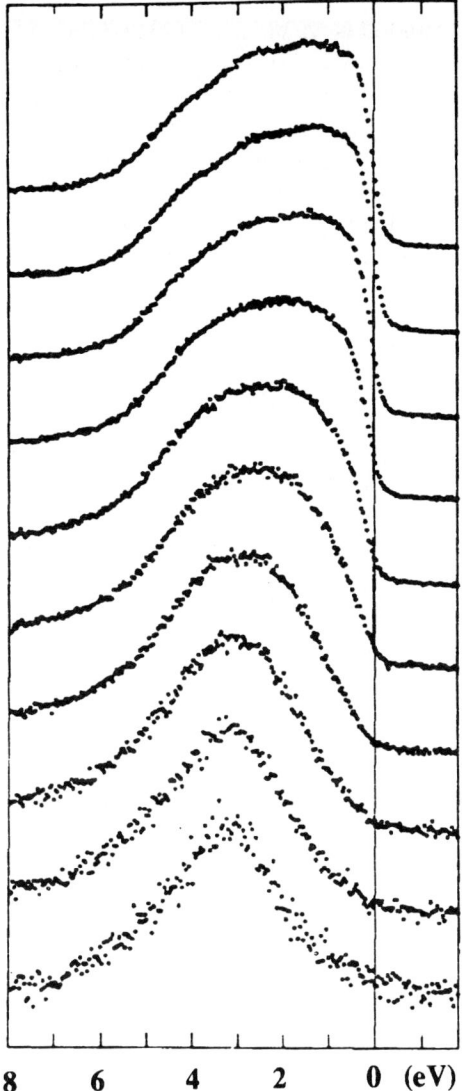

Fig. 3. Valence band photoemission spectra of Pd clusters deposited on carbon. Cluster volume increases by factors of 2 between adjacent spectra. (Reproduced by permission from [5]).

number and increasing localization, similar as for surface sites in bulk metals. The effect is exemplified in Figure 3 for the case of Pd clusters on carbon, which present a narrow, symmetric, atom-like VB at the lowest coverage.

In Figure 4, the same effect is shown for a few transition metals. In addition to VB narrowing with decreasing nuclearity, an energy shift in the VB centroid

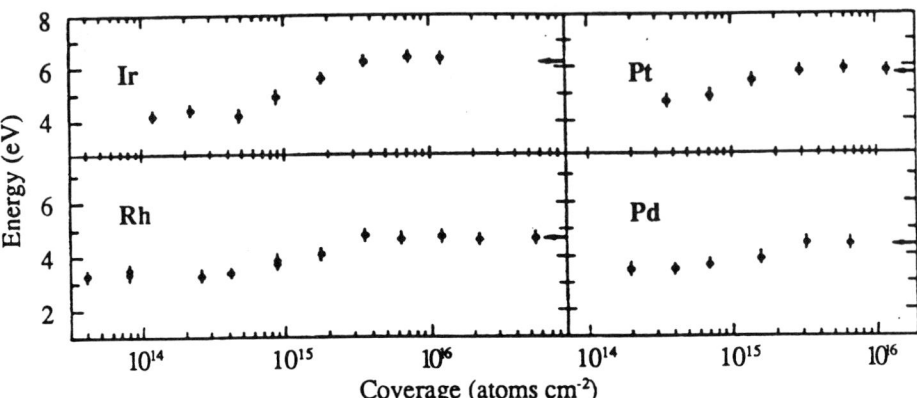

Fig. 4. Variation in measured valence band width of various transition metals as a function of metal coverage. (Reproduced by permission from [6]).

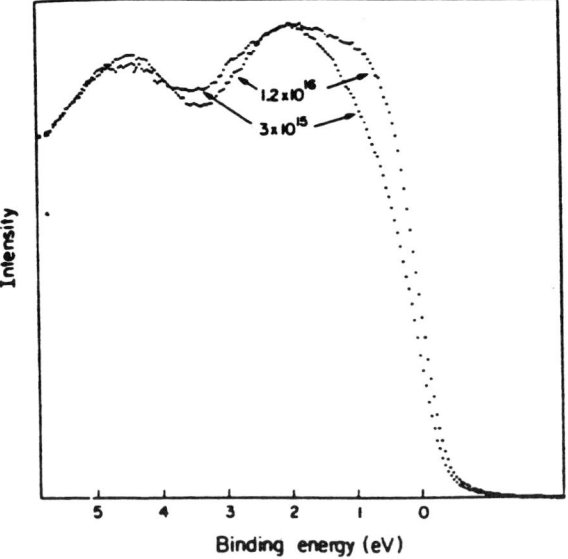

Fig. 5. Valence band photoemission spectra of Pt clusters at two coverages. Average particle size at 3×10^{15} atoms/cm^2 is 18 Å, while particles are severely coalesced at the higher coverage. (Reproduced by permission from [4]).

has been reported, in the same direction of core lines (i.e. to higher BEs) but somewhat smaller in module, because a hole in the VB is more delocalized.

In addition to the above effects, a variation in the density of states at the Fermi level with mean cluster size has been found for different TM clusters [4]. Figure 5 shows the effect in the case of Pd clusters deposited on carbon at two different coverages [4].

The variation has been interpreted as due to a different d-orbital occupation, i.e., as an initial state effect. These results show the possibility offered by XPS of following the entire process of metal nucleation, from isolated atoms to the bulk state.

The existence of a Fermi edge in the VB of a small nuclearity cluster does not constitute, *per se*, a conclusive evidence for the onset of a true metallic behaviour. A standard criterion for the definition of a metal is that the spacing of the electronic levels at the Fermi energy must be smaller than $k_B T$, where k_B is the Boltzmann constant and T is the absolute temperature. This criterion is, however, difficult to be experimentally tested. The complementary information to the number of occupied d orbitals in a TM cluster as a function of nuclearity is the number of d vacancies. This information can be extracted by means of X-ray Absorption Spectroscopy (XAS). The final state corresponding to the XAS edge occurs at the minimum energy required to excite an electron from the core hole to the first unoccupied state, which, for a transition metal cluster, is generally in the d-band, near the Fermi level. The final state is therefore, a core hole plus an extra electron in the valence band. This state is not accessible in XPS, where the cluster will be necessarily ionized. It has been proposed that the final state at the XAS edge corresponds to a completely screened state in an XPS measurement [4]. A comparison between corresponding XPS and XAS shifts with cluster size may afford a measure of the extent of extra-atomic screening. Mason reported such a comparison for Pd/C and Cu/C [4]. The equivalence, within experimental errors, in both XPS and XAS shifts for the two systems has been taken as a strong evidence favouring a complete extra-atomic relaxation effect at the smaller reported cluster size.

5.3. XPS of Molecular Clusters

A detailed account of part of the existing literature on the subject of photoemission studies on pure molecular clusters up to 1985 has been given by Guczi [16].

XPS studies have been reported on pure, unsupported molecular clusters of Fe [17], Co [17–19], Ru, Rh [17, 20], Os [17, 21–23], Ir [17, 23, 24], Pt [24, 25], Au [26–30] and bimetallic clusters containing Pt_3Au [25], Pt_3Sn [31], Ru_2Rh [20], Mo_2W, MoW_2 entities [32]. Most examined molecules are metal carbonyls. One of the few exceptions is represented by XPS studies on a series of Au triphenylphosphine clusters, reported by several authors [25–30]. A few, representative structures of investigated compounds are reported in Figure 6 [28].

For these compounds, a distinction can be made between centred and not centred structures, respectively corresponding to the presence or absence of a central Au atom in the metal cage. The distinction is clearly reflected in Au 4f spectra of $Au_{11}L_7X_3$ (L = triphenylphosphine; X = Cl, I) shown in Figure 7,

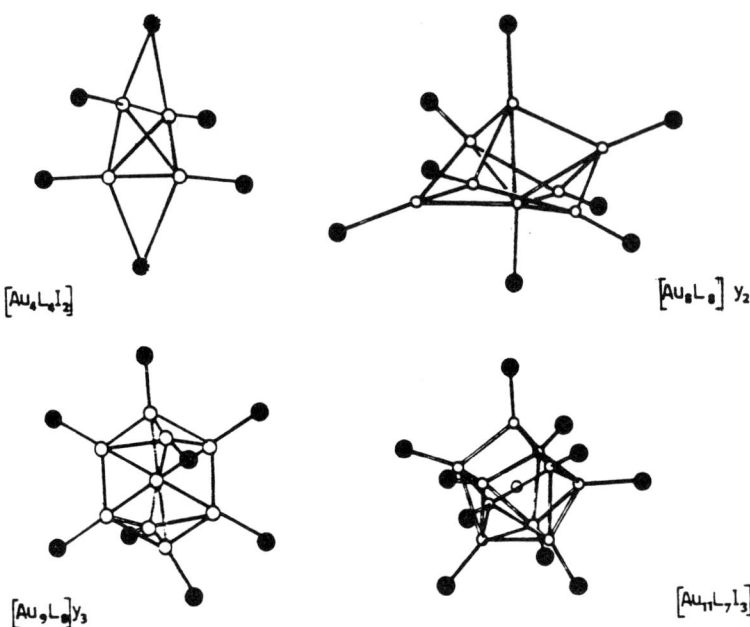

Fig. 6. Geometries of a series of Au molecular clusters. (L = triphenylphosphine). From [28].

where an accurate curve fitting gave separated BEs for the central Au, Au-P and Au-Cl bonded atoms [29].

Withdrawal of charge from the metal core exerted by attached triphenylphosphine and X groups occurs mainly at the expenses of the central, bare Au atom and of the Au atoms bound to halogens. The total charge transferred from the central Au atom to the halogens has been estimated to correspond to the loss of one s electron, from a comparison with AuLX Au4f BE shifts with respect to bulk Au [29]. In the same paper, the authors compare the valence bands of the molecular clusters $Au_{11}L_7X_3$ deposited on carbon to those of bare gold clusters, grown on carbon by controlled evaporation. The purpose of the comparison is to check for the presence and extent of d-d overlap in small gold clusters. A $5d$ splitting 0.5 eV larger than the free atom value has been found for $Au_{11}L_7X_3$ clusters, indicative of d-d interaction between neighboring atoms. The $5d$ splitting value of 2 eV closely approaches the corresponding value for the bare Au clusters analyzed in [29], containing, on the average, 20 atoms, thus suggesting comparable d-d overlap for the two systems.

On this experimental basis, the authors conclude that this class of molecular metal clusters presents a good means to study the effect of increasing nuclearity on the development of the energy spectra into the final band structure.

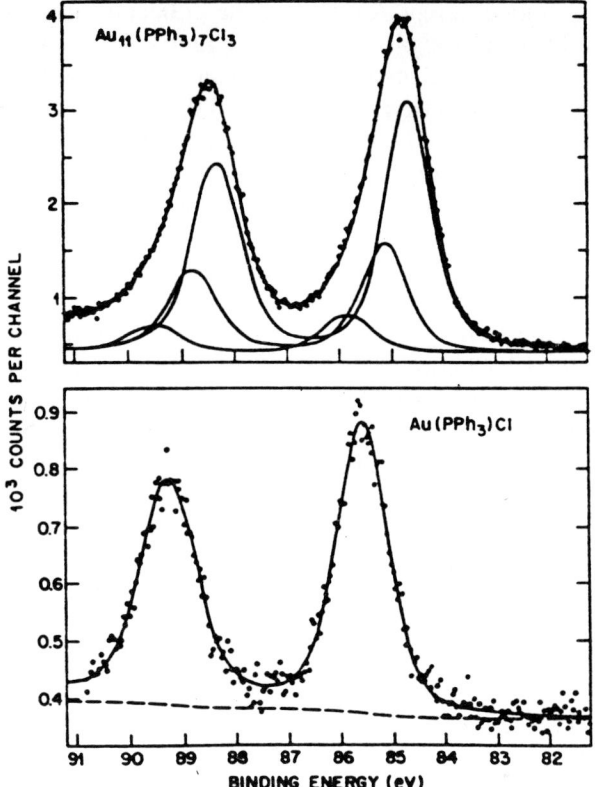

Fig. 7. Au4f XPS spectra of $Au_{11}L_7Cl_3$ and AuLCl (L = triphenylphosphine). (Reproduced by permission from [29]).

An XPS study on $Au_{55}(PPh_3)_{12}Cl_6$ (PPh_3= triphenilphosphine) has been recently reported [33]. The molecule can not be obtained in the form of a crystal and is supposed to have a cuboctahedral structure, with 24 surface unbound Au atoms and a naked core of 13 Au atoms, all the remaining metal atoms being either P- or Cl-bound at the surface [34a]. Au4f and VB spectra have been interpreted in terms of two components only, one encompassing all surface ligated and unligated Au atoms, and a core component, shifted to higher BEs [33]. Quantitative accord with the nominal cluster stoichiometry is, however, poor, since the expected surface-to-bulk atomic ratio is 3.23 while the value deduced from curve-fitting is 1.70. Moreover, the reported Au 4f curve fitting into an overall ligated and unligated surface component and a core component is in strong contradiction with earlier XPS results for $Au_{11}L_7X_3$ [29]. showing relevant chemical shifts from bare gold to Au-P and Au-Cl sites. XPS data on the same cluster have been reported by a different group, as a part of a thorough study of the electronic structure and properties of $Au_{55}(PPh_3)_{12}Cl_6$ [34b]. In the paper, the contributions to the composite

Au 4f peak of four different Au sites in the cuboctahedral Au_{55} structure have been resolved and assigned. The relative ratios of the Au 4f components reproduce the expected values from Au_{55} stoichiometry and structure. The four components for the cluster are found at 84.0, 84.4, 85.3 and 86.3 eV. The peak component at 84.0 eV is assigned to the unbonded gold surface atoms. The BE value obtained for the core Au 4f component, 84.4 eV, is larger than in bulk gold. The authors find this result consistent with the Mössbauer isomer shifts of bulk gold and Au_{55} [34c] and with the results from EXAFS experiments [34d], which showed a smaller inter-gold distance than in the bulk. Because of the relative peak intensities and BE separation of the Au-P and Au-Cl components [28, 29], the least intense Au 4f peak component should correspond to Cl-coordinated gold atoms. The ΔBE of 1.9 eV between Au core site peak and the peak due to Cl-coordinated Au should correspond to the loss of one 6s electron by the gold cluster to the chlorine atom. With nearly complete metallic screening, the negative charge on the chlorine will then be about one. The shift between the Au-P and Au-Cl components is 1.0 eV, which would then correspond to the loss of about one half 6s electron by the cluster to each phosphorus atom in the PPh_3 ligands. As for the valence band, reported in Figure 8, the experimentally observed separation between the $5d_{5/2}$ and $5d_{3/2}$ components of Au_{55} is 2.4 eV, [34b] closely approaching the bulk value.

The separation is distinctly different from that in smaller molecular gold clusters with nuclearities in the range of 4 to 13, where the corresponding values were reported to be $\cong$ 1.9 eV [28, 29].

The general characteristic features of X-ray photoemission spectra of transition metal molecular clusters in the solid state have been extensively discussed, in the case of carbonyls [17,19], often in a comparison with the bonding mode of CO to a molecular species and to a metal surface. All spectra show remarkable resemblance, each presenting an intense band in the 0–7 eV BE region, which can be attributed to ionization peaks of predominantly metal character, and two relatively weak bands, in the 8–13 eV region. These last are mainly composed of the $5\sigma + 1\pi$ and 4σ molecular orbitals of CO. A differential BE shift of the 5σ and 1π orbitals, which are differently involved in the bonding to the metal atom, has been indicated by several authors as the cause of the near degeneracy of the two valence levels in the carbonyl photoemission spectra [17]. A mononuclear carbonyl complex already reproduces the BEs of levels not involved in the bonding of the photoemission spectrum of adsorbed CO [17]. In a polynuclear carbonyl, the increased delocalization of the metal valence levels and the 5σ-derived orbital results in a photoelectron spectrum which, in the opinion of the authors in [17], is almost identical to the spectra of adsorbed CO (see Figure 9).

Multielectron excitations are present in the inner and outer ionizations of a carbonyl as satellite peaks on the high-BE side of the main ionization line. Multielectron excitations may generally be considered simply as eigenstates

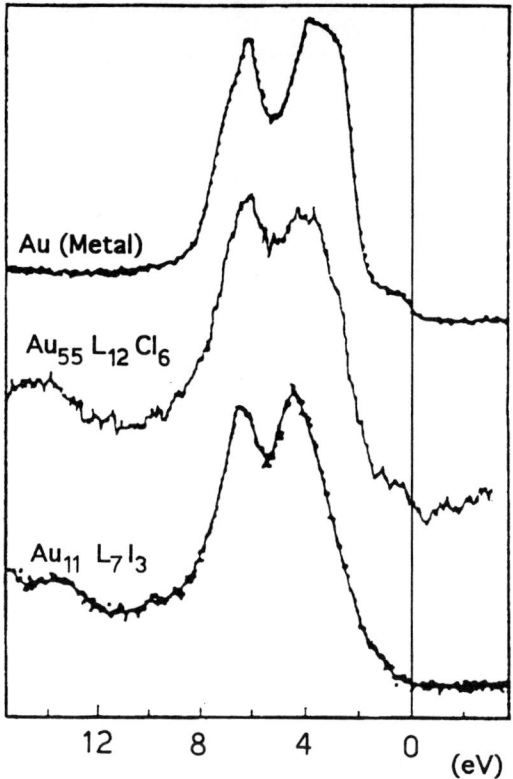

Fig. 8. XPS spectra of the valence band of gold and gold clusters. From top to bottom: bulk Au, $Au_{55}L_{12}Cl_6$ and $Au_{11}L_7Cl_3$ (L = triphenylphosphine). (Reproduced from [34b]).

of the $(N - 1)$-electron ion [1, Vol. 1, p. 183], created as a direct consequence of the screening of the hole. The origin of satellite peaks lies in the excitation of outer valence electrons of the molecule within the time scale of the photoemission experiment. This excitation leaves the ion in a higher energy state, therefore ionizations will be found at somewhat higher BEs, usually a few eVs, and are termed 'shake-up peaks'. A detailed study of core as well as VB shake-up peaks affords a deep insight into the bonding mechanism between the metal core and the ligands, although it requires a detailed theoretical analysis of the bonding [19].

Experimental BE values for the metal atoms in carbonyl clusters are found to be larger than the values for the corresponding bulk metals. In a few cases, they are very close to the values displayed by the corresponding metal oxides, as for $Ir_4(CO)_{12}$ and IrO_2 [35]. This behaviour has been attributed to electron withdrawal from the metal core by the attached CO ligands, and to a balance of initial (i.e., electronic charge redistribution between orbitals) and final state effects (i.e., screening of the hole created by the photoemission process),

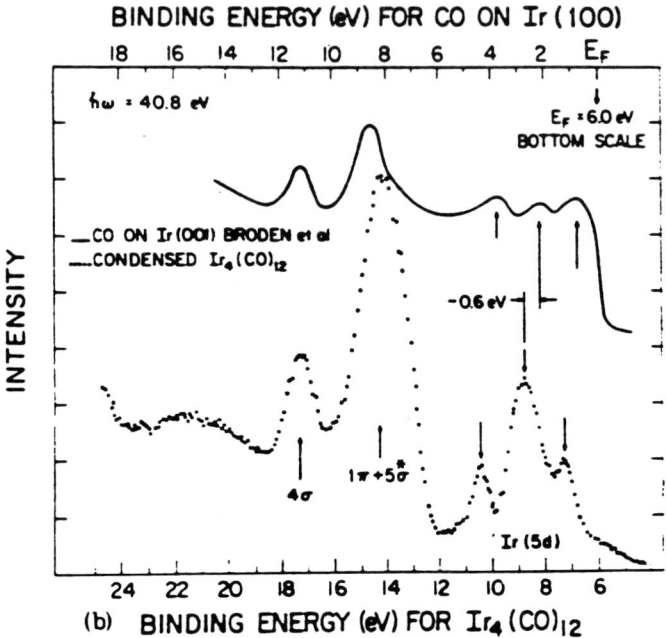

Fig. 9. Valence band photoemission spectra of $Ir_4(CO)_{12}$ and CO on Ir(001) taken at 40.8 eV photon energy. (Reproduced by permission from [17]).

related to the transition from a bulk metal to a cluster. The effect may, e.g., hamper the assignment of metal oxidation states in species derived from the metal cluster after heat and/or gas treatment, which may reproduce catalytic reactions.

A guide-line to follow changes in the oxidation state or in the environment of a particular supported cluster by means of photoemission, may be obtained by the application to the specific class of metal clusters of one of the existing models to estimate experimental XPS BEs [36]. In the simplest version, suited for large molecular systems with various attached ligands, these models allow to calculate the BE for the most intense metal ionization peak from parameter values associated to particular ligated atoms or atomic groups and from the metal oxidation state [36]. The parameter values are generally obtained through a best-fitting analysis of experimental data on several compounds. In this way one can predict the chemical shift in BE values to be associated to some bond rearrangement in the cluster after reaction, and compare the predicted amount with the overall instrumental resolution to check for experimental detectability of a particular reaction [21, 22].

5.4. XPS of Supported Molecular Clusters

Photoelectron spectroscopies have made a significant contribution to the structural electronic characterization of supported molecular metal clusters, as testified by the number of original and review papers published so far in the field [4–6, 8, 16]. The interest in a characterization by photoemission of these species, as inferred from the existing literature, lies mainly in their potential use as catalysts. A selection of topics in the field includes:

- *Nature of the supported cluster*: oxidation states of the metal atoms; determination of the range of existence of a particular surface species as a function of the temperature of the process; surface modification of a supported cluster as determined by reaction-induced changes in the relative quantitative ratio's of the elements;
- *Cluster-support interaction*: physisorption and chemisorption of a cluster on a substrate; role played by the nature of a support on cluster reactivity, also in terms of surface pretreatments;
- *Cluster-induced catalytic reactions*: assessment of cluster integrity; estimate of the actual dispersion on a substrate from XPS relative quantitative measurements. These aspects have been often investigated as a function of different initial preparation conditions and induced surface reactions;
- *Models of catalysts, based on supported molecular clusters*: identification of bonding sites of ligands on single crystals and specially grown surfaces.

Despite the potential interest of such XPS studies, the number of investigations dealing with supported molecular metal clusters, mostly carbonyl clusters, is limited [35, 37–59]. References to photoelectron spectroscopic studies of metal carbonyl clusters supported on oxides and characterized in catalytic reactions are rare [35, 38–47, 53, 56, 59]. The main reason which may be envisaged for this limited use of the technique is that some major possible drawbacks connected to the experimental setup of the technique itself, and/or to the molecular clusters as investigated samples have to be considered in order to plan a photoelectron spectroscopic experiment. From inspection of the existing literature in the field, two relevant problems may be extracted:

(1) Molecular metal clusters are generally sensitive to the high vacuum conditions, necessary in order to excite photoemission from conventional X-ray anodes. Some classes of molecular clusters, particularly metal carbonyls, are also sensitive to light, so that photodecomposition processes may occur, in the most unfavourable cases, within the acquisition time of a single spectrum, which is typically few minutes. Vacuum-induced decomposition may be reduced by cooling the sample to low temperature, typically liquid N_2 temperature, but it cannot be completely avoided. A fast acquisition of photoelectron spectra is, therefore, always desirable and often required. The actual sensitivity to monochromatic photons of a particular carbonyl cluster

has been found to depend, inter alia, on the nature of the substrate [29, 48, 49]. This means that a careful determination of BE's of a particular molecule could be possible by finding a suitable substrate. Moreover, photodecomposition during spectra acquisition may, in favourable cases, be circumvented in the case of pure solid compounds. The technique consists in collecting spectra from freshly deposited sample, by allowing a continuous evaporation or sublimation of the specimen from a source (which can be differentially pumped) against a cooled sample holder. Formation of bonds between cluster and substrate as a result of breaking of metal-ligand bonds, may modify the thermodynamics of the photodecomposition process, besides affecting the electronic structure of the species. As a result, a supported molecular cluster may be much more stable to X-rays than the pure species. We experimentally found a much higher stability of $Os_3(CO)_{12}$ chemisorbed on SiO_2 with respect to both pure and simply physisorbed samples [52, 53].

In summary, the stability of the supported molecular cluster has to be preliminarily tested under the experimental conditions for a photoemission experiment.

(2) A univocally accepted method for referencing BEs in the case of supported metal clusters is still lacking.

The problem has been the object of a long debate in the literature [3]. It has been emphasized that a metal support will unavoidably alter the electronic nature of the metal cluster, while if the support is insulating, the cluster BEs can assume different values, depending on the support [3]. Even more puzzling is the intermediate situation, represented by semimetallic substrates, among which carbon is one of the more commonly used.

Connected to this problem is the phenomenon of charging under photoemission. The procedure, widely used in the case of supported species, of referencing BEs by assigning a particular value to a substrate peak, does not solve all the underlying problems. This referencing procedure, though, may afford reproducibility of BE values within a related series of experiments as, e.g., in subsequent in situ heating or gas treatments on a supported cluster. The procedure is mainly intended to compensate for charging. Charging is a general phenomenon in photoemission. Its effects, however, are reflected in a photoemission spectrum only if electron screening is not effective within the time scale of the photoemission measurement. This generally happens in the case of insulating and poorly conducting species (either substrates and supported samples), hampering a correct assignment of BEs for all elements, particularly because of the intrinsically variable amount of the actual charge in different experiments.

A second effect which is due to charging is peak broadening, which prevents a meaningful curve-fitting of composite peaks and, in the most serious cases, leaves doubts about the actual number of atoms of the same element in different electronic states present in a broad peak. Of course, broadening of core metal peaks in clusters with reduced size may be a true effect, which, in the absence of cluster inhomogeneity, is attributed to lifetime broadening or reduced screening, as discussed in the previous section.

We will enter the problem of BE referencing in some more details, because of its basic relevance to the assessment of cluster core metal BEs.

Three major aspects of the problem have been debated in the literature: (i) actual size of the cluster (bulklike vs. non-bulklike dimensions); (ii) residual positive charge on the cluster, following photoemission; (iii) effect of the support.

Both the cluster size and the nature of the support are involved in the initial as well as in the final state processes. The presence of a residual positive charge in the core-ionized state of a cluster only affects its final state. Although this charge can be neutralized by electrons from the substrate, it may still be reflected in the photoemission spectrum; the crucial point is the relative time scales of the photoemission measurement vs. the cluster neutralization. The photoemission time is determined by the cluster core-hole lifetime, typically 10^{-15} sec. The time scale needed to neutralize the charge deficit induced by the photoemission by charge transfer from the support may be different. The main factor which determines the rate of charge transfer in a supported cluster may be found, in analogy with the already reported case of adsorbates on a surface [60], in the 'resistance to the support' [61]. As pointed out by Wertheim and DiCenzo [5], the factors to be considered are the contact area, the interfacial barrier, the density of states at the Fermi energy in the substrate and the Coulomb potential of the cluster. The actual value of the resistance is a function of the type (ionic, covalent, van der Waals), and strenght of the bond formed between adsorbate and support, which may vary from weak physorption to strong chemisorption. It is expected that strong chemisorbed clusters are well screened in their final states, while physisorbed clusters are isolated. An intermediate situation, represented by a weak cluster chemisorption, requires a sophisticated quantum-mechanical treatment, because of the probability of occupation of different final states for the positive charge [60]. To experimentally assess the existence of these different final states, a uniform cluster size distribution is needed, together with a high-resolution photoelectron apparatus. In the case of molecular metal clusters, the discussion of BE referencing is further complicated by the difficulty in knowing in detail the nature of the interaction with the substrate, mainly because of the presence of attached ligands. These last introduce different polarization channels in the support-to-cluster charge-transfer mechanism, offering a local response to the core hole. However, they provide the possibility of having size-selected clusters for detailed experiments.

A few papers have been published, dealing with a comparison between naked and ligated supported cluster properties, from this viewpoint [23, 29, 62]. We already commented in Section 2 on a high resolution XPS study of bare and molecular Au clusters [29]. Size distribution in the bare clusters is a crucial parameter in this comparative studies, as shown, among others, by

Fritsch and Légaré, who have reported a comparative XPS study of $Ir_4(CO)_{12}$ and naked Ir clusters, grown from the vapor [62]. Both systems were deposited on pyrographite. They obtained the Ir 4f relevant lineshape parameters for $Ir_4(CO)_{12}$ and for a series of supported Ir clusters, the largest of which had developed a clear Fermi edge, and used these to fit experimental curves in terms of Doniach-Sunjíc lineshapes [63]. A Gaussian broadening (purely instrumental) was convoluted with the peaks. The final asymmetry index and half-width at half-maximum of the Lorentzian curve vary with increasing Ir cluster nuclearity. The actual trend is indicative of a spurious effect, attributed by the authors to inhomogeneity in the size distribution, since, when the screening would be qualitatively the same, i.e., when the smaller Ir cluster obtained and $Ir_4(CO)_{12}$ are compared, the two Ir 4f signals exhibit a very different and asymmetric lineshape, while a symmetric peak is expected for small, uniform clusters.

Several different ways of avoiding charging or limiting its extent have been proposed, as extensively discussed in the literature [64]. The use of substrates made from perfectly planar inorganic oxides, like oxidic layers grown under ultrahigh vacuum on top of a metal, has been proposed a few years ago as a possible solution to the problem of differential charging, which results in different sample areas developing a different extent of charging under photoemission. A few studies have demonstrated the lack of any build-up of charge under illumination with X-rays of rhodium carbonyl clusters supported on Al_2O_3 grown in situ on carefully cleaned Al [48, 49]. An important preliminary point here is that the experimentally found chemical behaviour of Rh clusters on specially prepared planar aluminas and on powder with high surface area, prepared in a more traditional way, is similar. This justifies the use of the special substrate for a comparative study. By supporting $Rh_4(CO)_{12}$ and $Rh_6(CO)_{16}$ on in situ grown graphitic C or Al_2O_3 the authors were able to carefully determine the O 1s, C 1s and Rh 3d BE's, avoiding the problem of differential charging. In a preliminary investigation [48]. it was shown that by application of conventional XPS, the terminal, edge-bridging and face-bridging carbonyls in supported Rh carbonyl clusters can be readily distinguished from each other by inspection of the C 1s and O 1s core levels. The relative abundance of terminal vs. bridging CO's was shown to vary, depending on the stage of the decomposition process for both $Rh_4(CO)_{12}$ and $Rh_6(CO)_{16}$. This variation was interpreted as being due to both an increased electronic charge on the metal frame upon CO removal and the tendency to keep the largest possible degree of coordination on each metal atom. The trend in the BE's displayed by Rh3d, combined with the absence of changes in the observed O 1s and C 1s shake-up satellite intensities upon decomposition, were interpreted in terms of a non-aggregating reduction of the carbonyls into supported metal particles, evidencing that production of highly dispersed metal clusters from molecular carbonyls is at hand. In a subsequent paper [49], evidence was given to the formation of a rhodium dicarbonyl species on Al_2O_3. The species has been extensively characterized in terms of BEs of all the constituent elements and its instability under the conditions of the experiment has been pointed out. A significant result obtained is a Rh3d BE value which definitively assigns the supported Rh dicarbonyl to a highly oxidized rhodium species, most probably Rh^{3+}. This could result from interaction of the $Rh(CO)_2$ entity with three different O-Al surface groups. This dicarbonyl species is of particular importance in catalytic studies on metal carbonyl clusters also because the corresponding Os species was reported to be produced from supported $Os_3(CO)_{12}$ [65]. Both species were postulated by different groups to be responsible for catalytic reactions implying Rh or Os carbonyl clusters as the precursors [66].

The level of information on bonding site obtained in [48, 49] is quite comparable to that reached in chemisorption studies on single-crystal surfaces.

5.5. Outlook for the Future

Conventional XPS has been shown to supply truly unique information on pure and supported molecular clusters. Although a detailed knowledge of several small-nuclearity species has been reached, a general progress in the interpretation of photoemission spectra of pure molecular metal clusters is mainly connected to advances in the quantum mechanical treatment of large systems. Progress in the field of theoretical computation of metal clusters with attached ligands are becoming progressively easier, mainly because of the recent advances in the theoretical description of surface adsorption reactions in terms of a molecule-cluster reaction [67–69].

An increased signal intensity, coupled with monochromaticity of the photon source, may strongly deepen the characterization of supported, highly dispersed catalytic species or catalyst precursors derived from molecular clusters, like carbonyl clusters on inorganic oxides. The increasing availability to the scientific community of the advanced excitation source represented by synchrotron radiation, giving improvement of several orders of magnitude in signal intensity with respect to X-ray anodes sources and continuously tunable, monochromatic, highly polarized radiation, is opening up a new chapter also in catalytic studies by photoemission, because of the real possibility of doing time- and polarization-dependent as well as spin-resolved experiments [7]. When dealing with a pure or supported molecular cluster, however, several problems are facing both the experimentalyst and the theoretician and a successful experiment has to be selected and planned in great detail, as we tried to point out through this chapter.

References

1. C. R. Brundle and A. D. Baker (eds.), *Electron Spectroscopy*, Vol. 1, Vol. 2, Vol. 3, Vol. 4, Vol. 5., Academic Press, New York (1977–1984).
2. M. Cardona and L. Ley (eds.), *Photoemission in Solids, Topics in Applied Physics*, Vol. 26, Vol. 27, Springer, Berlin (1978, 1979).
3. W. F. Egelhoff, Jr., *Surf. Sci. Reports* **6** (1987) 253.
4. M. G. Mason, *Phys. Rev. B* **27** (1983) 748 (the article contains a number of references to photoelectron studies on pure and supported clusters).
5. S. B. DiCenzo and G. K. Wertheim, *Comments Solid State Phys.* **11** (1985) 203.
6. M. G. Mason, *Cluster Models for Surface And Bulk Phenomena*, NATO-ASI Series B, Vol. 283, Plenum Press, New York (1992), pp. 115–129.
7. G. Margaritondo, *Introduction to Synchrotron Radiation*, Oxford University Press, New York (1988), pp. 139–181.
8. R. C. Baetzold and J. F. Hamilton, *Progr. Solid St. Chem.* **15** (1983) 1.
9. S. T. Lee, G. Apai, M. G. Mason, R. Benbow and Z. Hurych, *Phys. Rev.* **B23** (1981) 505.
10. (a) L. Hedin and A. Johansson, *J. Phys.* **B2** (1969) 1336; (b) U. Gelius, *Physica Scripta* **9** (1974) 133.
11. C. D. Wagner, *Anal. Chem.* **44** (1972) 967.
12. B. Johansson and N. Mårtensson, *Phys. Rev.* **B21** (1980) 4427.
13. P. Ascarelli, M. Cini, G. Missoni and N. Nisticò, *J. Phys. (Paris) Colloque* **38** (1977) 125.
14. (a) M. Cini, *Surf. Sci.* **62** (1978) 148; (b) M. Cini and P. Ascarelli, *J. Phys. F (Metal Phys.)* **4** (1974) 1998.
15. F. Parmigiani, E. Kay and P. S. Bagus, *J. Electron Spectrosc. Relat. Phenom.* **36** (1985) 257.
16. L. Guczi, in *Metal Clusters in Catalysis*, B. C. Gates, L. Guczi and H. Knözinger (eds.), Elsevier Science Publishers, Amsterdam (1986), pp. 209–219 and references therein.
17. E. W. Plummer, W. R. Salaneck and J. S. Miller, *Phys. Rev.* **B18** (1978) 1673.
18. S. F. Yang, A. A. Bakke, H.-W. Chen, C. J. Eyermann, J. L. Hoskins, T. H. Lee, D. Seyferth, H. P. Withers, Jr. and W. L. Jolly, *Organometallics* **1** (1982) 699.
19. H.-J. Freund, F. Greuter, D. Heskett and E. W. Plummer, *Phys. Rev.* **B28** (1983) 1727.
20. Y. Sasaki, A. Tokiwa and T. Ito, *J. Am. Chem. Soc.* **109** (1987) 6341.
21. R. Zanoni, V. Carinci, H. Abu-Samn, R. Psaro and C. Dossi, *J. Mol. Struct.* **131** (1985) 363.
22. R. Zanoni and J. Puga, *J. Mol. Struct.* **240** (1990) 89.
23. P. Légaré, Y. Sakisaka, C. F. Brucker and T. N. Rhodin, *Surf. Sci.* **139** (1984) 316.
24. G. Apai, S.-T. Lee, M. G. Mason, L. J. Gerenser and S. A. Gardner, *J. Am. Chem. Soc.* **101** (1979) 6880.
25. M. Arfelli, C. Battistoni, G. Mattogno, D. M. P. Mingos, *J. Electron Spectrosc. Relat. Phenom.* **49** (1989) 273.
26. C. Battistoni, G. Mattogno, F. Cariati, L. Naldini and A. Sgamellotti, *Inorg. Chim. Acta* **24** (1977) 207.
27. P. M. Th. M. van Attekum, J. W. A. Van Der Velden and J. M. Trooster, *Inorg. Chem.* **19** (1980) 701.
28. C. Battistoni, G. Mattogno, R. Zanoni and L. Naldini, *J. Electron Spectrosc. Relat. Phenom.* **33** (1984) 107.
29. G. K. Wertheim, J. K. Kwo, B. K. Teo and K. A. Keating, *Solid St. Commun.* **55** (1985) 357.
30. C. Battistoni, G. Mattogno, D. M. P. Mingos, *J. Electron Spectrosc. Relat. Phenom.* **33** (1984) 107.
31. M. C. Jennings, G. Schoettel, S. Roy and R. J. Puddenphatt, *Organometallics* **10** (1991) 580.
32. B. Wang, Y. Sasaki, S. Ikari, K. Kimura and T. Ito, *Chem. Lett.*, in press.

33. M. Quinten, I. Sander, P. Steiner, U. Kreibig, K. Fauth and G. Schmid, *Z. Phys. D* **20** (1991) 377.
34. (a) G. Schmid, R. Pfeil, R. Boese, F. Bandermann, S. Meyer, G. H. M. Callis and J. W. A. van der Velden, *Chem. Ber.* **114** (1981) 3634; (b) R. C. Thiel, R. E. Benfield, R. Zanoni, H. H. Smit, M. W. Dirken, *Struct. Bond.* **81** (1993) 1; (c) H. H. A. Smit, Ph.D. Thesis University of Leiden, The Netherlands (1988); H. H. A. Smit, P. R. Nugteren, R. C. Thiel, L. J. de Jongh, *Physica* **B153** (1988) 33; H. H. A. Smit, R. C. Thiel, L. J. deJongh, G. Schmid, N. Klein, *Solid State Commun.* **65** (1988) 915; (d) M. C. Fairbanks, R. E. Benfield, R. J. Newport, and G. Schmid, *Solid State Commun.* **74** (1990) 431; M. A. Marcus, M. P. Andrews, J. Zegenhagen, A. S. Bommannavar and P. Montano, *Phys. Rev.* **B42** (1990) 3312; P. D. Cluskey, R. J. Newport, R. E. Benfield, S. J. Gurman and G. Schmid, *Materials Research Society*, Spring 1992 Meeting Proceedings, in press.
35. R. Zanoni, R. Psaro, C. Dossi, L. Garlaschelli and R. Della Pergola, *J. Cluster Sci.*, **1** (1990) 241.
36. R. D. Feltham and P. Brant, *J. Am. Chem. Soc.* **104** (1982) 641.
37. M. Kawai, M. Uda and M. Ichikawa, *J. Phys. Chem.* **89** (1985) 1654.
38. K. Tanaka, K. L. Watters, R. F. Howe and S. L. T. Andersson, *J. Catal.* **79** (1983) 251.
39. G. F. Meyers and M. B. Hall, *Inorg. Chem.* **23** (1984) 124.
40. G. F. Meyers and M. B. Hall, *Inorg. Chim. Acta* **129** (1987) 153.
41. M. J. Kelley, A. S. Fung, M. R. McDevitt, P. A. Tooley and B. C. Gates, *Mater. Res. Soc. Symp. Proc.*, Vol. 111 (1988) p. 23.
42. S. L. T. Andersson, K. L. Watters and R. F. Howe, *J. Catal.* **69** (1981) 212.
43. H. Knözinger, Y. Zhao, B. Tesche, R. Barth, R. Epstein, B. C. Gates and J. P. Scott, *Faraday Disc. Chem. Soc.* **72** (1981) 53.
44. X.-J. Li, J. H. Onuferko and B. C. Gates, *J. Catal.* **85** (1984) 176.
45. J. R. Budge, B. F. Lucke, J. P. Scott and B. C. Gates, *Proc. 8th Int. Congr. Catal.*, Berlin (West), 2–6 July, Verlag Chemie, Frankfurt (1984), Vol. 5, p. 89.
46. P. G. Gopal and K. L. Watters, *Proc. 8th Int. Congr. Catal.*, Berlin (West), 2–6 July, Verlag Chemie, Frankfurt, (1984), Vol. 5, p. 75.
47. H. F. J. van't Blik, J. B. A. D. van Zon, T. Huizinga, D. C. Köningsberger and R. Prins, *J. Am. Chem. Soc.* **107** (1985) 3139.
48. G. Apai and B. G. Frederick, *Langmuir* **3** (1987) 395.
49. B. G. Frederick, G. Apai and T. N. Rhodin, *J. Am. Chem. Soc.* **109** (1987) 4797.
50. C. Furlani, R. Zanoni, C. Dossi and R. Psaro, *The Physics and Chemistry of Small Clusters*, NATO-ASI Series, Plenum Press, New York (1987), pp. 775–780.
51. C. Dossi, R. Psaro, R. Zanoni and F. S. Stone, *Spectroch. Acta, Part A* **43** (1987) 1507.
52. R. Zanoni and R. Psaro, *Spectroch. Acta, Part A* **43** (1987) 1497.
53. C. Dossi, A. Fusi, E. Grilli, R. Psaro, R. Ugo and R. Zanoni, *Catalysis Today* **2** (1988) 585.
54. C. Dossi, A. Fusi, R. Psaro, R. Ugo and R. Zanoni, *Structure and Reactivity of Surfaces*, Elsevier Science Publishers, Amsterdam (1989), pp. 375–384.
55. C. Dossi, A. Fusi, E. Grilli, R. Psaro, R. Ugo and R. Zanoni, *J. Catal.* **123** (1990) 181.
56. A. S. Fung, P. A. Tooley, M. J. Kelley, D. C. Koningsberger and B. C. Gates, *J. Chem. Phys.* **95** (1991) 225.
57. T. K. Sham, Z. F. Liu and K. H. Tan, *J. Chem. Phys.* **94** (1991) 6250.
58. T. K. Sham, J. Hrbekand and K. H. Tan, *Surf. Sci.* **236** (1991) 259.
59. R. M. Sanyal, D. K. Ghorai, D. R. Dutta, S. K. Adhya, B. Sen and B. Viswanathan, *Appl. Catal.* **74** (1991) 153.
60. S. L. Qiu, X. Pan, M. Strongin and P. H. Citrin, *Phys. Rev.* **B36** (1987) 1292n.
61. K. Schonhammer and O. Gunnarsson, *Solid State Commun.* **23** (1977) 691.
62. A. Fritsch and P. Légaré, *Surf. Sci* **145** (1984) L517.
63. S. Doniach and M. Sunjíc, *J. Phys. C* **3** (1970) 285.
64. D. Briggs, *Handbook of X-ray and Ultraviolet Photoelectron Spectroscopy*, Heyden & Sons Ltd., London (1977), p. 121.

65. R. Psaro, C. Dossi and R. Ugo, *J. Mol. Catal.* **21** (1983) 331.
66. R. Psaro and R. Ugo, *Metal Clusters in Catalysis*, Elsevier Science Publishers, Amsterdam (1986), pp. 427–496.
67. H.-J. Freund and M. Neumann, *Appl. Phys. A* **47** (1988) 3.
68. G. Pacchioni, P. Bagus and F. Parmigiani (eds.), *Cluster Models for Surface And Bulk Phenomena*, NATO-ASI Series B, Vol. 283, Plenum Press, New York (1992).
69. D. Salahub and N. Russo (eds.), *Metal-Ligand Interactions: From Atoms, to Clusters, to Surfaces*, NATO-ASI Series C, Vol. 378, Kluwer Academic Publishers, Dordrecht (1992).

6. APPLICATION OF MÖSSBAUER EFFECT SPECTROSCOPY TO CLUSTER RESEARCH

R. C. THIEL, H. H. SMIT and L. J. DE JONGH
Kamerlingh Onnes Laboratory, Leiden University,
P.O. Box 9506, 2300 RA Leiden, The Netherlands

6.1. Introduction

During the course of the last 35 years, Mössbauer Effect Spectroscopy (MES) has been well established as a powerful measurement technique in both solid state physics and chemistry [1, 2, 3, 4, 5]. Because of the existence of so many excellent reviews of MES, we will only give a brief sketch of the basics needed to understand the use of the technique, in particular as applied to cluster research.

In this chapter we will examine the use of MES for the study of small particles and clusters, which until now has been relatively limited. The advantages and disadvantages of MES will be examined, and the few molecular cluster compounds having fairly large metal atom cores and which have been thoroughly studied, $[Au_9L_8]^{n+}(NO_3)_3$ (with $n = 1, 3$), $Au_{11}L_7(SCN)_3$, and $Au_{55}L_{12}Cl_6$, will be used as illustrative examples. Here and in what follows, we will use the abbreviation L = PPh$_3$ unless there is a chance of confusion.

6.2. Mössbauer Effect Spectroscopy

6.2.1. ELEMENTS OF MES

Before beginning with a discussion of the specific application of MES to cluster research, it would seem useful to give the basics of the technique for those not familiar with the method.

The Mössbauer Effect, sometimes called nuclear resonant fluorescence, makes use of a transition between a low lying nuclear level and the nuclear ground state, with the concomitant emission or absorption of a photon. This is quite similar to atomic and molecular absorption processes, except that with MES the photon is a γ-ray with an energy generally of the order of a few tens of keV. In principle, the width of the lorentzian emission or absorption line is given by the lifetime of the excited state, and for the Mössbauer excited state it may be of the order of tens to hundreds of peV. In the case of ^{197}Au MES, which we will use as an example and for which the decay scheme is shown

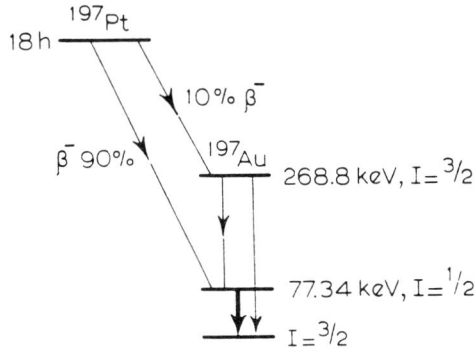

Fig. 1. The nuclear decay scheme of ^{197}Pt to ^{197}Au, showing the Mössbauer level, nuclear spins, lifetimes, and transition energies.

in Figure 1, the excited state has an energy of 77.3 keV and the line width is 240 peV. As indicated in the figure, the ground state has a nuclear spin of 3/2 and the excited state a nuclear spin of 1/2.

In order to be able to perform spectroscopy, it is necessary to have a range of energies available; for example, to obtain an absorption spectrum we need a beam of photons covering a certain range of energies, out of which the resonant energies can be absorbed and then isotropically re-emitted.

Although the use of synchrotron radiation as a source of γ-ray photons for MES has advanced rapidly during the past few years, it has as yet not been used for cluster research. For this reason we will limit ourselves here to the normal situation, where recourse has had to be taken to the use of a radioactive isotope, such as ^{197}Pt, to populate the excited state of the nucleus, in this case ^{197}Au, which then emits the needed γ-ray on its return to the ground state. The Mössbauer source then contains both the radioactive precursor as well as the Mössbauer isotope, either in the excited or in the ground state.

If we consider absorption spectroscopy, it is the absorber which will be studied and which must contain the Mössbauer isotope in its ground state; in the case of gold, the only stable isotope is ^{197}Au, our isotope of interest. Because the emission and absorption lines in a solid have approximately the same – fixed – energies, use is made of the Doppler effect to give one of these two lines, usually the emission line, a continuous range of energies. In moving the source toward and away from the absorber with a velocity ranging from about -15 to +15 mm/s, one is able to scan the region of interest for a normal gold absorption spectrum. This is shown in a typical experimental arrangement in Figure 2.

For MES measurements on ^{197}Au, the detector indicated in this figure is usually a high purity Germanium solid state detector, followed by electronics to separate the Mössbauer γ-rays from other unwanted γ's also emanating from the source.

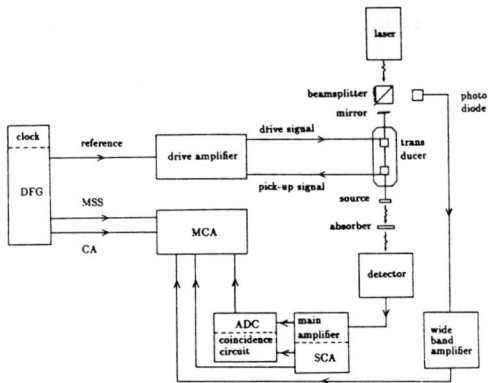

Fig. 2. A schematic experimental arrangement for MES.

When the ground state nuclear volume is different from that of the excited state, as it is for ^{197}Au, and at the same time there is a difference of electronic density within the nuclear volume present in the source and the absorber, the absorption line observed will be shifted from the zero-velocity position. This so-called Isomer Shift (I.S.), or in chemical literature Chemical Shift (C.S.), shown in Figure 3a, can thus provide information about the chemical surroundings of a given Mössbauer site, as reflected in the local electron density within the nuclear volume of Mössbauer isotope in the absorber, relative to that in the source [2]. For ^{197}Au MES, we will be comparing the density in the source, i.e. gold in metallic platinum, with the density in the nuclear volume due to the 6s-, 6p-, and 5d-electrons in the absorber.

This electron density difference between source and absorber can be thought of as being composed of four components: the actual difference in total density of s-electrons, the difference in total relativistic density of 6p-electrons, the contribution due to the difference in screening of the 6s-electrons by the 6p-electrons, and the contribution due to the difference in screening of 6s-electrons by 5d-electrons. The first contribution is the average density of all s-electrons within the nuclear volume, regardless of the energy and the spin. This should not be confused with the density of states at the Fermi level, which plays a major role in the NMR Knight Shift, the electronic specific heat, and the magnetic susceptibility, as discussed in other chapters of this volume. For gold, the contribution due to the 6p screening is almost completely cancelled by the relativistic density of the 6p-electrons. The main contribution comes from the 6s-electrons, which corresponds to +8.0 mm/s per 6s-electron added to the gold atom, while the 5d-electrons have a screening effect of -1.6 mm/s per 5d-electron added to the gold atom [6]. In gold, the second order Doppler shift [2], which leads to a temperature dependent shift of the centroid, is more than an order of magnitude smaller than the usual fitting errors of ±0.02 mm/s, and can be disregarded. Note that

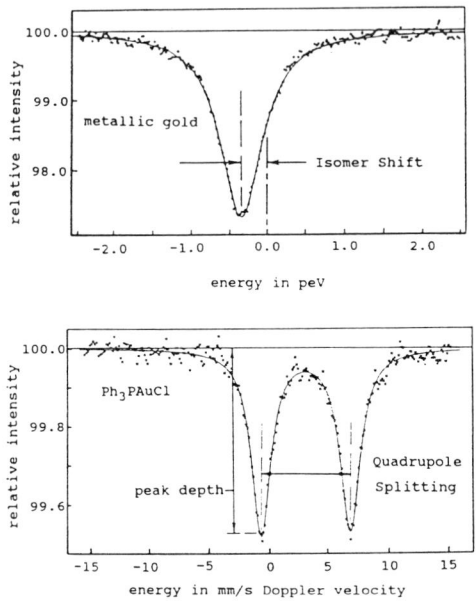

Fig. 3. (a) A representative ^{197}Au Mössbauer absorption singlet from bulk gold, showing the I.S. The energy scale is given in peV. (b) A representative absorption doublet. Here the energy scale is expressed in terms of the more usual Doppler velocity in mm/s.

while there is room for interpretation as to which effects are responsible for the electron density seen by the I.S., the I.S. itself is measured directly.

If one (or both) of the nuclear levels carries a quadrupole moment and at the same time the nucleus finds itself in an electric field gradient (EFG), the originally single absorption line will exhibit Quadrupole Splitting (Q.S.) [2]. Only the spin 3/2 ground state of gold has a quadrupole moment, so the nuclear quadrupole interaction leads to the splitting of the original absorption singlet into a symmetric doublet, as shown schematically in Figure 3b.

In gold compounds, it is usually the non-cubic distribution of the electrons of the Mössbauer ion itself, often expressed in terms of occupation of the molecular orbital levels [7], which leads to an EFG at the nuclear site. The Q.S., taken together with the I.S., also provides information as to the local chemistry of the site, since there is a strong correlation between the Q.S. and the I.S. of gold compounds. This is shown in Figure 4, which has been extracted from refs. [7, 8, 9, 10]. Note that Au^I, etc. are nominal valence states, and should not be taken literally.

Here again, while the origin of the Q.S. may be open to interpretation, i.e. the origin of the EFG present at the nucleus, the actual value of the Q.S. is obtained directly from the spectra, via the fitting procedure, and can be considered a physical fact.

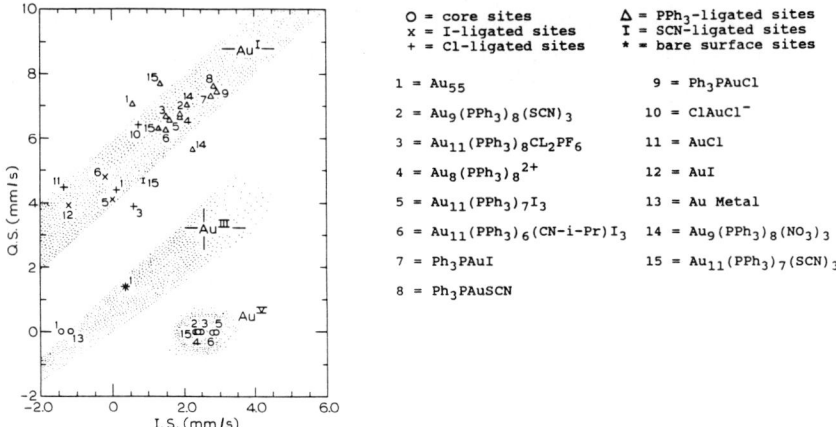

Fig. 4. Q.S. versus I.S. for several gold cluster compounds and for bulk gold. Included is an indication of the location of the various valence and/or hybridization regions.

Since gold is a diamagnetic element and no measurements in external magnetic field will be presented here, we need not consider the possible nuclear hyperfine splitting (hfs) due to the presence of a magnetic field.

As a consequence of the high photon energies and the very narrow line widths, resonant absorption and resonant emission can only take place under those circumstances where there is both conservation of energy and of recoil momentum, i.e. in zero phonon processes, without recoil. Although a classical description is actually incorrect, it gives a useful picture for describing the thermal behaviour of the gold clusters discussed below: one considers the recoil to be taken up by the lattice as a whole. The fraction of events which are recoilless, the recoil-free fraction or f-factor, can be written as

$$f = \exp\{-k\langle x^2\rangle_T\}, \tag{1}$$

where k contains all the nuclear parameters, and $\langle x^2\rangle_T$ is the thermal average of the mean square amplitude of the emitting or absorbing atom. The mean square amplitude can be written in the Debye approximation as

$$\langle x^2\rangle_T = \frac{3\hbar^2}{4mK_B\Theta_D}\left(1 + 4\frac{T^2}{\Theta_D^2}\int_0^{\Theta_D/T}\frac{x\,dx}{e^x - 1}\right), \tag{2}$$

where m is the atomic mass, $\hbar$ and k_B both have their usual meanings, and Θ_D is the Debye temperature [1]. The more strongly bound the Mössbauer atom is in its surroundings, usually a lattice or a glass-like matrix, the larger the f-factor. In our case, with small ligated particles, the Mössbauer atom is in a 'lattice' of neighbors, which in turn are in a lattice of clusters or in an amorphous matrix. The Debye–Waller factor from x-ray diffraction is a more familiar name for the same effect, which in MES literature bears the name Lamb–Mössbauer factor. The f-factor manifests itself in the observed absorption intensity, or peak depth, of the lorentzian peak, as shown in Figures 3a and 3b. The peak intensity decreases strongly as the temperature is increased above θ_D. Below $\theta_D/2$, the temperature dependence of the peak intensity becomes relatively weak.

If there are multiple sites present in the absorber, which can be intrinsic to the sample or may be due to impurities located in or on the sample, one will see a superposition of spectra, one from each distinct site. If the bonding of all the sites to the lattice (or matrix) is the same, then the relative intensities of the sites will be proportional to the site occupations. If the bonding of the individual sites differs, but one knows the site occupations, the f-factors for the individual sites can still be extracted directly.

Anisotropic surroundings of a (surface) atom will lead to an anisotropic f-factor, which in turn *may* cause asymmetry in a quadrupole doublet, the so-called Goldanskii–Karyagin Effect [2].

6.2.2. MES SPECIFIC TO CLUSTERS

MES has a number of very strong points that have made it the popular technique which it has become. Among these is the fact that generally a large amount of information can be extracted from the spectra, which can often be compared quite directly to physical models, preferably with a minimum of assumptions. Although use of computer fitting of the spectra is essential, in most cases the analysis is relatively straightforward, but does require adequate knowledge of the system in which the Mössbauer isotope in located. Furthermore MES is a local probe, in the sense that the parameters obtained are site specific. And finally, Mössbauer absorption measurements require only a few tens of milligrams of sample, and can be performed on powders and without the use of an external magnetic field (as contrasted to NMR). In fact, Mössbauer emission spectroscopy, where the sample contains the radioactive isotope, requires only a few milligrams, but is seldom utilized because of the need for often sophisticated radiochemistry. For some exceptions, see [11] and the last section of this chapter [12, 13].

On the other hand, there is a serious limitation to the use of MES, especially in research on compounds of large molecular clusters. The point is that there are only a few isotopes suitable for simple MES which combine useful physical characteristics with adequate availability. One requires a not-too-high γ energy and a line-width small enough to be able to utilize the I.S., the Q.S., and the hfs, as well as that these quantities are themselves large enough to be useful observables. And one also requires a Mössbauer source which is either commercially available or which can be simply prepared, and which has a long enough half-life to be practicable. Of these, there are to date none which have been built into metallic cluster compounds containing more than a few metal atoms [14], with the exception of ^{197}Au. This has limited the use of MES in research on molecular compounds with large metallic cores until now to the study of gold clusters [7, 9, 10, 15, 16, 17, 18]. ^{197}Au is in fact also a rather exotic isotope, with a high γ energy and low θ_D, requiring cooling of both source and absorber to below 100 K. The source, prepared by thermal neutron irradiation of the isotope ^{196}Pt naturally present in platinum metal, has a half-life of only about 18 hours, which necessitates proximity to a reactor as well as the use of specialized cryogenic equipment for rapid manipulation of samples.

In addition to studying gold clusters by MES, it would be very interesting to be able to study platinum clusters as well. Unfortunately, ^{195}Pt, the only platinum Mössbauer isotope, is an example of a Mössbauer isotope having physical characteristics which make it unattractive for cluster research: relatively broad lines, giving poor site resolution, a very short half-life, making measurements even more difficult, and probably the worst characteristic, no nuclear quadrupole moment [20], making site resolution completely impossible.

A very recent development, which will be treated briefly in the last section of this chapter, is the realization [12, 13] that neutron activation of the natural abundance of 25% ^{196}Pt contained in platinum clusters might be carried out without damaging the cluster. Preliminary results indicate that this is indeed possible, at least for the large cluster type having a core of 309 platinum atoms [21]. In this case, one does Mössbauer emission spectroscopy, using a gold foil as absorber.

In a sense, there is another 'limitation' to the use of MES in clusters, as mentioned above, namely the requirement of adequate knowledge of the system containing the Mössbauer isotope. Due to the necessity of least squares fitting of the spectra and the fact that, given enough fitting parameters, one can in principal fit any spectrum, the introduction of physical restraints to reduce the number of fitting parameters becomes essential. This in turn requires a minimum of preliminary knowledge of the system. In order to be able to make a physically meaningful analysis, it is essential to have the results of as many other measurements on the system as possible.

Although ^{197}Au Mössbauer spectroscopy on molecular cluster compounds with large cluster cores is the central topic of this chapter, it should be mentioned that there is an extensive literature on ^{57}Fe Mössbauer absorption spectroscopy and ^{57}Co Mössbauer emission spectroscopy on iron catalysts deposited on various substrates [11, 19, 22, 23].

6.2.3. Previous MES Measurements on Gold Particles

Over the past two decades there have only been a few MES measurements carried out on small gold particles. As early as 1970, Schroeer *et al.* [24]

attempted to demonstrate the effect of low frequency cutoffs on the recoil-free fraction of small gold colloids in gelatine.

In an elegant but none-the-less inconclusive study of the temperature dependence of ^{197}Au MES on gold colloids, Viegers [8, 25] was able to give an indication that the center-of-mass motion of the entire colloid might be responsible for a large discrepancy in the measured f-factors at 4.2 K when compared to that of bulk gold. It will be obvious that colloids have the serious disadvantage of having relatively large size- and shape-distributions. It is just this point which complicated those results.

An improvement in the results of MES measurements on small particles came with the advent of the cluster compounds. The first ones, containing only a few gold atoms, such as $Au_4(PPh_2CH_2PPh_2)_3I_2$ and $Au_5(PPh_2CH_2PPh_2)_3(PPh_2CHPPh_2)^{2+}$ [7], were measured by MES at 4.2 K. Similar measurements have been performed on $[Au_8L_8](PF_6)_2$ and $[Au_9(PAr_3)_8](PF_6)_3$ clusters [16], where the cluster surfaces were coordinated by (for example) PPh_3, and where the clusters could be either neutral or ionic.

MES measurements at 4.2 K on $Au_{11}L_7X_3$ clusters, with X = I, Cl, SCN, and a variety of other ligands [7, 17, 26, 27], have also been reported. MES measurements on Au_{13} clusters have been mentioned in the literature [7, 17], but without quoting results or presenting spectra. These cluster compounds, as well as those mentioned in the previous paragraph, form crystals, and x-ray diffraction (XRD) has shown that they are all basically derivable from an icosohedral structure, which only becomes complete for the $Au_{13}(dppm)_6(NO_3)_4$ cluster[16].

Finally, an Au_{55} cluster, ligated by 12 PPh_3 and 6 Cl ions, was also investigated by MES at 4.2 K[28]. Although this compound doesn't form crystals, there was some evidence that the cluster structure is cuboctahedral.

The above mentioned investigations were all especially aimed at an understanding of the chemistry of these cluster compounds. They provided a rich background against which the temperature dependent MES measurements performed in our own laboratory could be placed. As will be discussed below, we have made an extensive study of the $Au_{55}(PPh_3)_{12}Cl_6$ cluster compound, which we will abbreviate as Au_{55}, as well as additional measurements on the related compounds $Au_{55}(PPh_2C_6H_4NaSO_3 \cdot 2H_2O)_{12}Cl_6$, a water soluble version abbreviated as Au_{55}^* [29], and $Au_{55}(P[C_6H_4CH_3]_3)_{12}Cl_6$, a methylated version abbreviated as Au_{55}-tolyl [30].

Before beginning with a discussion of the measurements on Au_{55} and an analysis thereof, it is necessary to give a short description of this molecular cluster system: the cluster core consists of two shells around a single central gold atom, a first shell of twelve gold atoms, and a second shell of 42 surface atoms, all arranged in a cuboctahedral stacking, i.e. cubic close-packed, as sketched in Figure 5a. The gold atoms at the corner surface sites, shown schematically in grey, are coordinated by a phosporus ion which forms part

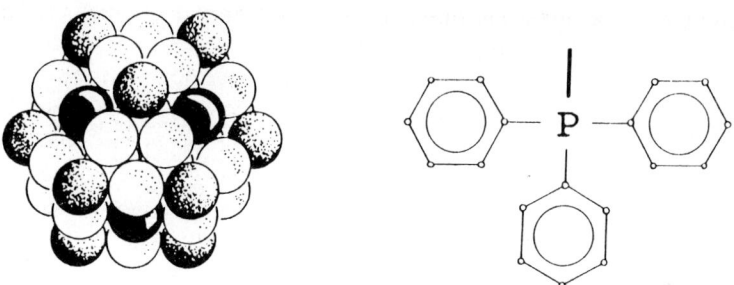

Fig. 5. (a) A schematic representation of the four sites of the bare Au$_{55}$ cluster. (b) A flattened representation of the PPh$_3$ ligand.

of the triphenyl phosphine ligand shown in a flattened view in Figure 5b. In fact, the three rings attached to each of the phosphorus atoms branch symmetrically outward, forming 'cones' which serves as an insulating jacket around the cluster core, thus keeping the metal cores from coalescing. The gold atoms shown in black in Figure 5a, at the center of the square faces, are coordinated by chlorine atoms.

At the time that our first Mössbauer measurements were being analyzed, the cuboctahedral structure was only a postulate, and several other possibilities also had to be considered, such as icosahedral, antiprismatic cubic, and hcp. The explanation of the thermal behaviour to be discussed in the following section was one of the strongest pieces of evidence for the cuboctahedral structure, which is now generally accepted.

6.3. Our Mössbauer Results

6.3.1. THERMAL BEHAVIOUR

In systems containing a number of physically inequivalent sites, Mössbauer spectroscopy can often allow the determination of the properties of the individual sites. Au$_{55}$, with three different surface sites and a fourth core site, proved to be an ideal candidate for studying the thermal properties of two-shell, cuboctahedral cluster compounds.

Because the I.S. and the Q.S. for ^{197}Au are dependent mainly on the local surroundings, they can be used as a label for the site: a plot of Q.S. versus I.S. gives a good picture of the local site 'chemistry' [8, 9, 10], as shown in Figure 6, with the three most common nominal valence state assignments and local symmetries grouped into well defined regions and indicated here by the speckled regions. This information can provide a unique site label for the study of the behaviour of the four sites individually. As explained above, from the study of the thermal evolution of the Mössbauer spectra, the Debye–Waller factors of the different contributions to the spectrum may be obtained

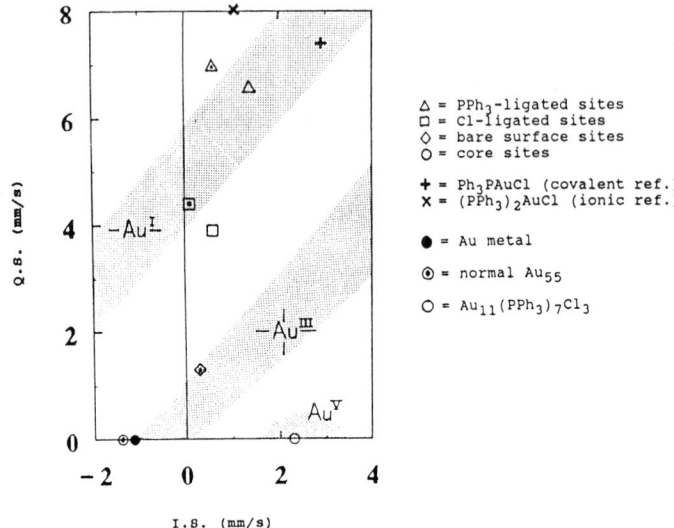

Fig. 6. The Q.S. versus I.S. values of LAuCl, L_2AuCl (as reference materials), and the various sites of $Au_{11}L_7Cl_3$, and $Au_{55}L_{12}Cl_6$ with L = PPh_3. The symbols used are discussed in the text.

and thereby the vibrational mean squared amplitudes of the individual sites may be estimated.

When considering the thermal behaviour of small particles, the usual approach is to start with the continuum model with the necessary corrections, such as the Born procedure for the counting of modes using different cut-off frequencies for transverse and longitudinal modes [35], separation into bulk modes and surface modes [36], introduction of low frequency cut-off due to the particle size [37, 38], which comes down to assigning a θ_D^{bulk} and a θ_D^{surf} to a given particle (or ensemble of particles), perhaps together with a θ_0^{bulk} and a θ_0^{surf} as low frequency cutoffs.

Initial analysis of measurements done at 4.2 K [17] made use of this idea, leading to 13 free parameters for the fitting procedure: an I.S.-value for each site and a Q.S.-value for each site except the core site with Q.S. = 0, four line-widths, and two relative line intensities, one for the core site (i.e. for the bulk modes) and one for the surface sites (i.e. for the surface modes, the relative site occupations of the surface sites being known). Unfortunately, in [17] a superposition of lorentzian lines was used for their fits, which is an acceptable approximation for very thin absorbers and no overlap of the lines, but is unacceptable for the case discussed here. This provided these authors with an erroneous extra two unjustified degrees of freedom in their fit.

On the other hand, analysis of MES measurements as a function of temperature between 1.25 and 60 K on the Au_{55} system [9], employing the transmission integral method [1, 8], the only correct method when lines

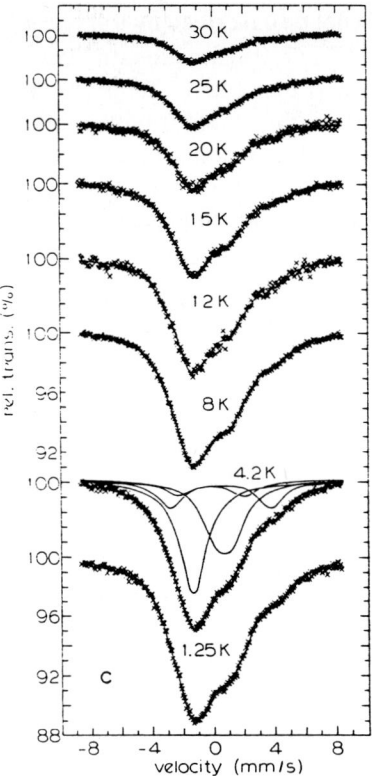

Fig. 7. Mössbauer spectra at several temperatures between 1.25 and 30 K, with transmission integral fit; the sub-spectra at 4.2 K are indicated [9].

strongly overlap, gave very satisfactory results using one unsplit Mössbauer absorption line for the 13 core site atoms and three Q.S. doublets for the three distinct surface sites, using the natural line-width for all sites, plus variable absorption intensities for all four sites, i.e. only 11 free parameters. The positions of the four sites in a Q.S. versus I.S. plot is shown in Figure 6, together with the positions of the three sites of Au_{11} for reference. The fitted spectra in the temperature range stated are shown in Figure 7; the sub-spectra shown for 4.2 K are lorentzian simulations made using the parameters obtained from the transmission integral fit, and are meant only as a guide to the eye.

From the temperature dependence of the fitted line intensities, and from the known relative site occupations, the Mössbauer f-factors for the four sites of Au_{55} could be calculated [9]. These, in turn, could be related to effective Einstein (or Debye) temperatures [39] θ_E's (or θ_D's) associated with the vibrations of the atoms on the individual sites.

Unexpectedly, the relative intensities corresponding to the three surface sites were not in the same ratio as the site occupations. Therefore the f-factors of the surface sites could not be described by a single $\theta_E^{\text{surface}}$, meaning that the use of a θ_E^{bulk} and a $\theta_E^{\text{surface}}$ [37, 38] was insufficient to explain these results. An additional problem was that the f-factors for all four sites were also considerably lower at 1.25 K than that for bulk gold at the same temperature, similar to what was observed with colloidal gold [8, 25].

A further very important result of this analysis, also clearly visible in Figure 7, is that the temperature dependence of all the sub-spectra together in the temperature range studied is determined predominantly by the motion of the center of mass of the whole Au_{55} cluster. This can be concluded from the observed uniform decrease of the total intensity with increasing temperature, without any visible change in the general shape of the spectrum. In effect, this means that the f-factors for the individual sites must be multiplied by an f-factor due to the motion of the center of mass of the whole particle [9]. See also [40], where this concept was originally developed, and [8, 25], where it was tested on colloids. The use of such an inter-cluster f-factor, in addition to the usual intra-cluster f-factor, also resolved the problem of the apparent deficiency in the total f-factor at 1.25 K when compared to bulk gold.

Although the model calculations clearly showed that there is also strong anisotropy in the f-factor for two of the three surface sites, the Goldanskii–Karyagin Effect was *not* taken into account in the fits of the spectra. This would have necessitated use of extra free parameters, while any asymmetry present in the doublets would have been completely overshadowed by the statistics.

The eigen-frequencies and the degeneracies for the various gold sites have been calculated, using a simple brute force model of point masses and massless springs [15]. For this calculation, the ligands on the surface were completely ignored, since it is known that the energies of the vibrational and rotational modes for ligands such as those coordinating Au_{55} usually lie well above 50 K. The only free parameter used was the unknown spring constant for the motion of the center-of-mass of the cluster in the matrix of amorphously interlocked ligands, i.e. for the three inter-cluster translational degrees of freedom. The rotational degrees of freedom were also ignored. This single free parameter was determined by fitting the temperature dependence of the f-factors, as shown by the full curves drawn in Figure 8.

The calculated specific heat for Au_{55} at temperatures between 2 and 60 K, based on this model, is shown in Figure 9. Satisfactory agreement has been obtained for the f-factors and the specific heats of the cluster compounds Au_9 and Au_{11} as well[15]. Note that the full curve drawn in the figure corresponds to the specific heat calculated with *no free parameters*. It is in fact the 3 inter-cluster translational degrees of freedom usually ignored in treatments found in the literature, ie. the motion of the whole cluster, which govern the specific heat at temperatures below about 10 K.

That the surface-site f-factors are strongly sensitive to the details of the structure of the cluster should be less than surprising. The MES observation

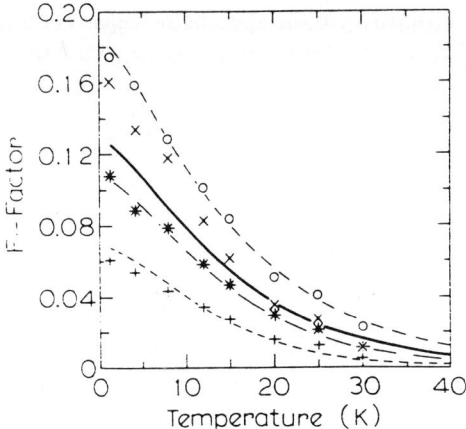

Fig. 8. The temperature dependence of the f-factors of the MES sub-spectra [15]. The fit is discussed below.

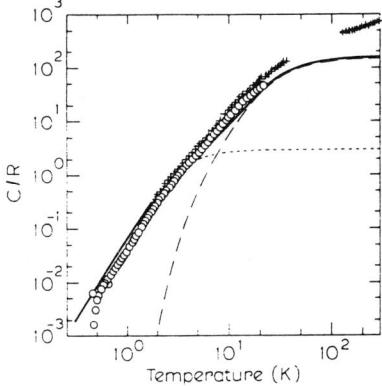

Fig. 9. The specific heat of Au_{55} at low temperatures, with the specific heat – calculated with no free parameters – included as a full line; the fine dashed line is the inter-cluster contribution, the coarse dashed line the intra-cluster contribution [15].

that one has different 'Debye temperatures' for the three distinct surface sites of the Au_{55} cluster casts doubt on the validity of the use of a continuum-model picture for discussing the site specific thermal behaviour. Indeed, for such small particle sizes, where the surface structure is so manifest, the use of the concept of surface modes becomes dubious, and is certainly inadequate to explain the observed temperature dependence of the f-factors. None the less, it has proven possible to describe the low temperature specific heat of Au_{55} rather well using such a continuum-model (see the chapter on specific heat), when the center-of-mass motion has been taken into account [41].

A linear term in the low temperature specific heat due to mobile electrons in a small metallic cluster must be sought in a temperature region in which the inter-cluster contribution to the specific heat has become small. Although it is doubtful whether a linear term should be expected for Au_{55} [43], a search for such a term would have to be carried out at temperatures low with respect to the θ_D^{inter} i.e. at temperatures well below 50 mK.

The anomaly seen in zero-field specific heat measurements [42] below 100 mK can be understood on the basis of Schottky anomalies [39] due to the Q.S. of the nuclear ground states of the three surface sites. The MES Q.S. values, used directly for the calculations [43], gave a nuclear contribution that is a factor five larger than that observed experimentally. This is probably due to thermal relaxation time effects, which can strongly influence the specific heat measurements [44].

6.3.2. Electronic Behaviour

Just as it proved possible to obtain site-specific information as to the thermal behaviour of the Au_{55} cluster, site-specific electronic behaviour could also be extracted from the Mössbauer spectra.

The I.S. value of bulk metallic gold is shown in Figure 6 by the filled circle. Mono-nuclear gold molecular compounds which have such negative I.S. values are all strongly ionic and often quite unstable as well. In the same figure, the Q.S. and I.S. values of Ph_3PAuCl (as a covalent reference material) are given by a vertical cross and of $(PPh_3)_2AuCl$ (as a partly ionic reference material) by a diagonal cross.

Open and centered triangles have been used to indicate the values for the PPh_3-coordinated sites of the Au_{11} and the Au_{55} clusters, respectively. The values for the Cl-coordinated sites are shown by open and centered squares. The values for the central site of Au_{11} are given by the open circle and the core sites of Au_{55} by the centered circle. The centered diamond is used for the values of the bare surface sites. The various surface sites of Au_{55} all have significantly smaller values of the I.S. than the surface sites of the Au_{11} cluster compound and all other smaller gold cluster compounds.

Note that the I.S. value of -1.4 ± 0.1 mm/s of the core sites of Au_{55} lies quite close to the -1.2 mm/s of bulk gold (given by the filled circle), in strong contrast to that of the central sites of Au_{11} and other small centered clusters, which are all greater than +2.0 mm/s (the I.S. of the central site could not always be extracted from the data [9, 16]). These very large positive values for the I.S.'s have been attributed [16] to a transfer of charge from the central site to the peripheral ligands, the central atom then having an effective valence of +5. This is an indication that the core sites of Au_{55}, which have nearest neighbor surroundings and an I.S. almost identical to bulk gold, also have electronic surroundings similar to bulk metallic gold.

But this also presents a difficulty. The I.S. observed for the 13 core atoms is more negative than the I.S. of bulk gold metal itself [33], despite having the same nearest neighbor coordination as bulk gold [9], while the decrease in Au-Au distance in Au_{55} of more than 3% relative to bulk gold, obtained from

EXAFS measurements [45, 46], would be expected to lead to a change in the I.S. of about 0.7 mm/s in the positive direction relative to the bulk, due to an effective compression of the $6s$ electrons into a volume about 10% smaller than in the bulk [48]. We should thus not expect a change of 0.2 mm/s in the negative direction, as is observed experimentally [9].

This effective loss of 6s electron density out of the central core can be understood [18] using the model of Citrin and Wertheim [49], based on earlier theoretical work by Desjonquères and Cyrot-Lackmann [50], which predicted a narrowing of the d-bands for the surface atoms in transition metals. As an example to demonstrate their model, which explained the surface atom core level shifts observed in the XPS spectra of metallic systems, they used a flat gold surface: due to the lower coordination of the surface sites, relative to the bulk metal, the electrons associated with these sites become somewhat more localized, i.e. by transforming some s-electron density into d-electron density, necessitating a small net flow of charge from the bulk metal to the surface layer. The same effect should also hold for the surface layer of the metal core of these cluster compounds. In fact, the ligands will have a very similar effect on the ligated surface gold atoms; on the other hand, more than half of the surface atoms of Au_{55} are unligated.

The 'bulk' metal in the Au_{55} cluster corresponds to the 13 core atoms, which have to supply charge to 42 surface atoms. In addition, the average surface coordination of 6.33 for the cluster is lower than the average of 8.5 over the (100) and (111) surface faces, the most stable in a flat gold surface [51]. Therefore more charge would need to leave each individual gold atom in the 13 atom core of Au_{55} than would leave a bulk atom of a normal macroscopic sample. This effect can provide a qualitative explanation for the huge drop of $6s$ electron density at the core sites.

A comparison of the inter-gold distances obtained by EXAFS on the Au_{55} clusters [45, 46] with those obtained on bare gold clusters [47] shows that the reduction of the inter-gold distance in Au_{55} is twice as large as in bare clusters of the same average size, relative to the bulk. Using the argument of [49] again, the loss of up to about 12 valence electrons to the ligand shell of the Au_{55} cluster (see the chapter on XPS in this volume) might well explain the difference in inter-gold distance between the bare and ligated clusters as being due to increased (core level) binding as a consequence of the decrease in Coulomb repulsion in the core. Recent voltametric measurements on Au_{55} [52] indicate that the cluster compound can accommodate up to 9 additional electrons before desintegrating.

In a recent publication [33], use was made of a simple empirical model [54], with proven predictive value, which was developed for predicting the I.S. values observed for Mössbauer isotopes imbedded in alloys or intermetallic compounds relative to those of the pure element. It was there assumed that the entire cluster core, including the ligated surface atoms, exhibits largely metal binding, i.e. that the model was applicable. The main idea behind the application of this model was to attempt to understand the observed value of I.S. for the bare surface sites. A further neglect of the effect of the ligand shell, as a first approximation, allowed a simplification of the model which left no free parameters. Although the model gave surprisingly good agreement between experiment and theory, more recent results on two versions of Au_{55} which have different ligands do show that the effect of the ligands on the I.S. of the surface atoms, although small, can certainly not be ignored, as discussed in the following section. Furthermore, it should be stressed that what is measured by the Mössbauer effect I.S. is mainly the result of s-electron density in the nuclear volume, and from this one can only infer a change of character from s-like to d-like.

There has been considerable interest in recent years in the question of a possible decrease in melting temperature of small metal particles, relative to the bulk, both theoretically [55, 56, 57, 58] as well as experimentally [59, 60]. Using the Lindemann criterium for melting, extrapolations from high temperature measurements have suggested that, for gold particles with a diameter of less than about 2 nm, the melting point would be 0 K [55, 57, 58, 60] i.e. gold particles of this size would always be in the form of liquid droplets. A recent theoretical study [61] has questioned these results, claiming a melting temperature for a fictitious bare Au_{13} fcc particle of about 680 K. This is not to be compared to the Au_{13} ligated clusters, which with their icosahedral structure are certainly not metallic in character, with much sharper d-band structure than metallic gold and a tangential average interatomic spacing larger than bulk gold. On the other hand, as we have already argued, there is reason to believe that the 55 atom core of Au_{55} does show metallic binding, with one inter-atomic distance, which is somewhat smaller than that of bulk gold. There is a substantial degree of delocalization of the s-electrons and development toward a metal-like d-band structure, with the cluster core having a 'diameter' of about 1.4 nm.

The presence of a MES spectrum consisting of a superposition of distinct lines of natural linewidth from the four different crystallographic sites also constitutes proof that gold core of the Au_{55} cluster is a *solid* up to temperatures of at least 30 K, on the time scale of the MES measurements: 10^{-5} s $\leq$ t $\leq 10^{-10}$ s, see [62, 63]. Surface melting on this time scale can also be refuted for Au_{55} for the same reason.

6.3.3. EFFECT OF LIGAND MODIFICATION

Finally we would like to discuss some recent data obtained on the two Au_{55} molecular cluster compounds containing modified ligands [34]. In Figures 10a and 10b, the modified ring structures of Au_{55}-tolyl and Au_{55}* are shown schematically. It is well established [64, 65] that, in the tolyl ligand, a modest transfer of electronic charge from the ring toward the phosphorus atom takes place, somewhat reducing the effective electronegativity of the atom with respect to its gold neighbor. The $NaSO_3$ substituent, on the other hand, can be

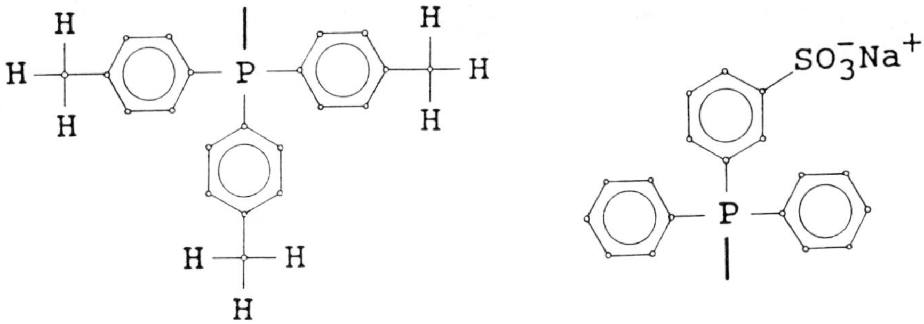

Fig. 10. (a) A flattened representation of the tolyl ligand. (b) A similar representation of the modified triphenyl phosphine ligand of the water soluble version.

expected to strongly attract electronic charge from the phosphorus atom, thus increasing its effective electronegativity. We can thus expect that there might be some slight modifications to the spectra observed for the four sites of the original Au_{55} compound as a consequence of these ligand substitutions.

Before examining the spectra, one might expect, on the basis of the apparent metallic bonding in the cluster core discussed above, that there would be a transfer of total charge on the cluster core to or from the modified ligands, since the chlorine ions already have a charge very close to one electron each, and should not take part in further charge transfer. This charge transfer would then be seen as a very small *increase* of the I.S. values of all four sites for Au_{55}-tolyl ligands, and a substantially larger *decrease* in the I.S. of all four sites for Au_{55}^*. If on the other hand the bonding of the cluster core were to be basically molecular, one could expect much larger changes in the I.S. values, but only for the gold atoms directly coordinated by the phosphorus atoms. Again the I.S. values would increase or decrease with charge transfer toward or away from the cluster.

The Mössbauer spectra of both new versions, taken at 4.2 K, are shown in Figure 11, together with that of Au_{55} for comparison. The Au_{55}-tolyl spectrum could be fit successfully by allowing only minor modifications of the I.S. and Q.S. values of the sites, and giving about the same site-specific f-factors as for the normal Au_{55}. For Au_{55}^* this was not the case: a satisfactory fit with only minor modifications of the I.S. and Q.S. values could only be obtained at the cost of drastic, and physically unrealistic, changes in the site-specific f-factors. Keeping the site-specific f-factors fixed at the values found for Au_{55} (and for Au_{55}-tolyl) resulted in an excellent fit, but with very large changes in I.S. and Q.S. values for two of the four surface sites. These latter values are listed in Table 1, together with the values for Au_{55}-tolyl and Au_{55}. It should be noted that the fits to the spectra in Figure 11 were done with the transmission integral method; the subspectra shown are lorentzian simulations to act as a guide for the eye only.

TABLE 1

The Q.S. and I.S. values of the various sites of Au_{55}, Au_{55}-tolyl, and $Au_{55}{}^*$ [34].

Compound	Core		Bare		Cl		P	
	IS	QS	IS	QS	IS	QS	IS	QS
Au_{55}-tolyl	-1.50	0.0	0.42	1.40	-0.20	4.63	0.21	7.04
Au_{55}	-1.44	0.0	0.28	1.40	0.11	4.20	0.70	7.12
Au_{55}^*	1.37	0.0	-0.82	1.06	1.46	0.71	0.74	6.93

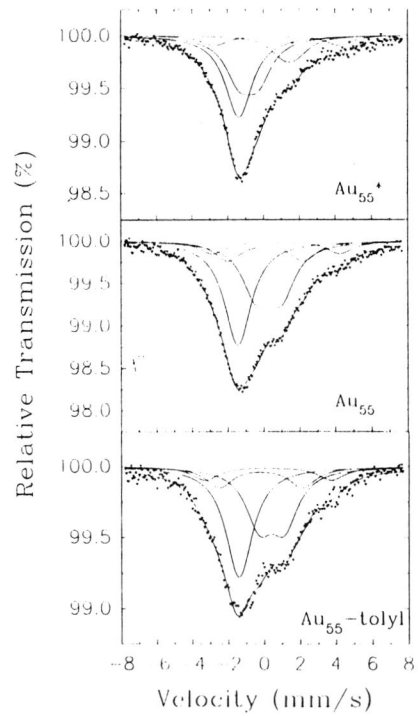

Fig. 11. The Mössbauer spectra obtained at 4.2 K on Au_{55}-tolyl and $Au_{55}{}^*$ are shown, together with that of Au_{55} for reference [34].

In Figure 12, the Q.S. versus I.S. values for all three varieties are presented. The filled figures represent the original Au_{55} compound, the open figures represent the methylated version Au_{55}-tolyl, and the centered figures represent the water-soluble $Au_{55}{}^*$ compound. As in Figure 6, the core site values are marked by circles, the bare surface site values by diamonds, the Cl-

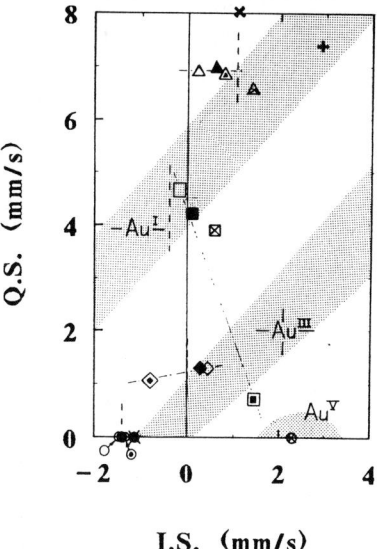

Fig. 12. The Q.S. versus I.S. values of the various sites of Au_{55}, Au_{55}-tolyl, and Au_{55}^* [34]. The symbols used are discussed in the text.

coordinated site values by squares, and the P-coordinated site values by the triangles.

A comparison between the positions of the Q.S. versus I.S. values of Au_{55}-tolyl relative to Au_{55} shows that, although the differences are not very large, the I.S. change for the two coordinated surface sites is negative, while it is positive for the bare surface sites. This means that neither of the extremes suggested above are actually entirely correct. A similar comparison for the Au_{55}^* with respect to Au_{55} shows a slight positive I.S. shift for the phosphorus-coordinated gold, a rather large negative shift for the bare gold surface sites, and a large positive shift for the chlorine-coordinated site. This latter is also accompanied by a huge drop in the Q.S., signaling a change from linear hybridization to planar hybridization of the gold orbitals. The reason for this drastic change is not yet understood.

So indeed, complete neglect of the influence of the ligands is a too crude approximation, especially when a comparison can be made between systems with different ligands, and can only be considered a starting point for a complete analysis. As stated above, although the interpretation of these results are still incomplete, the Q.S. and I.S. values are real physical results.

6.4. ^{197}Au MES on Platinum Clusters

As mentioned above in the section on MES specific to clusters, it has recently become clear that it is in some cases possible to neutron activate the ^{196}Pt occurring naturally in a platinum cluster (or colloid) [12].

In order to be able to carry out activation, it was necessary to develop a method of containment of the cluster material while it was undergoing the neutron irradiation [13], since handling loose powders having relatively high specific activity and a quite short half-life is a most unattractive proposition.

Another consideration in activation is whether or not the material to be studied will be damaged, either by the neutron and the hard γ-ray bombardment in the reactor, or by the neutron capture followed by high energy γ-emission [66]. It has turned out that a $Pt_{309}(Phen^*)_{36}O_{30\pm10}$ cluster did survive the mistreatment in the reactor without undergoing permanent damage [12, 13]. Any vacancy formation in the cluster cores seems to have been annealed out, due perhaps to the temperature in the reactor where the activation took place.

In Mössbauer emission experiments, one generally has to be concerned with the possibility of so-called 'after effects'. During the radioactive decay process which transforms the mother nucleus in the sample being studied into the Mössbauer nucleus, not only are the nuclear properties modified, but the electronic structure of the atoms are also subject to change, sometimes slight and sometimes drastic. In the case of the transformation of platinum into gold, β emission takes place, causing an increase of nuclear charge, and the gold ion thus formed is left lacking one outer electron, with respect to the platinum ion before the transition. If this deficiency is corrected rapidly enough (i.e. in less than about $10^{13} \pm 10^{14}$ s), both with respect to the ion as well as to its local surroundings, the surroundings will have come back into equilibrium, and no after effects will be seen in the Mössbauer spectrum. This is always the case for metallic systems, where electrons are readily available and can quickly redistribute themselves. On the other hand, in semiconducting and insulating materials [2, 3, 68], one may see several charge states present, due to only partial restoration of the charge deficiency of the ion and/or its surroundings.

Due to the fact that we have a transformation of one platinum atom to gold in each cluster measured, we have actually performed our measurements on clusters with metals cores of $Pt_{308}Au_1$. Estimates [12, 13, 20] also show that there is little chance ($\ll 10^{-6}$) of having two gold atoms formed in the same cluster. In fact, due to the insulating character of the ligands surrounding the clusters, the average times for hopping of electrons between clusters may be estimated from high frequency conductivity and NMR measurements [32, 67] to be of the order of seconds for temperatures around 4 K. Although this average time drops as the temperature rises, even at temperatures up to about 50 K, only a small fraction of the clusters will have recovered the lost electron within a time comparable to the Mössbauer time scale. This means

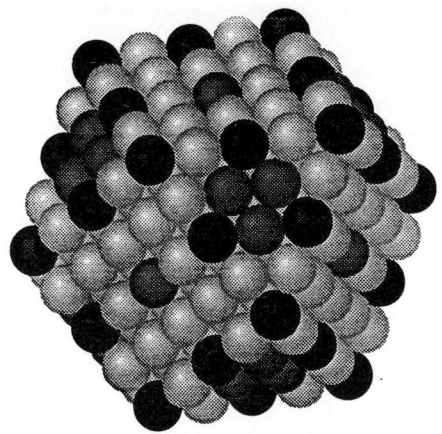

Fig. 13. An artist's conception of the cluster core of Pt_{309}, giving a view of the outside of the cluster. The black circles represent the 36 Phen*-coordinated Pt-sites, and the uncolored circles represent both the bare surface sites and the oxygen-coordinated sites.

that the large majority of the cluster cores on which we have measured are actually $Pt_{308}Au_1^+$.

Although no single crystal x-ray data is available for structure determination, high resolution electron microscopy [21] results confirm the size and cuboctahedral structure, and x-ray powder diffraction [21] has shown that the interatomic distance between the platinum atoms is the same as in bulk platinum metal. In Figure 13, an artist's conception of the $Pt_{309}(Phen^*)_{36}O_{28}$ cluster core is given, with the ligands removed for clarity. The black colored circles represent most probable location of the 36 surface atoms coordinated by the Phen*'s, which are themselves shown schematically in flattered view in Figure 14. The postulated sites not only have a high degree of symmetry, they also allow a maximum of room for these rather bulky ligands.

The location of the sites ligated by the 30 ± 10 oxygens, and thus the location of the bare surface sites as well, is less certain. Indeed, although it seems probable that the oxygens are present as O_2 molecules, connected to the platinum surface by one end of the oxygen molecule, even this remains uncertain. The light grey circles represent both the bare surface atoms, of which there are almost certainly 48 located on the ribs of the octahedron, and between 48 and 66 on the square and triangular square faces, depending on the actual location and form of the oxygen at the surface. The dark grey circles represent the ligated gold O_2 atoms of the cluster.

As has been mentioned above [10], the intra-cluster f-factors of Au_{55} depend strongly on the number of Au nearest neighbors, i.e. they are strongly site-dependent. The intra-cluster f-factors obtained for the Au_{55} clusters can be taken as reasonable estimates for the vertices, the edges, and the square-

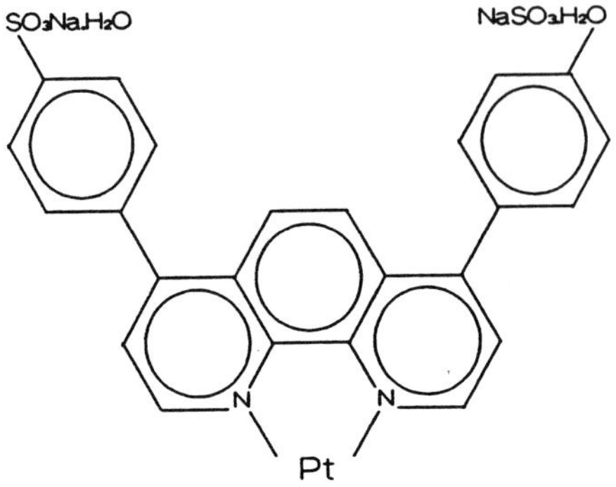

Fig. 14. A flattened schematic drawing of the Phen* ligand.

faces of the $Pt_{308}Au$ clusters. There is every reason to suppose that this also holds for the platinum atoms in the Pt_{309} cluster. Thus, the number of platinum nearest neighbors can be used to estimate the f-factors needed for the core sites, the vertex and rib sites, and the square face sites; an estimate for the sites on the triangular faces can also be obtained by interpolation.

Using the knowledge of the *known* site occupations and these estimates for the site f-factors, one can predict the relative intensities of the singlet due to the 147 central core sites, the quadrupole doublet due to the 36 Phen* ligated sites (assuming that the vertex and rib sites can be lumped together), and the quadrupole doublet due to the 48 bare rib sites. Looking at the flat square and triangular faces, one could assume that it would be possible to resolve 2 quadrupole doublets due to the unligated sites on these surfaces. None the less, we chose to lump all the unligated surface sites together, using 1 quadrupole doublet. If the oxygens are present in the form of O_2 molecules, as is expected from the chemistry of these clusters, it will *not* be possible to extract separate quadrupole doublets for these sites from our statistics, so they too have been lumped into one O_2 'site'. These considerations were used as the point of departure for fitting the spectra of Pt_{309}. In practice, it turned out that the quadrupole doublet with a relative spectral intensity corresponding closely with that expected for the 'lumped' unligated surface sites always collapsed into a singlet.

In Figure 15, spectra take at four temperatures between 1.8 and 40 K are shown, together with the transmission integral fit performed with one singlet and three quadrupole doublets. A lorentzian simulation of these lines is included as a guide to the eye.

One should note that the sign of the I.S. values in a ^{197}Au emission measurement is reversed with respect to gold absorber measurements against ^{197}Au in platinum metal. This has been taken into account in Table 2. Note also that the value of the I.S. for the core sites is then *exactly* the same as that for ^{197}Au in bulk metallic platinum! This, together with the identical

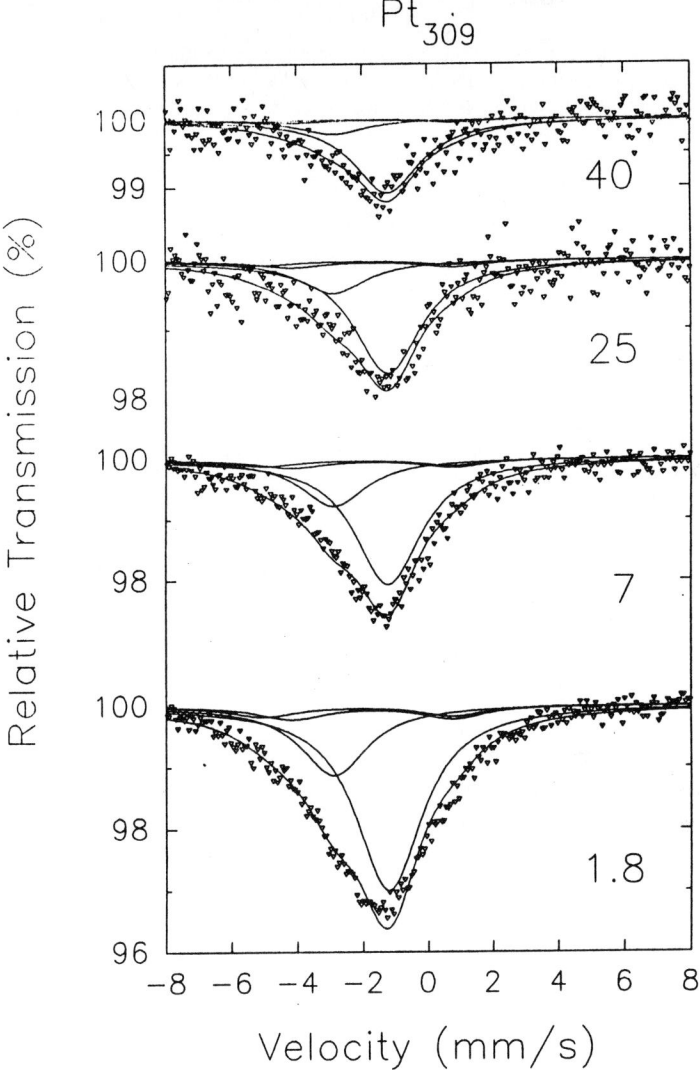

Fig. 15. Spectra of the $Pt_{308}Au_1^+$ cluster taken at 1.8, 7, 25 and 40 K are shown; included are the transmission integral fits and lorentzian simulations of the lines as a guide to the eye.

interatomic distance in the cluster core, the behavior of the low temperature specific heat, and the Korringa like relaxation times observed by NMR (see the specific heat and NMR chapters above) is a compelling argument that the central core sites in this cluster are indeed metallic-like.

Although in principle there are 5 distinct surface sites, these have been lumped together to form the 4 'sites' used for fitting, giving the Q.S. and I.S. values displayed in Table 2. The I.S. values have been referred to Au

TABLE 2

The average number of platinum nearest neighbors, the measured values for the relative intensities, and the Q.S. and I.S. values of the various sites of the $Pt_{308}Au_1$ cluster are given, as well as the expected site occupations. Gold foil is used as the I.S. reference.

Site	Site occu.	No. Pt n.n.	Ave. no. n.n.	I.S. mm/s	Q.S. mm/s	I %
Core	147	12	12	0.00 ± 0.05	0.0 ± 0.05	62
	48	7				
Unligated	48	8	7.71	1.7 ± 0.1	0.0 ± 0.1	23
	16	9				
Phen*	12	5	6.06	0.9 ± 0.4	5.7 ± 0.4	7
	24	7				
O_2	8	9	8.57	0.5 ± 0.4	5.4 ± 0.4	8

in platinum foil. Also included are the measured values of the f_{intra}'s of all the 4 sites, making use of the Θ_D of about 17 K found for the inter-cluster center-of-mass vibrations. This value was determined from fitting the temperature dependence of the spectra obtained between 1.8 and 40 K [13]. For convenience, the expected occupations of all the surface sites is given, with the total expected by lumping sites together.

It is worth mentioning that from the characterization of the I.S. value as a reflection of the 6s electron density in the nucleus [6] mentioned above, and from the accuracy with which we can measure the position of the singlet due to the central core sites, which account for more than 60% of the total spectral intensity, we are able to observe a change in the $6s$ electron density of the order of 0.0025 of an electron [18]. Seeing the same I.S. for ^{197}Au in Pt_{308} as for ^{197}Au in bulk metallic platinum within our error limits means that the local s-electron density of states in the metallic inner core of 147 atoms of the Pt_{309} cluster is the same as that of bulk platinum within 0.5%.

It should be emphasized that, since MES sees the total integrated s-electron density at the nucleus, it is not as sensitive to the appearance of quantum-gaps in the electronic energy level spectrum due to finite-size effects, as for instance NMR or the electronic specific heat. The reason is, obviously, that these latter quantities are determined by the electronic density of states *at* the Fermi level E_F (and are thus very sensitive to the appearance of quantum-gaps around E_F).

6.5. Conclusions

Although the use of MES in the study of giant metal-cluster compounds has until now been limited to only one (fairly exotic) Mössbauer isotope, namely ^{197}Au [69], it is clear that in this way local, site-specific information can be obtained which is unattainable by other techniques.

But for MES to be a viable technique for this type of study, it must be strongly integrated within a framework of results obtained from many other physical and chemical measurement techniques, e.g. EXAFS, XPS, NMR, specific heat, DSC, SIMS.

Within this framework, and with reasonable care in fitting the spectra, more-or-less model independent Mössbauer parameters can be obtained. In the final interpretation of the measurement results, naturally, recourse has to be made to models for the cluster system.

Making use of the convenient Q.S. to I.S. correlations, obtained from earlier chemical studies, as site-specific signatures, it has been possible to obtain direct information about the thermal properties of these giant metal-cluster compounds, which has helped in understanding the low temperature specific heat. The I.S. values observed for the central core sites have been instrumental in demonstrating, together with the XPS results, that the central core of the Au$_{55}$ variants and Pt$_{308}$Au$_1$ have $6s$ charge densities (implying similar $5d$ charge densities as well) which are close to and exactly the same as for gold in bulk gold and gold in bulk platinum, respectively. Further information concerning the effect of the ligands on the characteristics of the surface sites has also been obtained.

All in all, the fairly limited number of MES experiments carried out on Au$_{55}$ and Pt$_{309}$ have proven to be a rich source of information.

References

1. H. Wegener, *Der Mössbauer Effect und seine Anwendung in Physik und Chemie*, Mannheim (1965).
2. V. I. Goldanskii and R. H. Herber (eds.), *Chemical Applications of Mössbauer Spectroscopy*, Academic Press, New York (1968).
3. P. Gütlich, R. Link,and A. Trautwein (eds.), *Mössbauer Spectroscopy and Transition Metal Chemistry*, Springer-Verlag, Berlin (1978).
4. B. V. Thosar, P. K. Iyengar, S. C. Bhargava and J. K. Srivastava, *Advances in Mössbauer Spectroscopy*, Studies in Physical and Theoretical Chemistry, Elsevier Scientific Publ. Co., Amsterdam (1983).
5. D. L. Nagy, K. Lázár and Zs. Kajcsos, 'Applications of the Mössbauer Effect', *Proc. Intern. Conf. Appl. Möss. Spectr.*, J.C. Baltzer AG, Bazel (1989).
6. G. K. Shenoy and F. E. Wagner, *Mössbauer Isomer Shifts*, North-Holland, Amsterdam (1978).
7. J. W. A. van der Velden, Ph.D. Thesis, University of Nijmegen, The Netherlands (1983).
8. M. P. A. Viegers, Ph.D. Thesis, University of Nijmegen, The Netherlands (1976).
9. H. H. A. Smit, Ph.D. Thesis, University of Leiden, The Netherlands (1988).

10. H. H. A. Smit, R. C. Thiel, L. J. de Jongh, G. Schmid and N. Klein, *Solid State Commun.* **65** (1988) 915.
11. A. M. van der Kraan, *Hyperfine Int.* **40** (1988) 211.
12. F. M. Mulder, T. A. Stegink, R. C. Thiel, L. J. deJongh, and G. Schmid, *Nature* **867** (1994) 716.
13. F. M. Mulder, R. C. Thiel, T. A. Stegink, L. J. deJongh, P. Gubbens and G. Schmid, to be published.
14. R. V. Parish, L. S. Moore, A. J. J. Dens, D. M. P. Mingos and D. J. Sherman, *J. Chem. Soc. Dalton Trans.* (1989) 539.
15. H. H. A. Smit, P. R. Nugteren, R. C. Thiel and L. J. de Jongh, *Physica* **B153** (1988) 33.
16. F. A. Vollenbroek, Ph.D. Thesis, University of Nijmegen, The Netherlands (1979).
17. W. Bos, R. P. F. Kanters, C. J. van Halen, W. P. Bosman, H. Behm, J. M. M. Smits, P. T. Beurskens, J. J. Bour and L. H. Pignolet, *J. Organometallic Chem.* **307** (1986) 385.
18. R. C. Thiel, M. W. Dirken, and R. Zanoni, *Proceedings ICAME Budapest '89: Applications of the Mössbauer Effect*, J. C. Baltzer, Bazel (1990), p. 1729.
19. A. M. van der Kraan and J. W. Niemandsverdriet, in G. J. Long and J. G. Stevens (eds.), *Industrial Applications of the Mössbauer Effect*, Plenum Press, New York, London (1986), p. 609.
20. J. G. Stevens and V. E. Stevens (eds.), *Mössbauer Effect Data Index*, Adam Hilger, London (1972), p. 180.
21. G. Schmid, B. Morun, and J.-O. Malm, *Angew. Chem. Int. Ed. Engl.* **28** (1989) 778.
22. W. N. Delgass, G. L. Hallen, R. Kellerman and J. H. Lunsford, in *Spectroscopy in Heterogeneous Catalysis*, Academic Press, New York (1979), p. 132.
23. F. J. Berry, in G. J. Long (ed.), *Mössbauer Spectroscopy Applied to Inorganic Chemistry*, Vol. 1, Plenum Press, New York (1984), p. 391.
24. D. Schroeer, R. F. Marzke, D. J. Erikson, S. W. Marshall and R. M. Wilenzick, *Phys. Rev.* **B2** (1970) 4414.
25. M. P. A. Viegers and J. M. Trooster, *Phys. Rev.* **B15** (1977) 72.
26. F. A. Vollenbroek, P. C. P. Bouten, J. M. Trooster, J. P. van den Berg and J. J. Bour, *Inorg. Chem.* **17** (1978) 1345.
27. F. A. Vollenbroek, J. J. Bour, J. M. Trooster and J. W. A. v.d. Velden, *J. C. S. Chem. Com.* (1978) 907.
28. G. Schmid, R. Pfeil, R. Boese, F. Bandermann, S. Meyer, G. H. M. Callis and J. W. A. v.d. Velden, *Chem. Ber.* **114** (1981) 3634.
29. G. Schmid, N. Klein, L. Korste, U. Kreibig and D. Schönauer, *Polyhedron* **7** (1988) 605.
30. The methylated version of Au_{55} Au_{55}-tolyl has been prepared by G. Schmid; ^{31}P NMR [31], electrical conductivity [31, 32], and MES [33, 34] results have been reported.
31. M. P. J. van Staveren, Ph.D. Thesis, University of Leiden, The Netherlands (1989).
32. M. P. J. van Staveren, H. B. Brom and L. J. de Jongh, *Physics Reports* **208** (1991) 1.
33. R. C. Thiel, R. E. Benfield, R. Zanoni, H. H. A. Smit, and M. W. Dirken, *Structure and Bonding 81: Structures and Biological Effects* (1993) 1–39.
34. F. M. Mulder, E. A. van der Zeeuw and R. C. Thiel, *Solid St. Comm.* (accepted for publication).
35. M. Born, *Atomtheorie des festen Zustandes*, Leipzig (1923).
36. E. W. Montroll, *J. Chem. Phys.* **18** (1950) 183.
37. S. W. Marshall and R. M. Wilenzick, *Phys. Rev. Letters* **16** (1966) 219.
38. R. C. Thiel, *Z. Phys.* **200** (1967) 227.
39. C. Kittel, *Introduction to Solid State Physics*, 5th edition, J. Wiley & Sons, New York (1976).
40. J. S. van Wieringen, *Phys. Lett.* **26A** (1968) 370.
41. J. Baak and H. B. Brom, *Proc. Intl. Symposium on the Physics and Chemistry of Finite Systems: From Clusters to Crystals*, Richmond, VA (1991) (in press).
42. G. Goll, H. v. Löhneysen, U. Kreibig and G. Schmid, *Z. Phys.* **D20** (1991) 329.

43. H. B. Brom, J. Baak, L. J. de Jongh, F. M. Mulder, R. C. Thiel and G. Schmid, Z. Phys. D (1992) (accepted for publication).
44. J. Baak, H. B. Brom, L. J. de Jongh and G. Schmid, Z. Phys. D (1992) (accepted for publication).
45. M. C. Fairbanks, R. E. Benfield, R. J. Newport and G. Schmid, Solid State Commun. **74** (1990) 431.
46. M. A. Marcus, M. P. Andrews, J. Zegenhagen, A. S. Bommannavar and P. Montano, Phys. Rev. **B42** (1990) 3312.
47. A. Balerna, E. Bernieri, P. Picozzi, A. Reale, S. Santucci, E. Burattini and S. Mobilio, Phys. Rev. **B31** (1985) 5058.
48. D. L. Williamson, in G. K. Shenoy and F. E. Wagner (eds.), *Mössbauer Isomer Shifts*, North-holland, Amsterdam (1978), Ch. 6b.
49. P. H. Citrin and G. K. Wertheim, Phys. Rev. **B27** (1983) 3176.
50. M. C. Desjonquère and F. Cyrot-Lackmann, J. Phys. **36** (1975) L45.
51. P. H. Citrin, G. K. Wertheim and Y. Baer, Phys. Rev. **B27** (1983) 3160.
52. J. J. Steggerda, J. G. M. van der Linden and J. E. F. Gootzen, Mat. Res. Soc. Symp. Proc. **272** (1992), 127-132.
53. F. van der Woude and A. R. Miedema, Solid State Commun. **39** (1981) 1097.
54. A. R. Miedema, F. R. de Boer and P. F. de Chatel, J. Phys. **F3** (1973) 1558.
55. Ph. Buffat and J.-P. Borel, Phys. Rev. **A13** (1976) 2287.
56. P. R. Couchmann and C. L. Ryan, Phil. Mag. **37** (1978) 369.
57. C. Solliard, Sol. St. Commun. **51** (1984) 947.
58. C. Solliard and M. Flueli, Surface Science **156** (1985) 487.
59. P. R. Couchmann, Phil. Mag. **A40** (1979) 637.
60. P. Labastie and R. L. Whetten, Phys. Rev. Lett. **65** (1990) 1567.
61. I. L. Garzon and J. Jellinek, Z. Phys. **D20** (1991) 235.
62. M. Blume and J. A. Tjon, Phys. Rev. **165** (1968) 446.
63. M. J. Clauser, Phys. Rev. **B3** (1971) 3748.
64. P. Sykes, *A Guidebood to Mechanisms in Organic Chemistry*, Longmans (1963).
65. N. S. Isaacs, *Physical Organic Chemistry*, Longman Scientific (1987).
66. A. Simopoulos and G. Vogl, Phys. Stat. Sol. (B) **59** (1973) 505–516.
67. D. Van der Putten, H. B. Brom, J. Witteveen, L. J. de Jongh and G. Schmid, Supplement to Z. Phys. **D26** (1993) 21–23.
68. H. Th. Le Fever, F. J. van Steenwijk and R. C. Thiel, Physica **86–88B** (1977) 1269.
69. At the moment that this manuscript was completed, the first MES experiments using [61]Ni on a series of nickel cluster compounds was just commencing at the University of Mainz.

7. SPECIFIC HEAT STUDIES ON METAL CLUSTER COMPOUNDS

H. B. BROM, J. BAAK and L. J. DE JONGH
Kamerlingh Onnes Laboratory, Leiden University,
P.O. Box 9506, 2300 RA Leiden, The Netherlands

7.1. Introduction

In what sense does the shape of a metal determines its electronic and vibrational properties? To get a feeling, let us compare a bulk piece of metal with a thin sheet of a few nm thick. The huge surface area of the latter will have its imprint on the electron density and on the vibrational spectrum. Due to the reduction of the number of neighbors, the surface atoms will be less strongly bound than in the bulk. The existence of a boundary is also the reason for a change in electron density of the conduction electrons at the surface atoms. In addition to these surface effects, quantum size effects (QSE) might also play a role in the above given example: if certain dimensions become of the order of a few nm's or less, the quantized nature of the energy levels will be more pronounced. For these reasons the low temperature lattice and electronic specific heat of small metal particles, which are the subject of this chapter, deviate from that of the bulk.

The history of the QSE goes back to the early days of quantum mechanics [1, 2, 3]. Halperin [3] has critically reviewed the status up to 1986. Although there is no doubt that quantum size effects exist, Halperin's review shows that unequivocal experimental proofs for the electronic QSE in small metal particles are hard to give. Experiments that have been claimed to show its presence [4], might be influenced by other effects as well. It is not easy to avoid contamination of the inherently huge surface, or to get rid of the influence of the substrate or to deal properly with the often present distribution in sizes.

Also on the theoretical side there are complications. Even for monodispersed samples the energy levels might be smeared. The samples will experience random surface potentials, that remove the orbital degeneracy and affect the energy levels in the particles. Dependent on a.o. the relative size of this effect different distribution functions (Poisson, Gaussian, unitary, orthogonal, symplectic) are appropriate [5, 6, 7], that lead to different results. Also the expectation that the susceptibility for an even numbered electron system will become zero at the lowest temperatures, is too naive. Due to spin-orbit coupling the groundstate will be a mixture of spin-up and spin-down wave functions and hence the total spin susceptibility for an even numbered electron system will be no longer zero.

Before discussing the specific heat data on small metal particles and especially on metal cluster molecules, we will present the theoretical framework for the lattice and electronic specific heat for these systems in the next section.

7.2. The Lattice Specific Heat

The total lattice specific heat contains the contribution of the surface modes together with that of the bulk. For large particles these contributions can be calculated in the continuum approximation. For particles of nm-size the continuum approximation is expected to break down and only the quantized limit will be appropriate.

For the metal cluster compounds apart from this intracluster contribution the vibrations of the cluster as a whole (inter-cluster vibrations) and those of the ligands attached to the core have to be taken into account. The ligands will not contribute to the specific heat at temperatures low compared to the vibrational frequencies involved (i.e. below 50 K [8]), while the inter-cluster vibrations can be calculated from the well known Debye formula [9].

$$\frac{C}{R} = 9\left(\frac{T}{\theta_D}\right)^3 \int_0^{\theta_D/T} \frac{x^4 e^x}{(e^x - 1)^2} \, dx \qquad (1)$$

where the parameters in the Debye temperature θ_D (the mass and sound velocity or spring constant) are to be related to the cluster as a whole, forming an effective medium.

As shown below, the intracluster vibrational frequencies can be calculated in the continuum approximation or in a molecular dynamics approach, where the atomic structure of the lattice is taken into account.

7.2.1. THE ELASTIC CONTINUUM APPROXIMATION

For the elastic continuum approximation applied to spherical particles (with N atoms each), we refer to the work of Nishiguchi and Sakuma [10]. The normal mode frequencies, ω's, are obtained by solving the Navier equation for an isotropic and homogeneous sphere, using the spherical coordinates r, θ and ϕ:

$$\rho\ddot{\mathbf{Q}} = (\lambda + 2\mu)\boldsymbol{\nabla}\boldsymbol{\nabla} \cdot \mathbf{Q} - \mu\boldsymbol{\nabla}\boldsymbol{\nabla} \cdot \mathbf{Q} \qquad (2)$$

where $\mathbf{Q}$ is the displacement vector with components Q_r, Q_θ and Q_ϕ, ρ the mass density and λ and μ the Lamé constants. As trial solutions we introduce:

$$\mathbf{Q}(r, \theta, \phi, t) = \cos(\omega t + \beta)\mathbf{A}(r, \theta, \phi) \qquad (3)$$

with β an arbitrary phase angle. The equation for $\mathbf{A}$ becomes:

$$\nabla\nabla \cdot \mathbf{A} - \mu \nabla \times \nabla \times \mathbf{A} = -\rho\omega^2 \mathbf{A}. \tag{4}$$

The solution can be found by introducing one scalar potential ψ_0 and two vector potentials ψ_1 and ψ_2 defined as $\psi_i = (r\psi_i, 0, 0)$:

$$\mathbf{A} = \nabla\psi_0 + \nabla \times \psi_1 + \nabla \times \nabla \times \psi_2. \tag{5}$$

The explicit expressions for the three components of $\mathbf{A}$ in terms of these potentials are:

$$A_r = \frac{\partial\psi_0}{\partial r} + \frac{\Lambda}{r}\psi_2 \tag{6}$$

$$A_\theta = \frac{1}{r}\frac{\partial\psi_0}{\partial\theta} + \frac{1}{\sin(\theta)}\frac{\partial\psi_1}{\partial\phi} + \left(\frac{1}{r} + \frac{\partial}{\partial r}\right)\frac{\partial\psi_2}{\partial\theta} \tag{7}$$

$$A_\phi = \frac{1}{r\sin(\theta)}\frac{\partial\psi_0}{\partial\phi} - \frac{\partial\psi_1}{\partial\theta} + \frac{1}{\sin(\theta)}\left(\frac{1}{r} + \frac{\partial}{\partial r}\right)\frac{\partial\psi_2}{\partial\phi} \tag{8}$$

where Λ is the angular momentum operator:

$$\Lambda = \frac{-1}{\sin(\theta)}\frac{\partial}{\partial\theta}\left(\sin(\theta)\frac{\partial}{\partial\theta}\right) - \frac{1}{\sin^2(\theta)}\frac{\partial^2}{\partial\phi^2}. \tag{9}$$

By substituting (5) into (2) we obtain:

$$((\lambda + \mu)\delta_{i,0} + \mu)\nabla \cdot \nabla\psi_i = -\rho\omega^2\psi_i. \tag{10}$$

The general solutions of (10) are given by:

$$\psi_i(r, \theta, \phi) = \sum_{l,m} K_i^{(l,m)} j_l\left(\frac{\omega_{l,m} r}{v_i}\right) Y_l^m(\theta, \phi) \tag{11}$$

where $j_l(x)$ and Y_l^m are the spherical Bessel function and the spherical harmonic function, respectively; $v_0 = \sqrt{\frac{\lambda+2\mu}{\rho}}$ corresponds to the longitudinal sound velocity and $v_1 = v_2 = \sqrt{\frac{\mu}{\rho}}$ to the transverse sound velocity.

If a stress-free surface is assumed, the boundary conditions written in terms of the radial components of the stress tensor σ become:

$$\sigma_{rr} = \left((\lambda + 2\mu)\frac{\partial}{\partial r} + \frac{2\lambda}{r}\right) Q_r$$

$$+ \frac{\lambda}{r}\left(\frac{\partial}{\partial\theta} + \cot(\theta)\right) Q_\theta + \frac{\lambda}{r\sin(\theta)}\frac{\partial Q_r}{\partial\phi}\bigg]_{r=R} = 0 \tag{12}$$

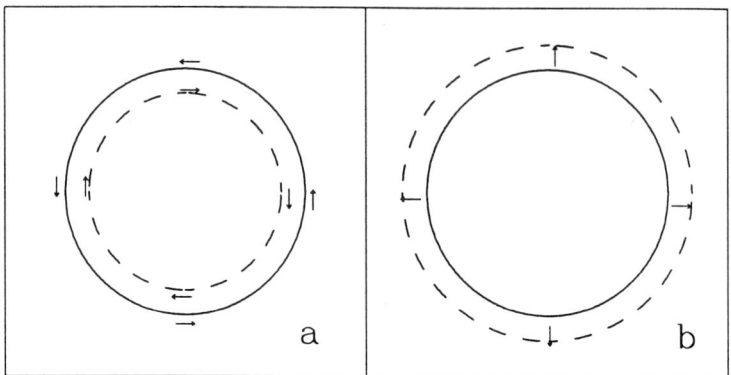

Fig. 1. The toroidal (a) and spheroidal modes (b) of a free spherical particle in the elastic continuum approximation.

$$\sigma_{r\theta} = \mu \left(\left(\frac{\partial}{\partial r} - \frac{1}{r} \right) Q_\theta + \frac{1}{r} \frac{\partial Q_R}{\partial \theta} \right]_{r=R} = 0 \qquad (13)$$

$$\sigma_{r\phi} = \mu \left(\left(\frac{\partial}{\partial r} - \frac{1}{r} \right) Q_\phi + \frac{1}{r \sin(\theta)} \frac{\partial Q_r}{\partial \phi} \right]_{r=R} = 0. \qquad (14)$$

R is the cluster radius. By substituting Equations (3–5) the boundary conditions can be written in terms of the scalar potentials ψ_i. After substituting the general solutions (11) for these scalar potentials and rearranging the results we obtain the two equations that determine the vibrational frequencies. The equation for the toroidal osciIIiations is given by:

$$j_{l+1}(\eta) - \frac{l-1}{\eta} j_l(\eta) = 0. \qquad (15)$$

The equation for spheroidal oscillations is given by:

$$2 \frac{v_t}{v_l} \eta \frac{j_{l+1}(\frac{v_t}{v_l}\eta)}{j_l(\frac{v_t}{v_l}\eta)} \left(\eta^2 + (l-1)(l+2) \left(\eta \frac{j_{l+1}(\eta)}{j_l(\eta)} - (l+1) \right) \right)$$

$$- \frac{1}{2} \eta^4 + (l-1)(2l+1)\eta^2$$

$$+ \eta \frac{j_{l+1}(\eta)}{j_l(\eta)} (\eta^2 - 2l(l-1)(l+2)) = 0 \qquad (16)$$

where $\eta = \omega R / v_l$. In Figure 1 these two vibrational modes are sketched.

The solutions of Equations (15) and (16) can be calculated using the bisection method. The spherical Bessel functions are given by:

$$j_0(x) = \frac{\sin(x)}{x} \qquad (17)$$

$$j_1(x) = \frac{\sin(x)}{x^2} - \frac{\cos(x)}{x} \tag{18}$$

$$j_n(x) = \frac{2n-1}{x} j_{n-1}(x) - j_{n-2}(x) \tag{19}$$

The recursion relation (19) is unstable for $x < 2l$, which can be handled by applying Miller's algorithm [11]. Since the zero's η of the equation which produces the toroidal oscillation frequencies are independent of the cluster parameters, this equation has to be solved only once. The zero's of the equation for spheroidal oscillations however depend on the ratio of the transverse and longitudinal sound velocities. Vibrational spectra and the corresponding specific heat of small lead particles, that are calculated in this way, are published by Nishiguchi and Sakuma [10] and Tamura et al. [12].

Since there are infinitely many l-values, each giving infinitely many ω's, a cut-off frequency has to be introduced. Most commonly the Debye cut-off method is used, which means that just the $3N$-6 lowest frequencies are used. A different cut off method was used by Tamura et al. [12]. By arguing that the wavelength should be larger than two times the interatomic distance they introduce an upperbound for l. After setting this upper limit the lowest frequencies are selected.

Once the vibrational frequencies are known, the specific heat contributions of the (Einstein) oscillators can be obtained from:

$$\frac{C}{R} = \frac{(\frac{\hbar\omega}{k_B T})^2 e^{\frac{\hbar\omega}{k_B T}}}{(e^{\frac{\hbar\omega}{k_B T}} - 1)^2}. \tag{20}$$

To calculate the total intra-cluster specific heat the contributions of all frequencies have to be summed.

7.2.2. THE MOLECULAR DYNAMICS APPROACH

If the number of atoms (N) in the particles (or core) is not too large ($N < 10^3$) a molecular dynamics approach [9] is feasible. The Newtonian equations of motion for the cluster atoms are given by:

$$m\frac{d^2 Q_i}{dt^2} + \sum_{j=1}^{3N} F_{ij} Q_j = 0 \tag{21}$$

where m is the atomic mass and Q_i are cartesian displacements coordinates and F_{ij} the force constant between atom i and atom j. Only nearest neighbour interactions are taken into account and the force constant between them is taken as an adjustable fitting parameter, see Figure 2.

By introducing the trial solution:

$$Q_i = A_i \cos(\omega t + \beta), \tag{22}$$

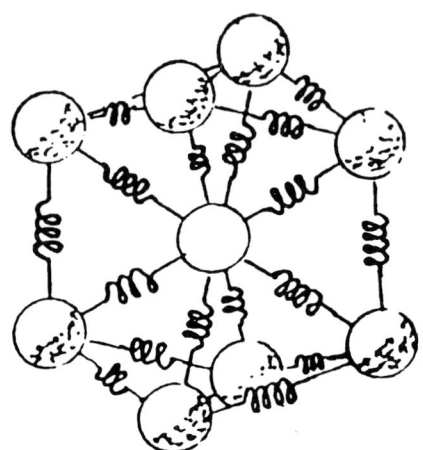

Fig. 2. Spring model used in the molecular dynamics approach.

β being an unimportant phase factor, we arrive at the eigenvalue problem:

$$\sum_{j=1}^{3N} F_{ij} A_j = m\omega^2 A_i. \tag{23}$$

The normal mode frequencies are obtained by numerical diagonalisation of the $3N \times 3N$ matrix F_{ij}. Six of the $3N$ eigenvalues are zero, corresponding to the 3 translational and 3 rotational degrees of freedom of the cluster as a whole.

The specific heat is calculated as in Section 7.2.1.

7.2.3. A SPHERICAL 309-ATOMIC PT-PARTICLE AS AN EXAMPLE

We illustrate the differences between the spring and continuum model using a hypothetical spherical 309-atomic Pt-particle as an example (such a 309-atomic unit is realized as the core in $Pt_{309}Phen^*_{36}O_{30}$, a metal cluster molecule to which we pay special attention). The diameter is about 2.1 nm. In Figure 3 the phonon density of states obtained from the spring model (dashed line) and the continuum model (drawn line) are given. For comparison we have also indicated the Debye density of states for bulk Pt (small dashes), normalized to the same volume.

From Figure 3 it can be seen, that for a 309-atomic Pt-particle the densities of states calculated in the continuum approximation and with the spring model are not identical, but at the lower end of the energy scale (below 50 K) become rather similar. Because the low temperature specific heat involves the lower levels only, it is hardly sensitive to the method used (see Figure 4).

If the data in Figure 4 are compared to the bulk specific heat of Pt one of the 'finger prints' of the QSE is visible, see Figure 5. Due to the spherical

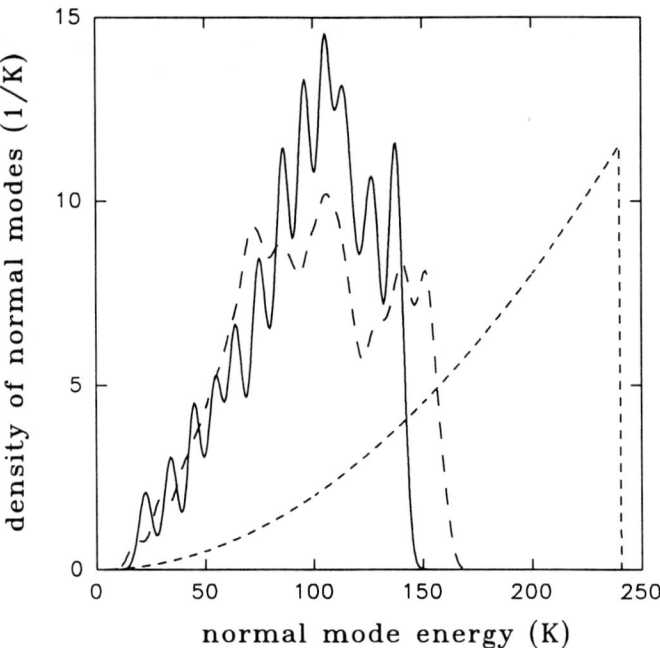

Fig. 3. The phonon density of states for the spring model (dashed line) and the continuum model (drawn) as appropriate for a spherical 309-atomic Pt-particle. The Debye density of states of bulk Pt, as calculated from the Debije temperature of 240 K, is indicated by small dashes.

boundary conditions, the degeneracy of the modes is enhanced compared to the bulk. The extra density of phonon modes, see Fig. 3, leads to an excess specific heat, ΔC, which almost follows a T^2 dependence, showing the importance of surface modes.

7.3. The Electronic Specific Heat

In this section results are derived for the electronic specific heat of isolated particles. To consider the particles in a specific heat sample as isolated is an oversimplification, both for metal cluster compounds and small metal particles. Metal cluster compounds form aggregates or crystals. The typical separation between the metal cores by the ligands is about 10 Å. Even if we neglect the influence of the ligands, the wave functions of neighboring cores will have some overlap. This should give a width in the energy level distribution due to dispersion in k-space. For the noncrystalline aggregates also the differences in the intermolecular distances will have a broadening effect. In the small metal particles the influence of the substrate forms another source for broadening. Depending on the broadening mechanism the resulting distri-

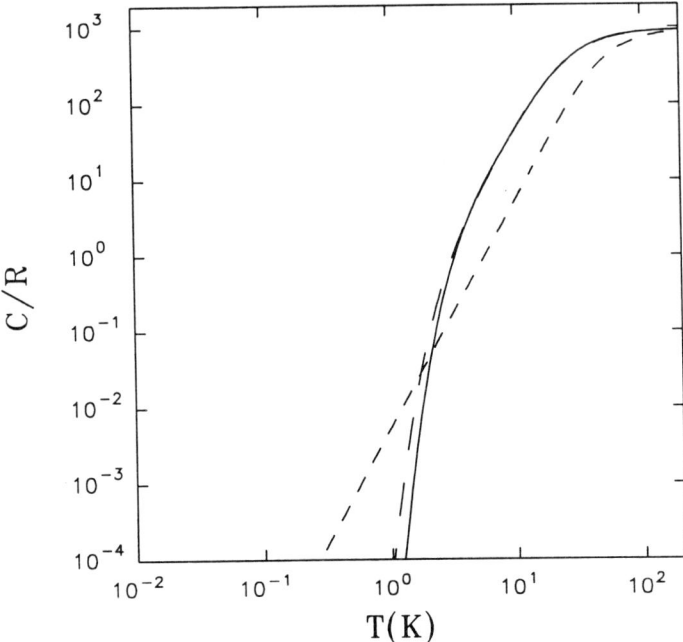

Fig. 4. C/R for a 309-atomic spherical Pt-particle as calculated in the spring (dashed) and continuum (drawn) model. The curves almost coincide. The Debye specific heat (small dashes) for bulk Pt is clearly different.

bution of the electronic energy levels might e.g. be Poisson-like, orthogonal, unitary or sympletic [3].

7.3.1. THE DEGENERATE ELECTRON GAS. SURFACE EFFECTS AND BULK STATES

For a degenerate electron gas the low-temperature specific heat is given by [13]

$$C_{el}/R = (dU/dT)/R = (\pi^2/3N_A)k_B T D(E_F) = (\gamma/R)T \qquad (24)$$

with U the internal energy per mole, N_A Avogrado's number and $D(E_F)$ the density of states per mole at the Fermi level. C_{el} is linear in temperature (for a free electron gas Equation (24) becomes $C_{el}/R = (\pi^2/2)(k_B T/E_F)$). The coefficient γ is related to the energy splitting δ at the Fermi level via

$$\delta \sim \frac{2}{D(E_F)} \frac{N_A}{N} = \frac{2\pi^2 R}{3} \frac{1}{N\gamma} \approx \frac{55}{N\gamma} \qquad (25)$$

with V the molar volume and d the diameter of the particle. For bulk Pd the value for γ is as large as $\gamma/R = 1.13 \times 10^{-3}$ K^{-1}. For Pt, $\gamma/R = 0.82 \times 10^{-3}$ K^{-1} and for Au, $\gamma/R = 8.8 \times 10^{-5}$ K^{-1}. For particle sizes of

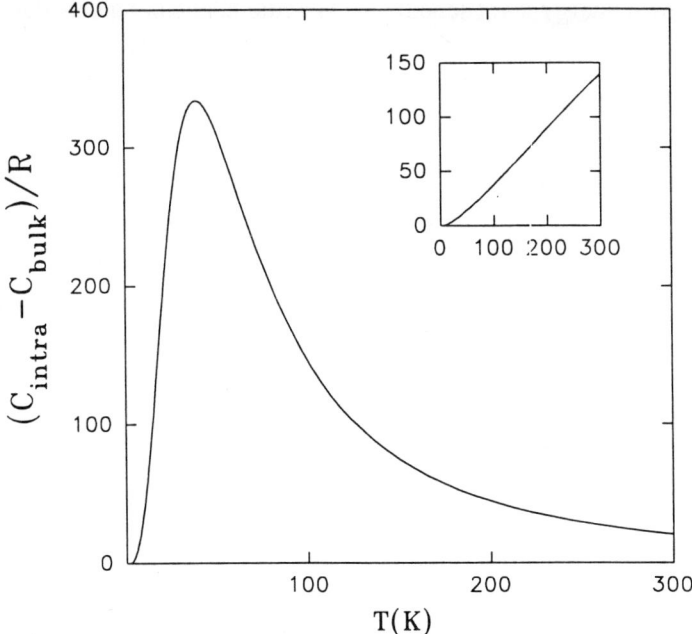

Fig. 5. QSE-Enhancement of the lattice specific heat of a 2.1 nm spherical Pt-particle (containing 309 atoms) compared to the bulk. In the insert $\Delta C/R$ is plotted versus T^2 for $T < 17$K. The almost linear dependence of $\Delta C/R$ on T^2 shows that especially the surface modes dominate at low temperatures. At the lowest temperatures C/R wil become negative, see Figure 4. This part is not shown here.

10 nm and larger, the average energy splitting will then be of the order of 1 K or smaller, so that for temperatures above 1 K the continuum approximation for the electronic level spectrum is justified, because $k_B T > \delta$.

Surface effects will modify the electron density at the outer atoms (see chapter about NMR). But, as long as the continuum approximation for these electronic states (which is the basis for the linear relation between C_{el} and T) remains valid, only the value of γ will be changed.

7.3.2. The Quantum-Size Limit

The continuum approximation breaks down, if the energy splitting around E_F becomes larger than $k_B T$. As an example, we will consider a 309 electron system with a diameter of 2.1 nm (about the diameter of the core of the Pt_{309} cluster considered before). From the measured γ-value of bulk Pt the average level splitting, δ, can be estimated with Equation (24) and Equation (25); the splitting is about 30 K. This implies that deviations from the linear contribution to the electronic specific heat are to be expected below 30 K.

For a known energy level scheme the specific heat can be calculated via the entropy

$$C/R = T\partial(S/R)/\partial T)_N \tag{26}$$

with $S = -\partial F/\partial T$. The free energy F is calculated from the partition function Z: $F = -RT \ln Z$. There is an additional twofold spin degeneracy per level.

As argued above, the electronic energy levels of the particles will most likely be broadened. Of the often cited distribution functions – Poisson, orthogonal, unitary and sympletic – only the Poisson distribution leads to a linear T-dependence of the specific heat at low temperatures: $C/R \propto kT/\delta$. For an even numbered electron system the proportionality constant is 5.0, for an odd system 3.3 [3, 7]. With the other distribution functions the specific heat has a stronger dependence on T.

7.4. Data and Discussion

Most of the thermodynamic studies on small metal particles have been performed on particles on a substrate. Also small metal particles were enclosed in the cages of e.g. SiO_2. For the metal cluster compounds, powders have been mixed with grease to make internal heat contact and contact with the heater and thermometer, or the compounds have been deposited on a substrate from solution.

7.4.1. SMALL METAL PARTICLES

The specific heat of small metal particles, e.g. of lead and indium [14], vanadium [15, 16], palladium [17] and platinum [4], but also of metaloxides like that of magnesium [18] has been measured and shows an enhancement of the lattice specific heat compared to the values of the bulk. The enhancement increases with decreasing particle size. For Pb [14] at a particle diameter of 2.2 nm and a temperature of 10 K the excess specific heat (compared to the bulk value) amounts to 300 mJ/moleK. The almost T^2 dependence of the enhancement strongly hints that surface modes are involved (the binding of the surface atoms is weaker than that of the bulk). The excess heat capacity compares well with the elastic continuum theory developed above [10]. For the Pb and In particles to get agreement between theory and experiment 'phonon softening' has to be introduced: the sound velocity (or Debije temperature) is lower than expected from the bulk. This again shows the importance of the surface layer. Size and temperature dependent particle sound velocities were also assumed for the palladium particles [17].

Concerning the electronic specific heat C_e, in the measurements of Stewart [4] on nanometer sized Pt-particles there are indications of the suppression of C_e around 1 K, although the matrix effects, that had to be subtracted, are large.

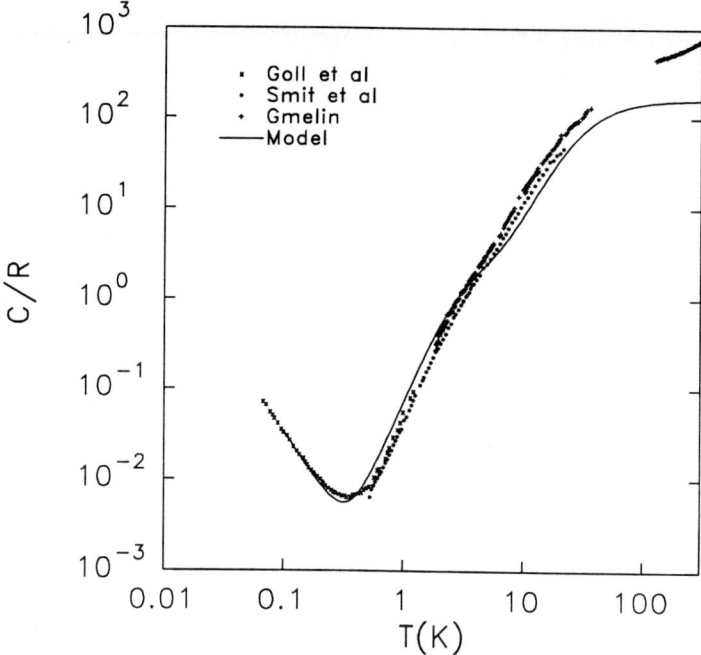

Fig. 6. Compilation of the low temperature specific heat data of the metal cluster compound Au_{55}. The drawn line is a fit discussed in the text.

7.4.2. METAL CLUSTER COMPOUNDS

A series of air stable metal cluster compounds with increasing diameter of the core are formed by $Au_{55}(PPh_3)_{12}Cl_6$, $Pt_{309}Phen^*_{36}O_{30}$ and $Pd_{561}Phen_{36}O_{200}$, abbreviated respectively as Au_{55}, Pt_{309} and Pd_{561}. If one starts counting the shells from the centre atom, then Au_{55} has completed the second shell of 42 atoms, Pt_{309} the fourth of 149 and Pd_{561} the fifth of 252. Till now it has not been possible to make such an air-stable series based on one single element, e.g. Pt only. We discuss the data in the same sequence of increasing core size.

Au_{55}. Specific heat data on Au_{55} are available from several groups [19, 9, 20, 21, 22]. A compilation is given in Figure 6. There is good overall agreement.

The intra-cluster contribution was calculated with the spring model with a spring constant for the intracluster vibrations of 30 N/m, about the bulk value of gold. From the Debye-Waller factor measured in a Mössbauer experiment on the same cluster the Debye temperature of the intercluster phonons was determined to be $\Theta_D = 15$ K. The two contributions together $C/R = C_{\text{inter}}/R + C_{\text{intra}}/R$ fit the data well above 0.5 K. The results of Goll

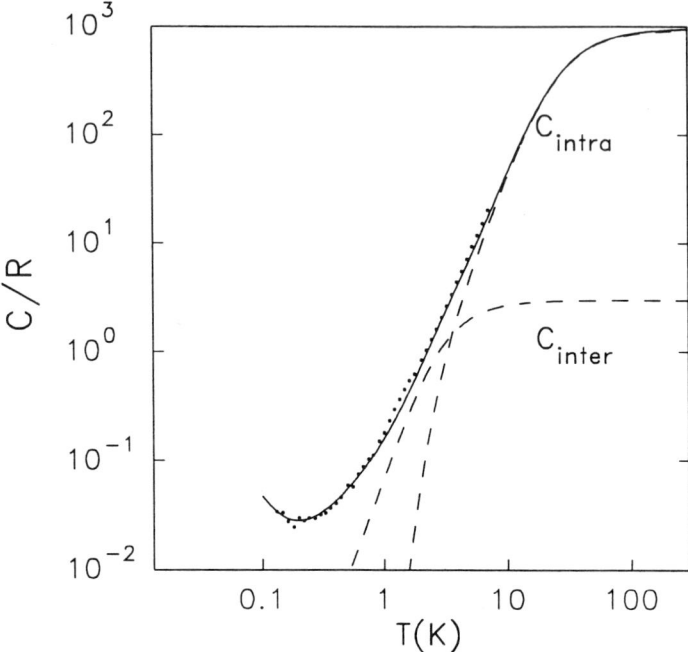

Fig. 7. C/R as a function of T for Pt_{309}. The inter- and intra-contributions are indicated by the dashed lines.

et al. [20], which extend down to 0.06 K, show the influence of a T^{-2} term at the lowest temperatures that normally arises from a Schottky-like anomaly.

Concerning the last mentioned contribution, $C/R = 3.4 \times 10^{-4} T^{-2}$, the electric quadrupolar splitting of the Au nuclei would give a similar T-dependence. However, the splittings as measured in the Mössbauer experiment would imply an even higher contribution to the specific heat. Because also the quadrupolar relaxation rate is estimated to be extremely low [22], another mechanism has to be found. The zero- and in-field dependence of C can be fitted with a hyperfine contribution of about one electronic spin per ten clusters coupled to a nuclear spin 3/2 (although higher nuclear spin values are not excluded) [22].

In the C/R units for Au_{55} as used in the figure, the γ/R term for bulk gold is 0.0048 at one Kelvin. Due to the size of the T^{-2} term, the presence of a linear contribution of the order of 0.001 can not be excluded, although it does not seem very likely, given the small size of the Au_{55} inner core.

Pt_{309}. In Figure 7 the experimental data for the Pt_{309} MCM are shown. The dashed lines are the calculated intra-cluster vibrational contribution (the low temperature outcome of the spring and continuum model are almost identical) and the estimated inter-cluster vibrational term (with $\Theta_D = 15$ K). The two

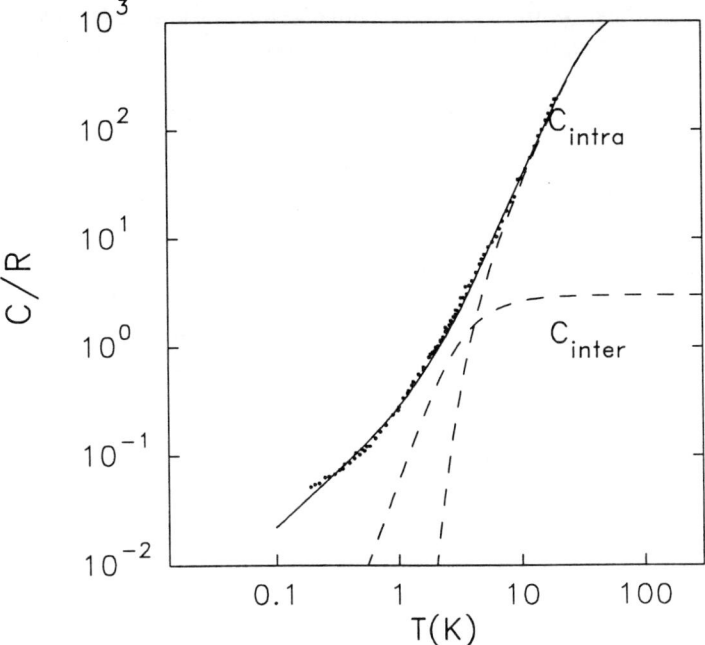

Fig. 8. Pd$_{561}$ Data and Fit. The experimental specific heat results are shown on a double logarithmic scale. The dashed lines represent the contributions to the specific heat due to intra-cluster vibrations, as calculated by the ECA method, and inter-cluster vibrations. The final fit (drawn line) includes a linear term (see text).

contributions are not sufficient to fit the data. Inclusion of an additional linear term improves the low temperature side of the fit appreciably. The final fitting expression is given by $C/R = 0.09T + 3.8.10^{-4}T^{-2} + C_{\text{inter}}/R + C_{\text{intra}}/R$. Below 2 K the intra-contribution can be neglected, while the Debye contribution of the inter-vibrations can be approximated by the familiar T^3-term: $C/R = 0.09T + 3.8 \times 10^{-4}T^{-2} + bT^3$. The T^{-2}-term, as in Au$_{55}$, might arise from hyperfine interaction. As in the small metal particles, the intra-contribution used in the fit is higher than calculated with the bulk value of the sound velocity, or the bulk spring constant. For bulk Pt the linear term is $0.25T$ (for a 309 atomic unit).

Pd$_{561}$. The cuboctahedral Pd$_{561}$ clusters of the Pd$_{561}$Phen$_{36}$O$_{200}$ compound consist of five shells of Pd atoms. The data are given in Figure 8. The lattice contributions are calculated with the MD and the ECA method. The vibrational spectra of both approaches differ in detail, especially at the higher energies [23, 21]. The calculated specific heats however are similar and compare well with experiment, see Figure 8. For the inter-cluster Debye temperature, the value $\Theta_D = 16$ K gives a good fit. The adjustable parameters in the calculations, i.e. the transverse sound velocity for the ECA method and the force constant for the MD method, were taken to be 1.63 km/s and 19 N/m respectively. Again, these values are below the values of the bulk. For bulk palladium the transverse sound velocity is 1.99 km/s and the spring constant 34 N/m.

The most important result is the observation that again the inter- and intra-contributions are insufficient to describe the specific heat at the lowest temperatures. In order to fit the experimental data over the whole temperature range, a linear term ($C/R = 0.22T$) has been added to the calculated specific heat. The solid line in Figure 8 represents the sum of the contributions of intra-and inter-cluster vibrations and the linear contribution. Like in Pt_{309}, the coefficient of the linear term amounts to approximately one third of the value of bulk value (for bulk Palladium $C/R = 561 \times (9.4 \times 10^{-3}T/R) = 0.64T$). Such a reduction of the bulk value has to be expected, because only the inner-shells (309 atoms) might show metallic behaviour, the outer atoms being partly chemically bound by ligand molecules.

In view of the NMR data on Pt_{309}, discussed elsewhere in this book, another scenario is even more likely. A non-exponential recovery of the Pt nuclear magnetization becomes visible below about 50 K, a temperature that is of the same order as the expected average level splitting δ. The recovery can be simulated if the electronic levels are distributed in a Poisson distribution,

$$P(\Delta) = \frac{1}{\delta} e^{-\Delta/\delta} \tag{27}$$

where Δ is the spacing between the adjacent energy levels and δ the average spacing. For systems with a high symmetry as for the cluster molecules, this is an allowed result of the broadening mechanism discussed earlier in Section 7.3. The Poisson distribution has as a unique feature that it gives a linear contribution to the low temperature specific heat. If we average between even an odd (see Section 7.2), $C/R \approx 4kT/\delta$. Using this formula for Pt_{309} an average energy splitting of about 40 K is found. For Pd_{561} this value becomes about 20 K. These values are somewhat higher than the average value for the energy splitting based on the bulk specific heat and the size of the core. But again, part of the outer layer of the core is non-metallic and hence the effective diameter of the metallic part is smaller than the diameter of the core, which we used for the estimate. If so, the quantum size effect would account remarkably well for the linear term in the low temperature specific heat for both Pt_{309} as Pd_{561}.

In Table 1 we summarize the parameters used in the expression $C = aT^{-2} + \gamma T + C_{intra} + C_{inter}$ for the low temperature specific heat of Au_{55}, Pt_{309} and Pd_{561}. The force constant K and inter-cluster Debye temperature Θ_D of Au_{55} were obtained by Mössbauer experiments. The ratio between the transverse and longitudinal sound velocities was fixed at the bulk value. All other parameters are obtained by fitting the experimental data. For comparison

TABLE 1

Fitting parameters for the low tempertaure specific heat in Au_{55}, Pt_{309} and Pd_{561}. The ratio given in the last column is assumed to be identical to the bulk value.

Cluster	a/R (K^2)	γ/R (K^{-1})	Θ_D (K)	K (N/m)	v_b (km/s)	v_t (km/s)	v_t/v_l
Au_{55}	0.34×10^{-3}	...	15	30	1.43	1.25	0.37
Pt_{309}	0.38×10^{-3}	0.09	15	20	2.00	1.20	0.53
Pd_{561}	...	0.22	16	19	2.25	1.63	0.43

we have also listed the average sound velocity v_b calculated from the low temperature Debye temperatures.

How do the specific heat data compare to data obtained via other techniques? The agreement with Mössbauer data for Au_{55} has been stressed already. NMR data have been the inspiration for invoking the Poisson distribution [25] of the electronic levels, that agrees with the linear term found in the low temperature specific heat. Susceptibility data are more difficult to compare with, because the Stoner enhancement that dominates the low temperature susceptibility data of e.g. Pd_{561} [24] has not to be reflected in the density of states around the Fermi level and hence in the electronic contribution to the low temperature specific heat. Transport measurements, as pointed out already, show the presence of wave function overlap between the cores of the metal cluster compounds (possibly by the assistance of the ligands) which is one candidate for the observed level broadening.

7.5. Summary

In small metal particles and cluster compounds quantization of the lattice vibrations is clearly visible. For core sizes of about 10 Å (Au_{55}) a spring model accounts well for the intra-cluster specific heat. For diameters of 20 Å and higher (Pt_{309}) a continuum approximation suffices for its calculation. The electronic QSE in the low temperature specific heat does not show up easily. In small metal particles there is no unambigous experimental demonstration. For the metal cluster compounds a Poisson distribution of the energy levels can account for the low temperature specific heat data. The latter interpretation is supported by NMR results, but its origin and validity remains to be investigated.

References

1. C. Schafer, *Z. Phys.* **7** (1921) 287.
2. M. Planck, *Theorie der Waärmtestrahlung*, 6th edition, Barth, Leipzig (1921).

3. W. P. Halperin, *Rev. Mod. Phys.* **58** (1986) 533.
4. G. R. Stewart, *Phys. Rev.* **B 15** (1977) 1143.
5. B. Mühlschlegel, D. J. Scalapino and R. Denton, *Phys. Rev.* **B 6** (1972) 1767.
6. R. Denton, B. Mühlschlegel and D. J. Scalapino, *Phys. Rev.* **B 7** (1973) 3589.
7. J. A. A. J. Perenboom, P. Wyder and F. Meier, *Phys. Rep.* **78** (1981) 173.
8. N. N. Greenwood and E. J. F. Ross, *Index of Vibrational Spectra of Organic and Inorganic Metallic Compounds*, Butterworth, London (1975).
9. H. H. A. Smit, P. R. Nugteren, R. C. Thiel and L. J. de Jongh, *Physica* **B 153** (1988) 33.
10. N. Nishiguchi and T. Sakuma, *Solid State Commun.* **38** (1981) 1073.
11. M. Abramowitz and I. Stegun, *Handbook of Mathematical Functions*, Dover Publ. Comp. (1965).
12. A. Tamura, K. Higeta and T. Ichinokawa, *J. Phys.* **C 15** (1982) 4975.
13. This relation holds only approximately, as e.g. the enhancement factor due to electron-phonon interaction is not taken into acount, see G. S. Knapp and R. W. Jones, *Phys. Rev.* **B 6** (1972) 1761.
14. V. Novotny, P. P. M. Meincke and J. H. P. Watson, *Phys. Rev. Lett.* **28** (1972) 901; V. Novotny and P. P. M. Meincke, *Phys. Rev.* **B 8** (1973) 4186.
15. G. H. Comsa, D. Heitkamp and H. S. Räde, *Solid State Comm.* **20** (1976) 877.
16. O. Vergara, D. Heitkamp and H. von Löhneysen, *J. Phys. Chem. Sol.* **45** (1984) 251.
17. G. H. Comsa, D. Heitkamp and H. S. Räde, *Solid State Comm.* **24** (1977) 547.
18. W. H. Lien and N. E. Phillips, *J. Chem. Phys.* **29** (1958) 1415.
19. E. Gmelin, unpublished results (1984).
20. G. Goll, H. van Löhneysen, U. Kreibig and G. Schmid, *Z. Phys.* **D 20** (1991) 329.
21. J. Baak, H. B. Brom, L. J. de Jongh and G. Schmid, *Z. Phys.* **D 26** (1993) S 30.
22. H. B. Brom, J. Baak, L. J. de Jongh, F. M. Mulder, R. C. Thiel and G. Schmid, *Z. Phys.* **D 26** (1993) S 27.
23. J. Baak, H. B. Brom, L. J. de Jongh and G. Schmid, in P. Jena *et al.* (eds.), *Physics and Chemistry of Finite Systems: From Clusters to Crystals*, Kluwer Academic Publishers, Dordrecht (1992), Vol. 1, p. 339.
24. D. A. van Leeuwen, J. M. van Ruitenbeek, L. J. de Jongh and G. Schmid, *Phys. Lett.* **A 170** (1992) 325.
25. D. van der Putten, H. B. Brom, J. Witteveen, L. J. de Jongh and G. Schmid, *Z. Phys.* **D 26** (1993) S 21.

8. NMR IN SUBMICRON PARTICLES

H. B. BROM, D. VAN DER PUTTEN and L. J. DE JONGH
Kamerlingh Onnes Laboratory, Leiden University,
P.O. Box 9506, 2300 RA Leiden, The Netherlands

8.1. Introduction

The properties of a metal will change if its size is no longer macroscopic. The size regime between large clusters (< 10 nm) and bulk solids (> 100 nm) is often referred to as mesoscopic condensed matter [1]. If the linear dimensions become comparable to characteristic length scales of the system new phenomena appear, that are not present in the bulk. As an example, if the sample dimensions are lowered to such an extent that the wave coherence length L_ϕ becomes comparable with the sample size, the transport properties will bear resemblance with a scattering states problem. If the sample size is reduced still further, the distance between the energy levels around the Fermi level will no longer be small compared to $k_B T$. In this final microscopic or cluster limit quantum size effects rather than intraparticle scattering will dominate the physical properties [1]. Because even between 1 and 10 nm the physics of a cluster changes drastically, one also might (as we will do below) let the mesoscopic regime start for particle sizes above 1 nm (roughly 50 atoms).

The nano-sized particles, with which we will mainly deal here, are often produced as hydrosols (method A). For the Pt-particles - we will use Pt as a generic example throughout this chapter - one might use H_2PtCl_6 as starting material [2, 3, 4, 5, 6]. The Northwestern group [2, 7] followed the standard techniques for chemical impreganation of a SiO_2 support to obtain samples of the order of 33–50 Å. The Lausanne group [4, 8, 9, 10] prepared an aqueous colloid of Pt particles of about 20 Å by citrate reduction of the Pt-hexachloride acid. After introducing SiO_2 or TiO_2 particles (about 200 Å diameter) in the solution and drying, SiO_2 (or TiO_2) – supported Pt particles were obtained. The Illinois group [5, 6] prepared a wider range of particle sizes (between 10 and 100 Å) and used alumina as support. In a first approximation the influence of the support can be neglected. The surface of the small metallic particles can be covered at will by various gas molecules, like oxygen or carbonyl-groups [8].

In a completely different approach (method B), compounds with large metal-atom cluster cores, are stabilized by ligand counterions: the so-called ligand-stabilized metal-cluster compounds [11]. An example is $Pt_{309}Phen^*_{36}O_{30}$ with a core of 309 Pt-atomes, packed in an fcc-structure. The difference with method A is that the metal-atom core always has a surface layer of ligand molecules.

Of the other preparation methods, we like to mention in particular the cluster beam approach. In this increasingly powerful method the particles are produced in a beam. One might even cover the surface with other atoms by a reaction with selected gases [12]. Finally the mass-selected clusters can be deposited in e.g. rare gas matrices and studied thereafter [13].

By the methods A and B the density of particles in a macroscopic volume can be made high enough to allow an NMR study. A peculiarity of method B is that frequently aggregates of particles are produced: the metal cores of the cluster molecules are only separated by the ligands, while the neutral metal cluster compounds are held together in a non-crystalline structure by Van der Waals forces. In case of ionic metal cluster compounds, crystalline ionic solids may be formed together with suitable counterions (recently a different class of crystalline compounds with unit cells containing up to 146 metal atoms has been synthesized [14]). These close-packed systems might have an appreciable conductivity because of the inter-cluster overlap of the wave functions. While NMR on a nucleus outside the metal core will probe especially the inter-cluster interaction, an intra-core NMR-nucleus will be a good probe for the degree of delocalization of the electrons inside the core.

In the next sections we present the general theoretical framework for this chapter in some detail, and shall thereby closely follow the approach taken by the Illinois [15, 5, 6, 16, 17, 18] and Lausanne groups [8, 9, 10]. Thereafter, we will summarize the insights obtained from the work on particles synthesized by method A. After a discussion of the NMR-data for the clusters (method B), the main conclusions will be formulated.

8.2. Surface and Quantum Size Effects

The reduction of the size of metallic particles has two aspects. Due to the reduction the ratio of surface-atoms to bulk atoms increases. Therefore we discuss below first the influence of a surface in a bulk material. Local density band structure calculations are a powerful method to account for these effects. Thereafter we address the so-called 'quantum size effect'. By the restriction in space, the continuum approximation for the energy level density of states breaks down, and the level spacings can become larger than the thermal energy $k_B T$. In a separate section we shall go into the relevant details of the ESR and NMR methods and show how these techniques are sensitive probes for the electronic properties.

TABLE 1

Calculated Knight shifts for bulk Pt-metal in the neighborhood of the metal surface, after Weinert and Freeman. The layers are counted inwards from the surface ($n = 0$). There is a large negative polarization of the core electrons due to the d moments, and a direct polarization of the valence s-electrons.

	Bulk ($n \ll 1$)	$n = 1$	Surface ($n = 0$)
Valence	1.3	1.3	2.4
Core	-5.4	-3.9	-3.0
Total	-4.1	-2.6	-0.6

8.2.1. SURFACE EFFECTS

In general, the band structure at the surface is expected to be different from that of the bulk due to the different coordination of the surface atoms. For a transition metal like Pt the d-band, which is already narrow in the bulk, is expected to become even more narrow (in a first approximation the highly correlated d-band will have a width proportional to the number of neighbors). The effect of the surface on the almost free s-electrons will not be that strong. By the narrowing of the d-band its density of states at the Fermi surface will decrease. Such a intuitive picture is indeed confirmed by the LDA calculations of Weinert and Freeman [19], as summarized in Table 1.

8.2.2. QUANTUM SIZE EFFECTS

As shown in many solid state text books [20], the energy levels and the allowed wave vectors of a single particle in a box with finite dimensions are quantized. For a system of N free particles in a volume V (density $n = N/V$) the Fermi energy E_F is given by

$$E_F = \hbar^2 \left(\frac{3\pi^2 n}{2m}\right)^{2/3} \tag{1}$$

and the energy separation ΔE

$$\Delta E = 2E_F/3N. \tag{2}$$

For most metals E_F is of the order of 5–10 eV. This implies that for metals of macroscopic dimensions the thermal energy $k_B T$ is much larger than ΔE and the discreteness of the energy levels is wiped out. For all practical purposes it is then allowed to consider the energy level structure around the Fermi level as continuous. Only if the sample becomes of nanometer size, one might expect the continuum approximation to break down. As can be seen from Equation (2) a rule of thumb to find the separation between energy levels is dividing E_F by the number of pairs of valence electrons (up and down spin) $N/2$. Returning to Pt as our paradigm ($E_F \approx 6$ eV) and considering its $6s$-and $5d$-electrons (10 electrons) as delocalized, then leads to an energy level splitting of the order of 50 K for a 309 Pt-cluster. Alternatively, one can use the density of states of bulk Pt, expressed per atom, 1.55×10^{-4} states/K atom, and multiply this number by 309. The resulting figure is about 20 K. If the energies of the electrons in such a cluster are calculated on basis of a model of free electrons confined to a sphere, electronic levels found with high degeneracies are found. Due to the fcc-symmetry of the core and the influence of the boundaries these degeneracies will be partly lifted. If the Fermi-level would lie within a highly degenerate level or multiplet, also Jahn–Teller like distortions will reduce the symmetry in the low temperature state [21]. In a more realistic approach one has to take into account the particular level schemes of the atoms and the packing arrangement [22]. From these calculations it transpires that at a cluster size of several hundreds of atoms a transition to the bulk structure takes place, and the energy level scheme approaches the solid state band structure.

The non-degenerate molecular orbitals that finally lead to e.g. the s- and d-bands, can at most contain two electrons with opposite spin. This means that (apart from the orbital contributions) the spin contribution to the magnetic susceptibility at $T = 0$ is either that of $S = 1/2$ or $S = 0$. Particles prepared according to method A will likely have a distribution in size and shape [23]. The resulting energy level distribution will be broadened compared to the idealized monodisperse particle picture, and the sharp distinction between even and odd will also be obscured. In addition the effects of the spin-orbit (L-S) coupling have to be taken into account. In a system with discrete energy levels the L-S coupling has a similar effect on the electronic susceptibility as in the case of singlet superconductivity. In both cases L-S coupling gives a non-zero susceptibility at $T = 0$ [3]. In the metal cluster compounds an additional factor is played by the inter-cluster wavefunction overlap. Aggregates of these materials have an appreciable conductivity at high frequency, i.e. if the electron hopping is confined to near or nearest neighbours [24]. Although the conductivity measurements are performed on pressed powders, the inter-cluster wavefunction overlap is likely only slightly modified in the loosely packed metal cluster compounds, as used in the NMR experiments.

8.3. ESR and NMR – Theory

In continuous wave (cw) ESR one measures intensity, position and shape of the ESR line. Via a saturation method an impression of the electron spin-lattice relaxation rate T_1^{-1} can be obtained as long as T_1 is not too short. If the line width is sufficiently small, pulsed ESR measurements are possible. Pulse

ESR gives in principle the same possibilities as pulse NMR (see NMR), e.g. the relaxation rates can be measured more accurately. The cw ESR method is widespread in chemistry, where it is mostly applied to non-conducting materials. In metallic systems the ESR lines are usually strongly broadened (except for the lighter alkali metals), which makes the technique of limited use [25].

Especially in cases of broad lines, NMR might reveal more details about the electronic spins than ESR. In solid-state NMR it is standard to measure four different quantities [26]: $K, \delta B, T_1^{-1}, T_2^{-1}$. The position of the NMR line is usually expressed as the relative field shift with respect to a well-known reference, and is denoted by the generalized Knight shift $K = \Delta B/B_{\text{ref}}$. A measure for the shape of the resonance line is the full width at half intensity δB. The energetic contact between the nuclear Zeeman system and the lattice is expressed by the nuclear Zeeman relaxation rate (T_1^{-1}). Lastly, the quantity T_2^{-1} is a measure for the mutual interaction between the nuclear spins. In metallic systems all these quantities are influenced by electronic parameters. Precise values of the parameters and dependences on e.g. the temperature, T, and magnetic field, B, contain a wealth of information about the electronic charge and spin distribution and about the electronic susceptibility.

8.3.1. ESR

Elliott [25] has shown theoretically that in alkali metals the scattering rate τ_s associated with the inverse linewidth $(\delta\omega_b^{-1})$ of an ESR line in the bulk should depend on the same scattering rate, (τ_R), that determines the resistivity:

$$\delta\omega_b \propto 1/\tau_s, \quad \tau_s \approx \tau_R(g-2)^{-2}, \tag{3}$$

with a proportionality constant of order one. For particle sizes (d) smaller than the spin diffusion length, which is in the micron regime, surface scattering will be a additional source for resistive scattering

$$\delta\omega_d \propto \left(\frac{1}{\tau_R} + \frac{v_F}{d}\right)(g-2)^2. \tag{4}$$

On the other hand, for submicron particle sizes the splittings between the energy levels will hamper the scattering. Like in superconductivity, the scattering mechanism has to overcome the (average) energy gap, δ, to be effective. For small enough particles the linewidth deduced from Equations (3) and (4) has to be multiplied with the electron-spin Zeeman-splitting ($\hbar\omega_z$), divided by δ [27]:

$$\delta\omega = \frac{\hbar\omega_z}{\delta} \cdot \delta\omega_d. \tag{5}$$

So, one might expect that even for those cases where in the bulk metal the observation of an ESR-line is hindered by too large a linewidth, in the small particle limit the observation of an ESR signal becomes possible. Indeed, ESR is observed in many elements for particle sizes below 100 nm. Changes in the distribution due to the size effect (see the NMR section) complicate this picture again, however. Also at low temperatures the decay of the longitudinal spin magnetization of the electrons in small metallic particles will be no longer exponential [28].

8.3.2. NMR

NMR in metals is well understood [29, 26]. In metals the delocalized s-electrons are strongly hyperfine coupled to the nucleus and are often responsible for the line-position and the relaxation rates. For simple metals with only s-electrons at the Fermi level, the relative shift of the line with respect to its atomic position, the Knight shift $K = \Delta B/B$, can be expressed in the total spin susceptibility of the electrons, χ_e^s, and the averaged wave function density at the nucleus evaluated at the Fermi level, $\langle |u_k(0)|^2 \rangle_{E_F}$:

$$K = \frac{\Delta B}{B} \approx \frac{8\pi}{3} \langle |u_k(0)|^2 \rangle_{E_F} \chi_e^s. \tag{6}$$

In general, the inner core s-electrons can be polarized due to electrons in other states (e.g. d-electrons) at the Fermi level. In Pt-metal this core polarization effect is very strong and is the reason for the unusually large negative Knight shift [19]. In such a more complicated case the Knight shift becomes the sum of individual contributions, which may even cancel each other:

$$K_{\text{tot}} = K_s + K_d. \tag{7}$$

Energy can be absorbed by the nuclear spin system if the energy quanta have the same splitting as the nuclear Zeeman splitting. If there is such a spectral density at the nuclear Zeeman splitting present from a energy reservoir at a temperature T, the nuclear levels involved in the energy exchange will finally be populated according to the appropriate Boltzmann distribution (for systems with nuclear spins $I > 1/2$ the situation can be more complicated). For metals the scattering between electronic spins is the origin of the spectral density and the relaxation rate is given by the well known Korringa relation [29], linking the relative Knight shift (K) to the nuclear-spin lattice relaxation rate (T_1^{-1}) as

$$T_1 T K^2 = \frac{\hbar^2}{4\pi k_B} \frac{\gamma_e^2}{\gamma_n^2}. \tag{8}$$

Different independent electronic contributions to the Knight shift always add up in the relaxation rate ($K_{\text{tot}}^2 = K_s^2 + K_d^2$)

$$\frac{1}{T_{\text{tot}}} = \frac{1}{T_s} + \frac{1}{T_d}. \tag{9}$$

In order to take into account the particle size effect for Pt, Slichter and coworkers [17] assumed the s-electrons as delocalized over the (spherical) volume of the particles. Because d-electrons are more tightly bound to the lattice sites, they were treated as in the bulk but with surface corrections. The effect of the d-band narrowing at the surface [17] is taken into account by the introduction of a healing distance λ: a few lattice spacings away from the surface the bulk d-densities are restored again (see Section 8.2.1).

$$K_d(r, R) = K_{db}[1 - \xi g(r, R)] \tag{10}$$

where ξ is the fractional reduction in the local density of states at the surface nuclei, r is the distance from the center and R the radius of the sphere. The function $g(r, R)$ was assumed to be (with lengths in Å):

$$g(r, R) = \begin{cases} 1 & r \geq R - 0.5 \\ \exp[(r + 0.5 - R)/\lambda] & r \leq R - 0.5 \end{cases} \tag{11}$$

The hybridization between the s- and d-bands will also lead to a distance dependence of K_s

$$K_s(r, R) = K_{sb} S(r, R)[1 + \eta g(r, R)]. \tag{12}$$

The parameter $S(r, R)$ takes the effect of the boundary conditions on the s-electron densities at the various nuclear sites into account. It is a text book problem [30] to calculate $S(r, R)$ for free electrons in a spherical square well.

For an odd electron system in the quantum size regime at $T = 0$, only one level, say the nth, has a spin. The r-dependence of the wave function is determined by the nth spherical Bessel function

$$\Psi_{R,E}(r) \propto j_n[r Z_{nk}/(R + 1.5)] \tag{13}$$

and $S(r, R)$ is simply given by $|\Psi_{R,E}|^2$. Different sites experience different shifts, so one has to make a weighted average for the final line shape. Because different final Knight shifts K correspond to different K_s and K_d contributions, the relaxation rates will vary over the resonance line.

The Lausanne group made a similar line shape analysis [8]. Differences in the Stoner enhancement for the Pt s and d electrons are taken explicitly into account, as is the orbital contribution to the Knight shift. The Stoner enhancement affects the relaxation rate and the Knight shift differently [31]. For clean particles [9, 10] the group stresses the reduction of the local density of states at the surface compared to the bulk due to the reduction in d-electrons. It was assumed that the spatial dependence in the density of states of the s-electrons only plays a minor role. Due to the differences in d electron density the relaxation rate is again a function of the line position.

In spin-spin relaxation processes one deals with those interactions that couple the spins without the transfer of energy to or from a heat bath. The omnipresent dipolar interaction, the pseudo-dipolar and pseudo-exchange coupling (via the electronic spins) are candidates. In the non-coherent limit [32] the ratio T_2/T_1 will be 2. If a given nuclear spin has Z instead of just 1 neighboring spins, the relaxation rate will be increased accordingly. Including the lifetime broadening effect [32], the final equation becomes

$$\frac{1}{T_2} = \frac{Z+2}{2T_1}. \qquad (14)$$

Inhomogeneities in the Knight shift, as expected for small particle sizes, modulate the shifts of neighboring nuclear spins. If the shifts are larger than the secular part of the interaction J between the nuclei, such a modulation leads to beats in the envelope of the spin-echo decay, that are directly linked to the strength of the coupling [33, 32]. Assuming a small gaussian spread in J, the Illinois group derived the following equation for the time dependence of the echo intensity $S(t)$:

$$S(t) = S_0 \exp(-2\tau_d/T_2)$$
$$\times \{B_0 + \exp[-(\tau_d/T_{2J})^2][B_1 \cos(J\tau_d) + B_2 \cos(2J\tau_d)]\}. \qquad (15)$$

A spread of J values shortens the relaxation rate T_{2J}^{-1} associated with the decay of the beats. As can be seen from the last equation, the beat phenomenon will die out rapidly for a large value of T_{2J}^{-1}, i.e. a large spread in J.

Independent of the particular model, some general statements can be made. In a magnetic field of about 10 Tesla, which is becoming more and more common in an NMR experiment, the electronic Zeeman splitting is about 13 K. Still, in a bulk conductor, the Korringa relation holds down to much lower temperatures. The reason for this is that in the scattering mechanism, mentioned in the section about ESR, only electrons at the Fermi level are involved.

In the case of submicron particles with a size distribution, the Knight shift of all particles with an even number of electrons will decrease with temperature, while those with an odd number will have a large spread in shifts ($\propto d^{-3}$).

In the quantum size regime, at sufficiently high level splittings and low enough temperatures, the electronic spins will be relaxed by crystal field modulations due to the phonons. In that limit the relaxation process is slowed down to such an extent that also other factors like impurity contributions or the interaction between electronic spins on neighboring particles [34] have to be considered.

8.4. Naked Clusters – Experiment and Discussion

In this section we will briefly summarize the experimental results on supported naked metal particles. A general review of experimental work has been presented by Halperin [3]. Although thereafter new and interesting results have been published, e.g. on small Cu particles [35], we will limit ourselves to the resonance results obtained on Pt particles. The reason for doing so, is to be able to make a comparison to the Pt-work in the metal cluster compounds.

8.4.1. ESR

ESR results for small Pt particles are reported by Gordon et al. [36]. The particles under study of average size of 2.1 nm were embedded in a gelatin matrix. The signal found had the free electron g-value and a width of 2 mT at 10 GHz (0.36 T) and 24 GHz. Its intensity was quite weak (corresponding to 0.04 unpaired electron per particle) and the temperature dependence was Curie-like. None of these features follows directly from the theoretical ideas developed above, e.g. constant linewidths are not expected from Equations (4) and (5), but are normally connected to impurity contributions. For that reason the authors also indicated possible other sources for the signal. Also in the metal cluster compounds of Pt_{55} and Pt_{309} up to now no ESR signal could be detected.

8.4.2. NMR

There is a wealth of NMR data on small platinum particles. One reason is the commercial interest in Pt as a catalyst, another the favorable properties of this nucleus: ^{195}Pt has a spin 1/2, an abundance of 33.8% and a nuclear moment, which is about 1/4 of that of a proton. The Illinois group has worked systematically from about 1980 on supported Pt-particles. The line shapes, relaxation rates and microscopic variations in the Knight shift were determined for different Pt-catalysts which surface to volume ratios from 4% up to 58% (particle sizes down to a few nm). A typical example of the line shape is shown in Figure 1.

As discussed in Section 8.2, in order to explain such a large spatial variation in the Knight shift K [5, 6], Slichter et al. proposed a model [17] in which the local density of states (LDOS) at the Fermi level of the $5d$-electrons is an exponentially decreasing function of distance when going from the center of the particle to the surface (see section on theory). The $6s$-LDOS was expected to exhibit large oscillations due to the geometry of the boundary conditions, which were taken spherical. Experimental evidence for the existence of these spatial Knight shift oscillations was given by Rhodes et al. [5, 6], who obtained anomalously large Knight shift gradients from the slow beats in the decay of the spin-echo intensities.

Bucher and van der Klink [8] observed similar line shapes, see Figure 2, although the width at the same surface to volume ratio is larger. Such a difference can be caused by the influence of molecules absorbed at the surface, which will reduce the effective size of the metal particle (see the discussion of the NMR line shape of the metal cluster compounds). Also here beat patterns in 'T_2' were observed.

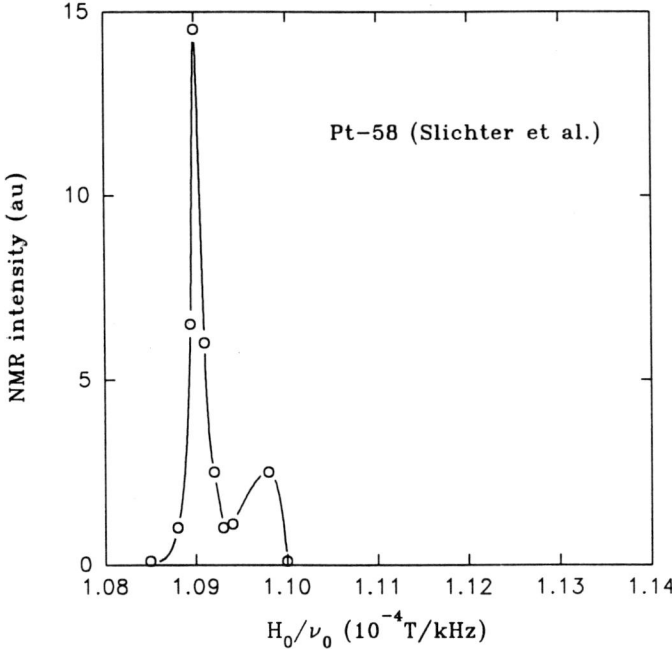

Fig. 1. The line shape of a typical Pt catalysts with a surface to volume ratio of 58% from the Illinois group. The typical particle size is a few nm.

The beat pattern appeared insensitive to the position within the line at which the measurement is performed. Because the observed beat frequency of 4 kHz is typical for Pt-alloys [33], this was seen as evidence for an almost constant (and metallic) density of states for the $6s$-electrons in their particles. Although the Knight shifts of the surface nuclei of the Pt catalysts used by Slichter *et al.* or the Pt particles prepared in Lausanne were almost zero, the nuclear relaxation times were rather short. In fact the values observed are in the ms-range, typical for a metal. For an explanation one has to realize that the positive contribution to K of the s-electrons has to be added to the negative one of the d-electrons and thus might cancel, while for the relaxation rates the various contributions always add up in the final rate. The overall picture that arises from these Pt-NMR experiments is that of particles that remain metallic down to the smallest radii (of about 20 Å). To describe the changes in the electron densities one has to introduce healing lenghts for the d-electrons. In addition the Illinois group favors a strong site dependence of the s-electron density.

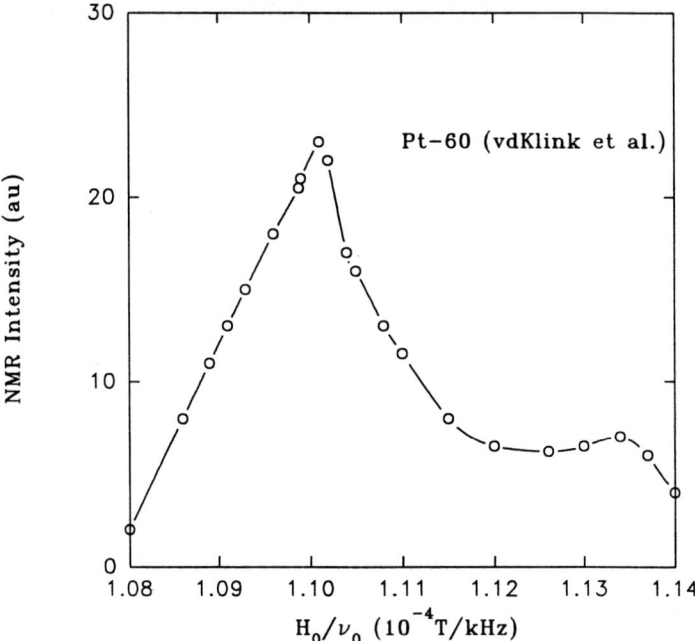

Fig. 2. Line shape of a naked Pt-cluster with a surface to volume ratio of 60% from the Lausanne group.

8.5. Aggregates of Metal Cluster Compounds – Experiment and Discussion

Although the physical properties of the small metal particles are supposed to depend strongly on their size, all the resonance data on naked clusters are, as far as we know, obtained on systems with a (small) distribution of particle sizes. With respect to this particle distribution problem, the metal cluster compounds are ideal candidates. The progress made in the synthesis of these materials now allows even a systematic study of physical properties as a function of particle sizes. Below, we will present representative data obtained via resonance on nuclei of the metal (Pt) core and on the ligand shell. For the latter we selected ^{31}P of the phosphine ligands used in the 55-clusters, which is directly bound to the surface of the core and ^{13}C-nuclei also present in the ligands of the larger cluster molecules. While the core resonance will reveal the electronic structure of the core itself, the ^{31}P and ^{13}C-NMR might be sensitive for inter-cluster charge transfer.

8.5.1. CORE-RESONANCE

We first consider the data obtained by ^{195}Pt NMR on $Pt_{309}Phen^{*}_{36}O_{30}$ [37, 38]. The 309 Pt atoms constituting the metal core have a cuboctahedral structure.

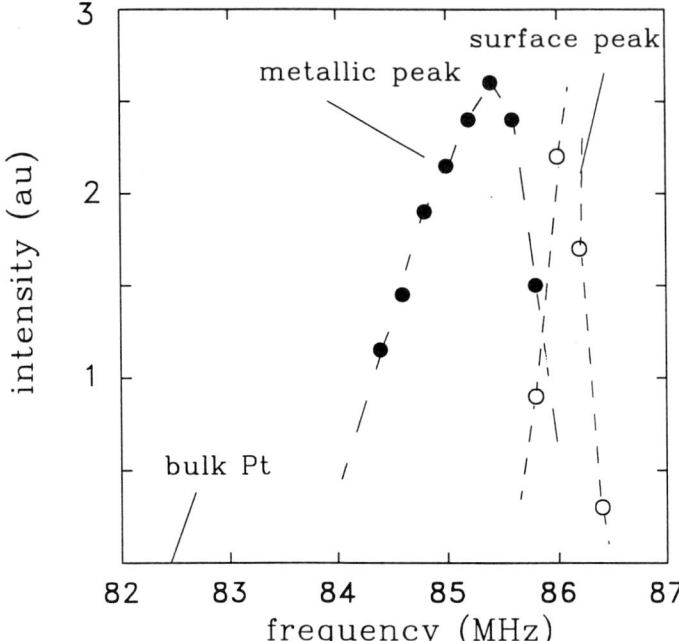

Fig. 3. ^{195}Pt NMR line shape at 77 K in a magnetic field strength of 9.39 T, measured in Pt$_{309}$Phen$^*_{36}$O$_{30}$. The Pt$_{309}$ core diameter is about 21 Å, with a surface to volume ratio of 110%. The decomposition of the line is based on the differences in relaxation rate. At least all 147 inner core Pt atoms with a diameter of about 16 Å seem to behave metallic, see text.

Using the molar volume of bulk Pt of 9.12 cm^3, the volume of the 309 core Pt-atoms can be represented by a sphere with a radius of 10.4 Å. The metal core is embedded in a dielectric matrix formed by the Phen* and oxygen molecules. The outer shell of the metal core has 162 atoms, the inner shells (from outside to inside) respectively 92, 42 and 12, surrounding a single atom in the centre. In metal cluster compounds with a core size below 55 atoms, Pronk et al. [34] performed measurements only at and below 4.2 K. The results found are consistent with those of Pt$_{309}$ and will not be further discussed here.

The NMR-experiments on Pt$_{309}$ were performed in the frequency range between 82–87 MHz. The applied magnetic field was 9.393 T. To cover the extremely inhomogeneously broadened resonance lines, the integrated spin-echo intensities taken with regular frequency intervals were put together [39]. Phase cycling was used to eliminate the effects of coherent noise. The typical pulse time for a 90° pulse was 2 μs. The delay time between the pulses was 30 μs. About 20 000 averages were taken per point. T_1-data were obtained by observing the recovery of the echo after a train of 90° pulses. In view of the non-exponential recovery at some line positions (initial time constant of a few ms followed by that of a few seconds) a logarithmic time increment above the usually employed linear time scale was preferred.

The line shape at 77 K as a function of the observation frequency ν is presented in Figure 3. The magnetic field strength was 9.39 T.

It clearly shows the absence of a bulk peak which would have been located at $\nu = 82.513$ MHz. Two separate peaks can be distinguished. The higher frequency peak at 86.0 MHz is located in the region where the chemical shift values of ^{195}Pt atoms in the insulating compounds are found. The full width at half height of the surface peak is about 0.4%. In view of its long T_1 the high frequency peak is attributed to the Pt-atoms on the core-surface, coordinated to ligand molecules. A similar observation is made by van der Klink et al. [8]. While hydrogen on a Pt surface has a minor effect on its metallicity, the presence of oxygen completely wipes out the metallic character. In the absence of delocalized electron spins the relaxation of the nuclei will be accomplished by localized paramagnetic spins (e.g. a free radical in one of the ligand molecules) or via nuclear spin diffusion. These processes are much less effective than those via conduction electrons and explain the reduced relaxation rate. (Note that for the small 'naked' platinum particles discussed above, the nuclei with $K = 0$, which are presumably also belonging to the surface atoms, T_1 is very fast. The short T_1 there is attributed to the cancellation of the positive Knight shift K_s, due to a non-zero density of states of the 6s-electrons, by the negative Knight shift K_d of the 5d-electrons (see Section 8.3.2). The relaxation rate is proportional to the electron densities i.e. to $K_s^2 + K_d^2$.

The lower frequency peak at 85 MHz, which will be discussed from now on, is attributed to platinum atoms in metallic environment. This is inferred from the short spin-lattice relaxation time, $T_1 T \approx 80$ msK, similar as for the Pt-catalysts and for bulk Pt. Down to about 65 K the recovery of the magnetization at all positions in the metallic line is exponential. Below 65 K the recovery becomes increasingly non-exponential with decreasing temperature. As an example the relaxation behavior at 4.2 K at 85.2 MHz is shown in Figure 4.

In order to account for the non-exponential decay, one has to assume a distribution of energy levels [38]. The form of the distribution function $P(\Delta)$ of the spacing Δ between two adjacent energy levels depends on the symmetry of the Hamiltonian. A random distribution of energy levels (Poisson distribution) gives

$$P(\Delta) = 1/\delta \, \exp(-\Delta/\delta) \qquad (16)$$

with δ the mean energy seperation between the energy levels.

The recovery of the magnetization is described by

$$M(t) \propto \int_0^\infty d\Delta P(\Delta)\{1 - \exp[-t/T_1(\Delta)]\} \qquad (17)$$

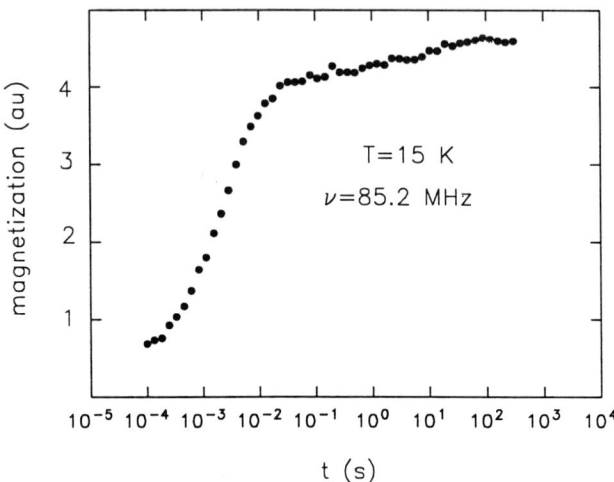

Fig. 4. Non-exponential spin-lattice relaxation rate of Pt in Pt_{309} at 15 K in a field of 9.33 T at a frequency of 85.2 MHz.

with $1/T_1(\Delta) \propto f(\varepsilon_0)[1 - f(\varepsilon_1)]W(\Delta) \propto e^{-\Delta/kT}$, in which $\Delta = \varepsilon_1 - \varepsilon_0$. $f(\varepsilon)$ is the Fermi–Dirac function. Because conservation of energy has to be fulfilled, the transition probability due to the electron-nucleus interaction, W, is assumed to depend (in first order) exponentially on Δ/kT. The Fermi level is assumed to be located between ε_0 and ε_1.

With the Poisson distribution, contrary to the orthogonal, unitary and simplectic distribution functions, the relatively rapid rise of the magnetization followed by a much slower increase can be simulated. For Pt_{309} additional evidence for a Poisson distribution is provided by specific heat data of the same sample. At low temperatures a linear contribution seems to be present [12]. From the distribution functions only that of Poisson can account for such a linear term. Furthermore, a value of about 40 K for δ was deduced, in fair agreement with the value of about 60 K obtained from the NMR experiments.

In Pt_{309} also the slow beats in the decay of the spin echo are observed, with a frequency of 4.8 kHz independent of the position in the metallic peak, see Figure 5. In 70%Au–30%Pt alloys a similar beat frequency of 5 kHz was found [33]. These beat patterns are characteristic for a distribution of Knight shifts between neighboring nuclei. T_2 does not follow the predictions of Equation (14).

As with the relaxation rates the lineshape can be explained in terms of electron spin densities. Since the Pt-atoms responsible for the high frequency peak do not demonstrate any metallic behavior, part of the Pt atoms have to be subtracted from the metal core. This yields a metallic particle whose volume is approximated by a sphere with a radius of 8 to 9 Å. From the location of the metallic peak it is concluded that the density of states is reduced to

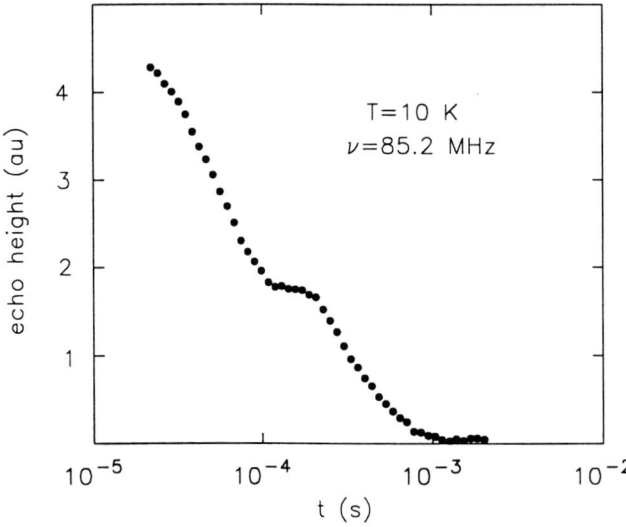

Fig. 5. Beats in the echo decay of ^{195}Pt in Pt$_{309}$ at 10 K in a field of 9.4 T.

a large extent with respect to bulk Pt. With a suitable choice of parameters the position and the width of the metallic peak can be reproduced by the model proposed by the Illinois group. Also the simpler model of Bucher and van der Klink [8], which has less parameters, fits the data. In Figures 6, 7 and 8 we show the results of our calculations for the Knight shifts, the line shape and the relaxation rates for two different scenarios. The drawn lines are obtained under the assumption that the s- and d-electron densities have a similar healing length, as in the Illinois model. In the calculation there is no other site-dependence in the s-electron density. If only the d-electron density is assumed to vary, with the s-electron density kept at the bulk value, the dashed lines result.

As can be seen from Figures 7 and 8, a good correspondence between experimental and calculated lineshapes and relaxation rates is already obtained if only the site dependence of the d electron density is taken into account.

The frequency of the beats in the echo-decay are independent of the line position. These slow beats arise from the exchange coupling between the nuclear spins, with a coupling constant that depends on the $6s$-LDOS. Consequently, the constant beat frequency is consistent with a constant $6s$-LDOS throughout the particle.

If we take the effect of the ligand atoms into account, the core experiment, as discussed here, shows that the about 200 fcc-packed inner-core Pt atoms of the Pt$_{309}$ cluster can be considered as metallic at room temperature. At about 50 K finite size effects become visible and are responsible for the non-

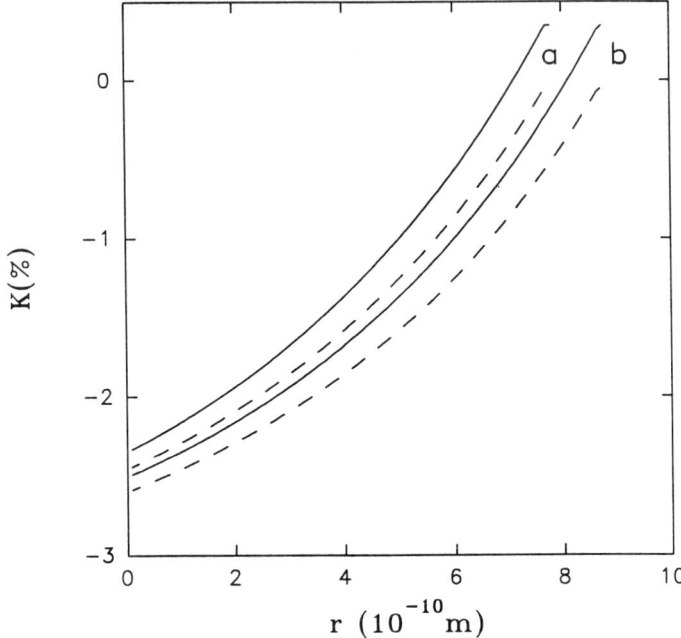

Fig. 6. The calculated distribution of Knight-shift values within a Pt-core with a diameter of 16 Å (set a) and 18 Å (set b). The drawn lines are obtained under the assumption that s- and d-electron densities have the same healing length. If only the d-electron density is assumed to vary, while the s-electron density is kept at its bulk value, the dashed lines result.

exponential magnetization recovery. We interpret this as an experimental observation of a transition from metallic to non-metallic behavior as the temperature decreases. Surface effects determine the line shape and the beats in the echo-decay (T_2).

8.5.2. LIGAND-SHELL RESONANCE

The Illinois-group [18] found that not only the relaxation time of the Pt-nuclei in the core, but even that of the ^{13}C of the carbonyl groups covering the Pt-particles can be described by the Korringa law. This shows that a non-zero conduction electron density exists at the ^{13}C-position. Apparently, such non-metallic 'surface probes' can also be used to investigate the nature of the metal-particle.

Here we first discuss the results of a NMR study on a phosphorous 'surface' atom of a representative member of the $M_{55}L_{12}Cl_x$ cluster compounds [11] (where M is a transition metal and L is an organic ligand molecule) [40]. For $Ru_{55}(P(t-Bu)_3)_{12}Cl_{20}$ the metal atom cores are coordinated partly by P-$(C_6H_3)_3$ groups that are bonded to the cluster via the phosphorus atom. ^{31}P is a suitable NMR nucleus, both because it has no quadrupolar moment

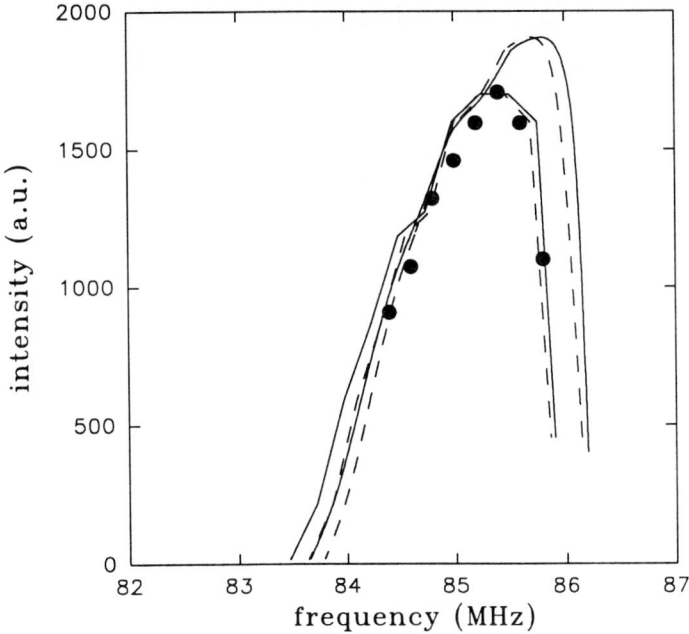

Fig. 7. The calculated lineshape and position for Pt-cores of a diameter of 16 and 18 Å. The smaller lines correspond to the particles with a 16 Å diameter. Dashed and drawn lines correspond to the two scenarios mentioned in Figure 6. Data (filled circles) are the same as in Figure 3.

(nuclear spin 1/2) and because of its relatively large NMR sensitivity. Also high resolution ^{13}C [41] data will be briefly mentioned.

^{31}*P-NMR*. Proton-decoupled ^{31}P CP/MAS NMR spectra at ambient temperature for the ligands in Au$_{55}$ have been recently measured [41]. The ^{31}P spectrum comprises a broad ($\sim$ 40 ppm or $\sim$ 6 kHz wide) feature at 50 ppm, with a small narrow component at 30 ppm. The latter is attributed to an impurity that is formed in the slow decomposition process of a metal cluster molecule [11], and the broad line is slightly shifted down-field from the ^{31}P resonance of PPh$_3$ in solution ($\sim$ −7 ppm). A hole burning experiment showed that the line is at least partly inhomogeneously broadened. The broad ^{31}P resonance provides evidence that the phosphorous is directly bonded to the metal core.

More insight in the binding to the metal core can be gained from the relaxation rate of these phosphorus nuclei. Figure 9 shows the ^{31}P longitudinal relaxation rate for the compound Ru$_{55}$(P(t-Bu)$_3$)$_{12}$Cl$_{20}$ as a function of T between 5 and 50 K at magnetic fields of 3.25 and 6.45 T. The relative accuracy of the data is estimated to be about 25%.

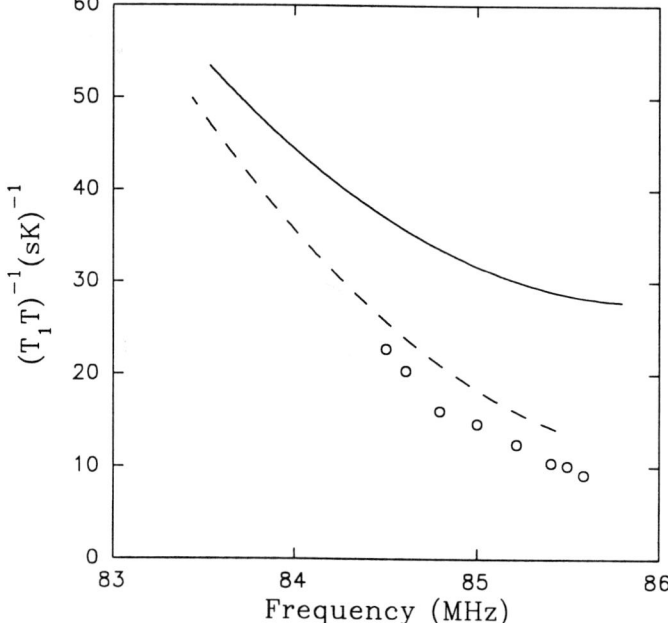

Fig. 8. The calculated and measured variation in $(T_1T)^{-1}$ as a function of frequency for the same two scenarios as in Figures 6 and 7. The particle diameter is fixed to 16 Å. In view of the uncertainties connected with the incorporation of the Stoner factor in the relaxation rates, the calculated relaxation rates shown are without Stoner enhancement. In the scenario where only the influence of the surface on the d-electron density is taken into account (dashed line), the experimental values (open circles) are reproduced reasonably well.

The combined temperature and field dependence of $1/T_1$ can be described as

$$1/T_1 \propto T^n B^{-m} \qquad (18)$$

with $m \approx 1.4$ and a slightly field dependent n of about 1.4 in a field of a few Tesla. A relaxation rate of the form of Equation (18) is rather unique and has only been observed in inorganic glasses at temperatures below about 100 K [42].

In these glasses the relaxation process is ascribed to an interaction between the nuclear spins and two-level (TLS) systems. Transitions between the TLS are possible by tunneling or hopping processes. The nuclear relaxation induced by the nucleus-TLS interactions is usually assumed to be a Raman process in which one TLS gets excited and another TLS gets deexcited. Because ^{31}P has no quadrupolar moment, the ^{31}P relaxation rate observed in the Ru$_{55}$-compound can only be due to magnetic interactions [42, 43]. The fit with Equation (18) suggests that the relaxation process is due to the interaction of the P-nuclei with electronic spins being involved in the two-level

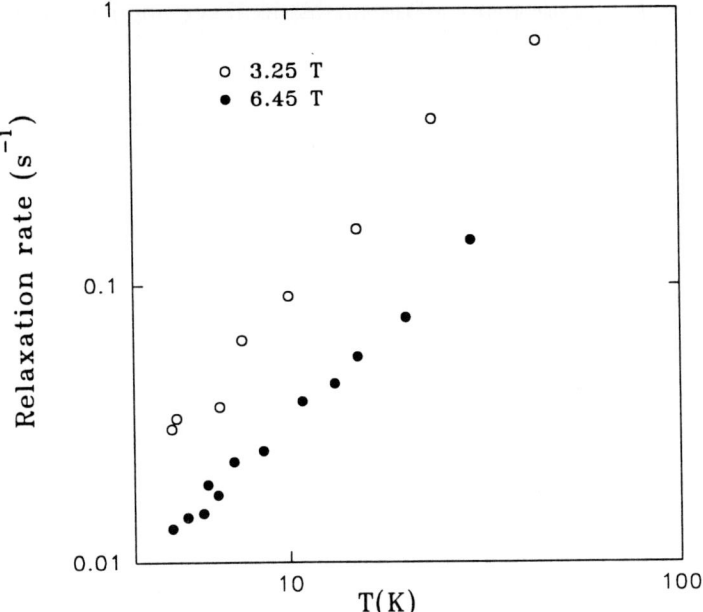

Fig. 9. ^{31}P relaxation rate in $Ru_{55}(P(t-Bu)_3)_{12}Cl_{20}$ as a function of T in 6.45 T and 3.25 T. Field- and temperature dependence of T_1 exclude a Korringa relaxation process.

system. It might well be possible that the relaxation channel is provided by the same carriers responsible for the the non-exponential relaxation in Pt_{309} and the observed inter-cluster hopping processes in the related cluster compounds Au_{55} and Pd_{561} (see below).

As far as the ^{13}C solid state NMR measurements on Au_{55} and Pt_{309} are concerned, the ligands behave like diamagnetic organic molecules, giving rise to relatively narrow lines. No pronounced Knight shifts or evidence of metallic-like relaxation were observed [41]. This result implies that there is hardly any unpaired spin density on the ligands.

8.5.3. RESULTS FROM OTHER EXPERIMENTS

Extensive studies of the frequency and temperature dependence of the conductivity have been performed on the related metal cluster compounds $Au_{55}(PPh_3)_{12}Cl_{16}$ and $Pd_{561}Phen_{36}O_{200}$. The observed strong suppression of the temperature dependence of the conductivity with higher frequency, and the scaling of the conductivity $\sigma(\omega)$ if expressed as function of $\omega/T\sigma(0)$ are consistent with the (almost) absence of a gap around the Fermi-energy. The here proposed Poisson distribution could also provide an explanation for these findings, as well as for the possible presence of a linear contribution in the specific heat data of the $Pd_{561}Phen_{36}O_{200}$ and $Pt_{309}Phen^*_{36}O_{30}$ compounds in addition to the intra- and inter-cluster phonon contributions.

However, the question why a Poisson distribution complies with the data is still an unresolved problem.

8.6. Summary

Using NMR as a nuclear probe for the electronic properties of small metal particles, metal clusters of only a few hundred atoms are found to be metallic, with important modifications due to surface effects. For the metal cluster compounds one needs to discriminate between the surface layer and the interior. If the interior is of the few hundred atom size, it behaves metallic, while the properties of the nuclei at the surface compare well to those in normal chemical compounds. In Pt-particles and clusters, which are used as a generic example, a gradual change of the d electron densities is sufficient to explain the room temperature linewidth and relaxation data. The low temperature T_1 data in Pt_{309}, which is equivalent to a naked metallic particle with a diameter of about 16 Å, are consistent with a Poisson distribution of the energy levels around the Fermi level. Also the data obtained by ligand NMR in metal cluster compounds with a core size of 55 metal atoms indicate an almost continuum of two level systems, that likely originate in the electronic level structure around the Fermi energy.

References

1. B. Mühlschlegel, *Z. Phys.* **D 20** (1991) 289.
2. I. Yu and W. P. Halperin, *J. Low Temp. Phys.* **45** (1981) 189.
3. W. P. Halperin, *Rev. Mod. Phys.* **58** (1986) 533.
4. J. J. van der Klink, J. Butter and M. Graetzel, *Phys. Rev.* **B 29** (1984) 6352.
5. H. E. Rhodes, P. K. Wang, H. T. Stokes, C. P. Slichter and J. H. Sinfelt, *Phys. Rev.* **B 26** (1982) 3359.
6. H. E. Rhodes, P. K. Wang, C. D. Makowka, S. L. Rudaz, H. T. Stokes, C. P. Slichter and J. H. Sinfelt, *Phys. Rev.* **B 26** (1982) 3569.
7. I. Yu, A. A. V. Gibson, E. R. Hunt and W. P. Halperin, *Phys. Rev. Lett.* **44** (1980) 348.
8. J. P. Bucher and J. J. van der Klink, *Phys. Rev.* **B 38** (1988) 11038.
9. J. P. Bucher, J. Buttet, J. J. van der Klink and M. Graetzel, *Surface Science* **214** (1989) 347.
10. J. J. van der Klink, in P. Jena *et al.* (eds), *Physics and Chemistry of Finite Systems: From Clusters to Crystals*, Vol. 1, Kluwer Academic Publishers, Dordrecht (1992), p. 537.
11. G. Schmid, *Structure and Bonding* **62** (1985) 51.
12. Proceedings of ISSPIC5, *Z. Phys.* **D 20** (1991).
13. W. Harbich, *J. Chem. Phys.* (1990).
14. D. Fenske *et al.*, *Angew. Chem. Int. Ed. Engl.* **32** (1993) 1303.
15. C. P. Slichter, *Surf. Sc.* **106** (1981) 373.
16. H. T. Stokes, H. E. Rhodes, P.-K. Wang, C. P. Slichter and J. H. Sinfelt, *Phys. Rev.* **B 26** (1982) 3575.
17. C. D. Makowka, C. P. Slichter and J. H. Sinfelt, *Phys. Rev.* **B 31** (1985) 5663.
18. J.-P. Ansermet, P.-K. Wang, C. P. Slichter and J. H. Sinfelt, *Phys. Rev.* **B 37** (1988) 1417.
19. M. Weinert and A. J. Freeman, *Phys. Rev.* **B 28** (1983) 2626.

20. N. W. Ashcroft and N. D. Mermin, *Solid State Physics*, Holt-Saunders Int. (1976), Chapter 2.
21. J. P. Bucher and J. J. van der Klink, *Helvetia Physica Acta* **61** (1988) 760.
22. J. Uppenbrink and D. J. Wales, *J. Chem. Phys.* **96** (1992) 8520.
23. J. P. Bucher and J. J. van de Klink, *J. Phys. Chem.* **94** (1990) 1209.
24. M. P. J. van Staveren, H. B. Brom and L. J. de Jongh, *Phys. Reports* **208** (1991) 1.
25. R. J. Elliott, *Phys. Rev.* **96** (1954) 266.
26. C. P. Slichter, *Principles of Magnetic Resonance*, Springer (1990).
27. A. Kawabata, *J. Phys. Soc. Jap.* **29** (1970) 902.
28. G. G. Khaliullin and M. G. Khusainov, *Sov. Phys. JETP* **67** (1988) 524.
29. J. Korringa, *Physica* **16** (1950) 601.
30. L. I. Schiff, *Quantum Mechanics*, 3rd edition, McGraww-Hill, New York (1968), pp. 83–88.
31. T. Moriya, *J. Phys. Soc. Jap.* **18** (1963) 516.
32. A. Abragam, *The Principles of Nuclear Magnetism*, Oxford University Press (1961). In paperback since 1983.
33. C. Froidevaux and M. Weger, *Phys. Rev. Lett.* **12** (1964) 123.
34. B. J. Pronk, H. B. Brom, A. Ceriotti and G. Longoni, *Sol. State Comm.* **64** (1987) 7.
35. T. Goto, F. Komori and S. Kobayashi, *J. Phys. Soc. Jap.* **58** (1989) 3788. H. Goto, S. Katsumoto, S. Kobayashi, *J. Phys. Soc. Jap.* **62** (1993) 1439.
36. D. A. Gordon, R. F. Marzke and W. S. Glaunsinger, *J. de Phys.* **38** C2 (1977) 87.
37. D. van der Putten, H. B. Brom, L. J. de Jongh and G. Schmid, in P. Jena et al. (eds.), *Physics and Chemistry of Finite Systems: From Clusters to Crystals*, Kluwer Academic Publishers, Dordrecht (1992), Vol. 2, p. 1007.
38. D. van der Putten, H. B. Brom, J. Witteveen, L. J. de Jongh and G. Schmid, *Z. Phys.* **D 26** (1993) S 21.
39. F. Hentsch, N. Winzek, M. Mehring, H. J. Mattausch, A. Simon and R. Kremer, *Physica* **C 165** (1990) 485.
40. M. P. J. van Staveren, H. B. Brom, L. J. de Jongh and G. Schmid, *Z. Phys.* **D 12** (1989) 451.
41. A. C. Kolbert, H. J. M. de Groot, D. van der Putten, H. B. Brom, L. J. de Jongh and G. Schmid, *Z. Phys.* **D 26** (1993) S 24.
42. G. Balzer-Jöllenbeck, O. Kanert and J. Steinert, *Sol. State Comm.* **65** (1988) 303.
43. M. Rubinstein, H. A. Resing, T. L. Reinecke and K. L. Ngai, *Phys. Rev. Lett.* **34** (1975) 1444.

9. MAGNETIC PROPERTIES AND UV-VISIBLE SPECTROSCOPIC STUDIES OF METAL CLUSTER COMPOUNDS

R. E. BENFIELD
*Chemical Laboratory, University of Kent,
Canterbury CT2 7NH, U.K.*

9.1. Introduction

Molecular metal cluster compounds have been known for many years; the essential features of the structure of $Fe_2(CO)_9$ were recognised during the 1930's [1]. The first crystallographic characterisation of a metal-metal bond unsupported by bridging ligands was achieved more than thirty years ago [2]. With the development of single-crystal X-ray diffraction as a powerful routine method, very many metal cluster compounds were structurally characterised, including many with interstitial main group atoms. The larger metal carbonyl clusters were found to resemble fragments of bulk metallic lattices (fcc, hcp, etc.) [3]. During the same time, it became apparent from theoretical and experimental (electron microscopy) studies that small, ligand-free metal particles could have geometries not available to continuous three-dimensional lattices: icosahedral, polytetrahedral, etc. [4, 5, 6, 7, 8]. Examples of these types of geometry were also then found among molecular cluster compounds, notably the extensive series of gold clusters based on icosahedral geometry [9].

Before 1980, the study of metal cluster compounds concentrated almost exclusively on their synthesis and structure. Detailed comparisons of metal-metal distances in clusters and bulk metals [3, 10] were made, and structural and bonding parallels drawn between ligands in metal clusters and molecules adsorbed on metal surfaces [11] – the 'cluster-surface analogy'. The aim was to exploit catalytic properties of cluster compounds. Electron-pair counting theories were evolved, to account for the cluster geometries in terms of their skeletal electron pair counts [12, 13].

However, almost no attention was paid to the physical properties of the metal cluster molecules. Only simple general ideas, such as the trend towards darker colours with increasing cluster size, had been noted [14]. As late as 1981, a major review on metal carbide clusters could include only a single paragraph on their physical properties, "because the requisite physical studies have not been made" [15].

The enormous electronic differences between low-nuclearity cluster molecules and bulk metals were well-recognised [10], but a general assumption was that a smooth transition from 'molecular' towards 'bulk' properties would occur with increasing cluster size. There was little experimental evidence either for or against this assumption. It had not been realised that these molecular compounds, representing arrays of exactly monodisperse metal clusters, of known size and geometry, separated from one another by envelopes of ligands [16], could be used as an ideal model system to study the important question: how many atoms are necessary in a metal particle to give bulk-like properties?

This was the situation when we commenced our physical studies of metal cluster compounds at the end of 1980 with experiments in Cambridge on the electron paramagnetic resonance (EPR) and magnetism of osmium carbonyl clusters. Chronologically, the first experiment was the observation of EPR from solid $[N(PPh_3)_2]_2^+[Os_{10}C(CO)_{24}]^{2-}$ at 77 K, but in this review it is more logical to discuss first the magnetic susceptibility measurements made during 1981–1983.

9.2. Magnetic Properties

9.2.1. Magnetic Susceptibility of Low-Nuclearity Clusters

The system chosen for this study was a series of electrically neutral osmium carbonyl compounds, containing between three and ten metal atoms. These were available from the synthetic group at Cambridge of Jack Lewis and Brian F. G. Johnson [17]. Beyond the general confirmation that most compounds of this type were diamagnetic and could therefore be taken to have a closed-shell electron configuration, magnetic susceptibility studies had not been made at all on these materials. The magnetic susceptibility seemed an ideal probe of whether these metal clusters indeed showed any trend towards metallic rather than simple molecular character [18, 19]. We were able to interest Prof. Mike Sienko in our ideas, and through him gained access to a very sensitive, computer-interfaced Faraday balance at Cornell. The results from these measurements, made over the temperature range 1.5–300 K, [20, 21] are summarised in Table 1.

As expected, the low-nuclearity clusters such as $Os_3(CO)_{12}$ and $H_4Os_4(CO)_{12}$ were found to be diamagnetic, with susceptibility independent of temperature; the larger cluster $H_2Os_{10}C(CO)_{24}$ was also diamagnetic at room temperature. In the high-temperature limit, the magnitudes of the diamagnetic susceptibilities increased with the number of metal atoms in the cluster, in accordance with the increased number of electrons present.

On subtracting the diamagnetic contributions of the individual atomic cores, using appropriate values for Pascal's constants for Os, C, O, H and I atoms, a residual temperature-independent paramagnetic component was

TABLE 1

Magnetic susceptibility data for osmium carbonyl clusters at 298 K [21].

Cluster	χ_m(obs) (10^6 emu mol^{-1})	χ_{ems} (10^6 emu mol^{-1})
$Os_3(CO)_{12}$	-170 ± 10	+146 ± 10
$H_2Os_3(CO)_{10}$	-188 ± 10	+114 ± 10
$H_4Os_4(CO)_{12}$	-200 ± 10	+219 ± 10
$Os_7(CO)_{21}$	-405 ± 20	+263 ± 20
$H_2Os_{10}C(CO)_{24}$	-580 ± 30	+326 ± 30
$Os_{10}C(CO)_{24}I_2$	-700 ± 40	+304 ± 40

revealed. The magnitude of this component, termed the *excess molar susceptibility*, χ_{ems}, increased with cluster nuclearity [21].

In the original analysis of these results [21], this excess molar susceptibility was interpreted as a Van Vleck temperature-independent paramagnetism resulting from the interaction of ground and excited states in the applied magnetic field. It was argued that the higher values of χ_{ems} in the larger clusters directly showed their smaller frontier orbital separation and were indicative of the development of an electronic band structure in these molecules. For $Os_3(CO)_{12}$, the HOMO-LUMO gap was estimated, using reasonable estimates of the relevant overlap integrals, to be about 3 eV. This is consistent with the UV-visible electronic spectrum of this cluster (see below), and with molecular orbital calculations. The frontier orbital structure of $Os_3(CO)_{12}$ is shown in Figure 1 [22]. For $H_2Os_{10}C(CO)_{24}$, the HOMO-LUMO gap was estimated to be only 1.0–1.5 eV.

As an additional point of interest, estimates of frontier orbital separation made by this magnetic method complement those made by spectroscopic measurements, because different selection rules apply [21]; see Figure 1.

Recently, similar ideas have been used to separate the diamagnetic and Van Vleck paramagnetic components of the magnetic susceptibility of C_{60}. As in the case of $Os_3(CO)_{12}$, the HOMO-LUMO transition in C_{60} is optically forbidden but magnetically allowed [25].

One difficulty encountered in [21] was the choice of appropriate values of Pascal's constants for the diamagnetic corrections. The value of -66×10^{-6} e.m.u./mol was used for the osmium atoms, this being the best available estimate of the appropriate value for osmium atoms in a (near) zero oxidation state. For the carbonyl ligands, it was not clear to what extent constitutive corrections should be made to the standard values for carbon and oxygen atoms, to take account of the different diamagnetic response of the π-electrons

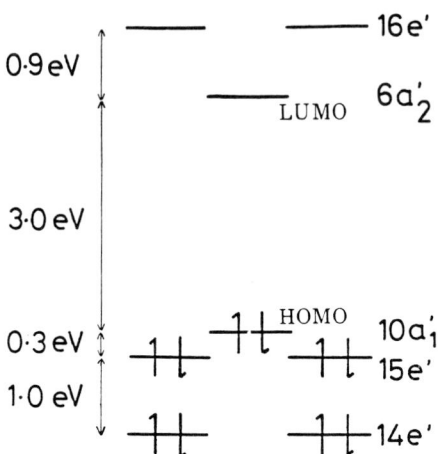

Fig. 1. Frontier orbital structure of $Os_3(CO)_{12}$ [22]. Constructed from results of X-alpha calculations [23], with energy spacings adjusted to give better agreement with electronic spectra [24] and other experimental measurements. The lowest energy allowed transitions in point group D_{3h} are: spectroscopic, $10a_1' \rightarrow 16e'$ and $15e' \rightarrow 6a_2'$; magnetic, $10a_1' \rightarrow 6a_2'$ (HOMO-LUMO).

in the carbon-oxygen multiple bonds, or whether the experimentally known value for the diamagnetic susceptibility of free gaseous carbon monoxide would be more appropriate. But the existence of large values of χ_{ems} in these molecules, and their correlation with cluster size, could not be eliminated by any reasonable choice of values for the Pascal's corrections.

Kimura and Bandow [26] related these results to their own measurements of the magnetic susceptibility of small (less than 100 Å in diameter) magnesium particles. They found that the temperature-independent paramagnetism is enhanced for smaller particles relative to the bulk metals. In fact, if the excess molar susceptibility χ_{ems} of the osmium carbonyl clusters is divided by the number of metal atoms per cluster to give the atom-normalised molar susceptibility χ_{ems}, the maximum value of χ_{ems} is obtained for the tetranuclear cluster $H_4Os_4(CO)_{12}$. They related the Van Vleck temperature-independent paramagnetism of the osmium cluster compounds to the orbital paramagnetism of the magnesium particles.

Recently, the analysis of the magnetic measurements discussed in [21] has been re-evaluated by Van Ruitenbeek and Van Leeuwen, who have considered the size effect of the orbital magnetic susceptibility of small metallic systems [27, 28]. They note that the Larmor diamagnetic and Van Vleck paramagnetic contributions cannot be separated in a such straightforward manner as was done in [21]. If this separation procedure is not applied, then no size dependence of the orbital susceptibility of the osmium clusters is found. The magnetism of the osmium valence electrons in the clusters is small, and the atomic diamagnetism is suppressed when the atoms are bound together to form a cluster.

Thus, our original interpretation of the magnetic measurements on these cluster compounds may need to be re-assessed.

9.2.2. MAGNETIC SUSCEPTIBILITY OF HIGH-NUCLEARITY CLUSTERS

Although the magnetic susceptibilities of the smaller osmium clusters were independent of temperature, the cluster $H_2Os_{10}C(CO)_{24}$ was found to behave differently [20, 21]. Below about 70 K, a temperature-dependent paramagnetic susceptibility was observed, conforming almost exactly with the simple Curie law, with a calculated magnetic moment of 0.62 Bohr Magnetons per cluster molecule. This paramagnetism is 'anomalous', because $H_2Os_{10}C(CO)_{24}$ is an even-electron molecule, with a structure conforming to that predicted on the basis of its skeletal electron pair count [12, 13], and so would have been expected to have a ground state electronic structure with all electrons paired.

The magnetic susceptibility of this decanuclear cluster was independent of the applied field strength up to 1.2 Tesla, demonstrating that the paramagnetism did not arise from ferromagnetic impurities, and this was confirmed by chemical analysis proving elements such as Fe and Co to be present in amounts below 10 ppm [20].

The paramagnetism of $H_2Os_{10}C(CO)_{24}$ is quite distinct from the behaviour of the lower-nuclearity clusters, which are diamagnetic with a temperature-independent paramagnetic susceptibility component, and from the Pauli paramagnetic susceptibility (also ideally temperature-independent) of bulk metals, which arises from the presence of itinerant electrons in fully developed metallic conduction bands. The term 'metametallic' was later introduced to describe distinctive properties of this type in metal clusters [29, 30, 31, 32]; the term 'protometallic' would probably have given a clearer idea of the concept.

This 'anomalous' weak low-temperature paramagnetism in $H_2Os_{10}C(CO)_{24}$ was taken as evidence [20] that a molecular cluster of ten osmium atoms could have magnetic properties similar to those of particulate metals, and displaying quantum-size effects [33, 34, 35, 36]. Similar Curie-type behaviour, also interpreted as arising from a quantum-size effect, had been reported in the previous few years from small particles of even-electron metals such as 12 Å particles of magnesium [37] and 22 Å particles of platinum [38].

At this stage, we had also commenced the use of EPR spectroscopy to study the paramagnetism in decanuclear osmium carbonyl clusters; the results of this work are summarised in the next section.

There was no reason to expect that this 'anomalous' paramagnetism in metal cluster compounds would be unique to osmium carbonyl clusters, and similar Curie-type paramagnetism was soon found in several even-electron carbonyl clusters of nickel [39] and platinum [39, 40]. These compounds were synthesised in the Milan laboratories of Giuliano Longoni and Alessandro Ceriotti. In all the osmium and platinum compounds, the magnetic moments observed per cluster molecule were low (1 Bohr Magneton or below), but larger moments (up to 8 Bohr Magnetons) were observed in the nickel carbonyl clusters [39].

The important question to be addressed was the origin of the paramagnetism in these high-nuclearity cluster compounds.

Molecular orbital calculations on the clusters were little help at this stage. Although in principle they could have shown the buildup of electronic band structure and the consequent physical properties, they were not powerful enough to deal with strongly spin-orbit-coupled multielectron systems. As an example, the extended Hückel method, which had been relatively successful for molecular clusters [41, 42], predicted an incorrect electron count for $H_2Os_{10}C(CO)_{24}$. The energy level separations and magnetic properties of this cluster clearly could not be predicted by this type of approach. Only later, with the development of computational methods such as local density functional theory, could the problem be approached in this way [43].

The explanation proposed at Leiden and Kent for the 'anomalous' paramagnetism [29, 30, 31, 32, 39, 44] was based on an analogy with the well-known magnetic properties of octahedral complexes of transition metal ions. The small metal clusters such as $Os_3(CO)_{12}$ have large frontier orbital (HOMO-LUMO) separations. For $Os_3(CO)_{12}$, the HOMO-LUMO gap was well-characterised by UV-visible spectroscopy and theoretical molecular orbital calculations to be about 3 eV. This is comparable with the frontier orbital separation in typical octahedral complexes of the Co^{3+} ion ($3d^6$), such as $[Co(NH_3)_6]^{3+}$. These complexes have a spin-paired t_{2g}^6 configuration and are diamagnetic with a Van Vleck temperature-independent paramagnetic susceptibility component [45]. However, as the cluster size increases, the frontier orbital separation decreases, as shown by the changes in the electronic (UV-visible-NIR) absorption spectra of the clusters (see below). When the HOMO-LUMO gap is sufficiently small, there must occur, in accordance with Hund's rules, a transition to a high-spin ground state electronic configuration; exactly as in the case of octahedral transition metal ion coordination complexes ML_6 [46].

The critical HOMO-LUMO separation will depend on the electron exchange energies and the pairing energy, i.e. the repulsion energy between two electrons spin-paired in the same molecular orbital [46, 47]. If the relevant molecular orbitals span more than one metal atom, the magnitude of the electron pairing energy will be more difficult to predict than in a simple ML_6 complex. The UV-visible spectra of close-packed decanuclear osmium clusters show clearly that the HOMO-LUMO energy gaps are in the range 1 eV or below. This is smaller than typical critical values for the transition to a high-spin ground state in coordination complexes [46, 48]. For example, $[CoF_6]^{3-}$, with a frontier orbital separation of 1.6 eV, is paramagnetic with a high-spin

TABLE 2

Magnetic behaviour of metal clusters, metal particles and bulk metals.

METAL	Low-nuclearity molecular cluster	High-nuclearity molecular cluster	Small metal particle	Bulk metal
Pt ; Os	diamagnetic/TIP	weak Curie paramagnetic	paramagnetic	Pauli paramagnetic
Ni	diamagnetic/TIP	strong Curie paramagnetic	superparamagnetic	ferromagnetic

$t_{2g}^4 e_g^2$ configuration [48]. Thus, in cluster compounds such as $H_2Os_{10}C(CO)_{24}$, a high-spin, paramagnetic electronic configuration is understandable.

The resulting magnetic properties are well understood in ML_6 complexes, where the unpaired electrons are essentially localised in d atomic orbitals on the central metal atom. However, the picture in metal cluster molecules is much more complex, because the degree of hybridisation of the metal atomic orbitals and the extent of electron delocalisation are not known. One effect of electron delocalisation might be to lower the observed paramagnetic moments to well below those expected classically for molecules with two unpaired electrons.

Further magnetic susceptibility measurements were made by the Cambridge group on a wider range of osmium cluster carbonyls, including compounds with decanuclear cluster units linked together to give up to 40 osmium atoms in a cluster. All the clusters with ten or more osmium atoms were found to be paramagnetic, showing a temperature-dependent susceptibility following Curie-law behaviour, with magnetic moments of the order of 1 Bohr Magneton [49].

In much higher-nuclearity cluster compounds, the molecular orbitals can be expected to form a quasi-continuous band structure with a very small HOMO-LUMO gap. This would resemble the band structure of a bulk metal sufficiently closely to lead to Pauli paramagnetism, causing a reversion to a temperature-independent magnetic susceptibility. For a discussion of the implication of these ideas within a molecular orbital framework, see [49].

It is quite reasonable for the critical nuclearity for the low-spin to high-spin transition to be different in different metals. In platinum carbonyl clusters, the onset of paramagnetism was found to occur at a nuclearity of six metal atoms [40]. Paramagnetic moments have been reported in ruthenium carbonyl clusters containing as few as four metal atoms [50].

The observation that the magnetic moments of the nickel carbonyl clusters are larger than those of the platinum and osmium compounds correlates well with the magnetic properties of the bulk and particulate metals (Table 2).

However, the more recent analysis by Van Ruitenbeek and Van Leeuwen of the size effect of the orbital magnetic susceptibility of small metallic systems [27, 28] has challenged this concept of a transition to a high-spin electronic configuration in large clusters. Instead, the atomic-like properties are predicted to evolve gradually towards a diamagnetism close to the bulk Landau value, with no paramagnetism appearing in the intermediate size regime. Although this work is based on the concept of free (non-interacting) electrons confined to potential wells of varying size and shape, thus ignoring correlations or spin-orbit coupling, it may require a new explanation of the observed paramagnetism of large cluster compounds.

One alternative explanation of the 'anomalous' paramagnetism found in the high-nuclearity clusters in the solid state is that disproportionation (electrical charge separation) might occur, to give a small population of odd-electron 'defect' species. This idea was originally ruled out [29], because it seemed less likely in cluster compounds than in the case of metal particles, in which electrostatic energy considerations are believed to favour charge neutrality [33, 51]. However, in view of the extensive measurements of electrical conductivity that have been made on these compounds [52], and the possibility of charge-transfer behaviour under UV-visible irradiation [53, 54] (see below), this idea may have to be reconsidered. Limited disproportionation would be consistent with the low values of the observed magnetic moments in the clusters.

Surface oxidation of the cluster samples may also be responsible for some of the observed paramagnetism [55], and more work is necessary to evaluate the importance of this effect.

9.2.3. EPR Studies of Decanuclear Osmium Clusters

The paramagnetism of $H_2Os_{10}C(CO)_{24}$ in the solid state was studied in more detail by EPR spectroscopy over the temperature range 4–300 K (Figures 2 and 3). The temperature-dependence of the EPR parameters gave considerable insight into the electronic structure of this cluster molecule [29, 30]. EPR also offered the prospect of helping to characterise the orbitals containing the spin density in more detail than could be achieved by static susceptibility measurements.

At temperatures below 100 K, a single resonance with Lorentzian lineshape was observed from polycrystalline $H_2Os_{10}C(CO)_{24}$ (Figure 2a). This was without any resolvable hyperfine splitting to nuclei such as ^1H or ^{189}Os. The g-factor of 2.280 ± 0.005 confirmed the strong orbital component expected in a cluster of low-valent $5d$ metal atoms [56]; unfortunately, there are no literature reports of EPR in colloidal osmium particles, which would have allowed an interesting comparison of g-value to be made.

The integrated intensity of the resonance followed the Curie law between 20 and 100 K (Figure 3). The linewidth varied linearly from 36 ± 1 G at 20 K to 94 ± 2 G at 120 K; taking the observed resonances to be homogeneously broadened with linewidths inversely proportional to the electronic spin-spin relaxation time T_2, this corresponds to a T_2 of the order of 10^{-9} sec. This

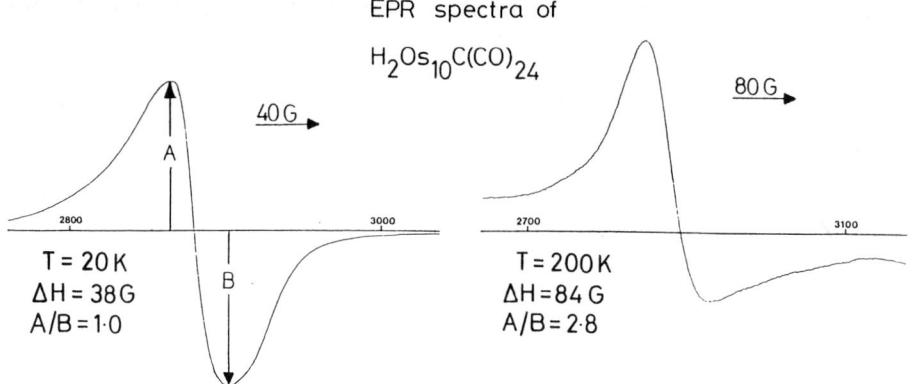

Fig. 2. X-band EPR spectra of polycrystalline $H_2Os_{10}C(CO)_{24}$ at 5 mW microwave power [29, 30].

is a much longer relaxation time than in bulk osmium metal, where T_2 can be estimated from the resistivity relaxation rate to be about 5×10^{-15} sec [21]. Clearly, the EPR from the conduction electrons in bulk osmium metal would have an extremely large linewidth and be observable only at very low temperatures; this has not yet been achieved experimentally.

This quenching of the electronic relaxation and narrowing of the EPR line in $H_2Os_{10}C(CO)_{24}$ compared to bulk osmium can be simply understood in terms of our view of the Os_{10} clusters as small metametallic particles with discrete energy levels; electronic spin transitions between these discrete energy levels are restricted in comparison to those in a bulk metallic sample where the electronic energy levels are continuous [56]. This gives a clear example of a quantum-size effect in the metal clusters [33, 34, 35, 36]. A similar narrowing of the EPR line in small silver particles has been reported [57].

Below 20 K, saturation effects at moderate microwave powers permitted electronic spin-lattice relaxation times T_1 in $H_2Os_{10}C(CO)_{24}$ to be measured; at 12 K, T_1 was about 10^{-3} sec. This further illustrates the difference in nature between the Os_{10} cluster and a bulk metal; in fully metallic cubic-symmetry systems, the two electronic relaxation times T_1 and T_2 are usually taken to be equal. Other high-nuclearity osmium carbonyl clusters were found to have electronic relaxation times T_1 and T_2 similar to those of $H_2Os_{10}C(CO)_{24}$ [58].

A very interesting observation was the progressive development of asymmetry in the EPR lineshape of solid $H_2Os_{10}C(CO)_{24}$ at temperatures above 100 K (Figure 2b). The parameter A/B, the ratio of the low- and high-field derivative peak amplitudes, reached 2.8 at 200 K, resembling the Dysonian lineshape associated with electrically conducting samples thicker than the microwave skindepth [59, 60]. In metal particles, the lineshape change from Lorentzian towards Dysonian takes place as the temperature is *lowered*

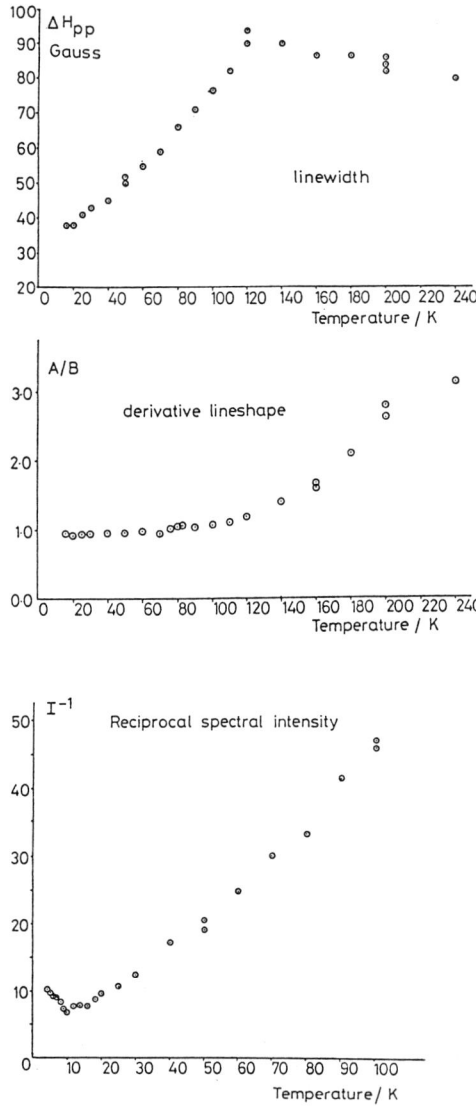

Fig. 3. Temperature-dependence of EPR spectral parameters of solid $H_2Os_{10}C(CO)_{24}$ [30].

[61]. Neither could the lineshape change in $H_2Os_{10}C(CO)_{24}$ be interpreted as showing an increase in electrical conductivity at higher temperatures, typical of semiconductors, because the EPR resonance reduced in intensity at higher temperatures. This is the opposite behaviour to that expected in semiconductors.

The probable cause of the asymmetric EPR lineshape above 100 K is intramolecular motion of the hydride ligands in the cluster. This perturbs the dielectric behaviour of the solid through localised charge transport, mixing the absorption and dispersion components of the Lorentzian derivative lineshape. Rapid molecular rearrangements of this type, with activation energies between 20 and 100 kJ/mol, are well known in metal carbonyl compounds [62]. In support of this idea, a similar EPR lineshape asymmetry has been observed in other Os_{10} clusters containing hydride ligands [63], but not in clusters without hydrides, such as $[Os_{10}C(CO)_{24}]^{2-}$ as its polycrystalline $[N(PPh_3)_2]^+$ salt [30].

At temperatures above 120 K, the behaviour of other parameters of the EPR spectrum of $H_2Os_{10}C(CO)_{24}$ also varied in a different way to that below 100 K [29, 30]. The linewidth reduced slightly, to 85 ± 4 G at 130 K, and then became independent of temperature to within ± 5 G up to 300 K (Figure 3). The spectral intensity no longer obeyed the Curie law, but reduced more than given by a simple reciprocal temperature-dependence; plots of reciprocal intensity against temperature followed the Curie–Weiss law with an extrapolated intercept on the temperature axis of 95 ± 10 K. These observations were not fully interpreted. The effect of the changing EPR lineshape on the other spectral parameters would have to be borne in mind in any fuller analysis of this Curie–Weiss behaviour.

We also observed EPR in several other decanuclear osmium carbonyl clusters in the solid state [30]. Polycrystalline $[Os_{10}C(CO)_{24}]^{2-}[N(PPh_3)_2]_2^+$ showed Curie-law EPR intensity variation up to 100 K. In contrast to the single resonance of $H_2Os_{10}C(CO)_{24}$, the lower crystal symmetry gave rise to three principal g-values. The central g-value of 1.92 was enormously shifted from the 2.280 of the isoelectronic parent compound. This observation was never satisfactorily explained, but the removal of the two protons from the decaosmium cluster may greatly perturb its frontier orbital pattern. As discussed above, no EPR lineshape asymmetry was observed from this cluster at higher temperatures.

An example of an axially symmetric substituted cluster is $[Os_{10}C(CO)_{24}AuPPh_3]^-[PPh_3Me]^+$. This material in polycrystalline form gave EPR below 50 K with $g_\perp = 2.28$ and $g_\parallel = 2.44$, consistent with the axial symmetry of the cluster (Figure 4).

The polycrystalline cluster $[Os_{10}C(CO)_{24}I]^-[N(PPh_3)_2]^+$ gave a resonance at $g = 2.14$, showing a six-line pattern characteristic of hyperfine coupling to a nucleus with $I = 5/2$. Although this was thought at the time to arise from the presence of manganese contamination, the possibility remains that this was really hyperfine couping to the ^{127}I nucleus, implying substantial spin density localisation on this atom.

Very interesting EPR spectra were obtained from solid $Os_{10}S_2(CO)_{23}$. At 150 K, a symmetric resonance was observed at $g = 2.23$, with linewidth 110 G. On cooling through 120–100 K, the linewidth diverged and there

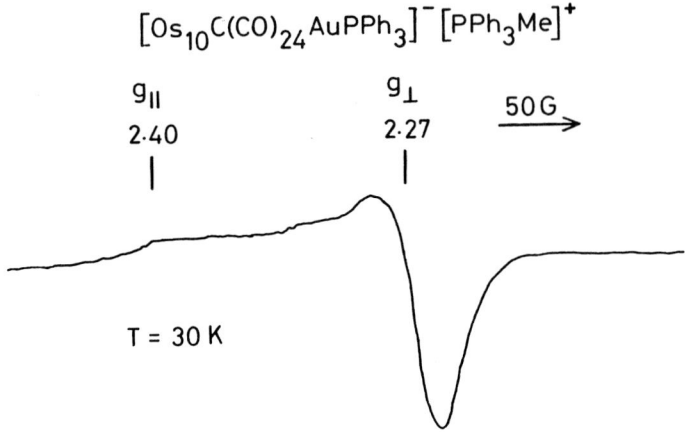

Fig. 4. EPR spectrum of polycrystalline $[Os_{10}C(CO)_{24}AuPPh_3]^-[PPh_3Me]^+$, showing axial symmetry.

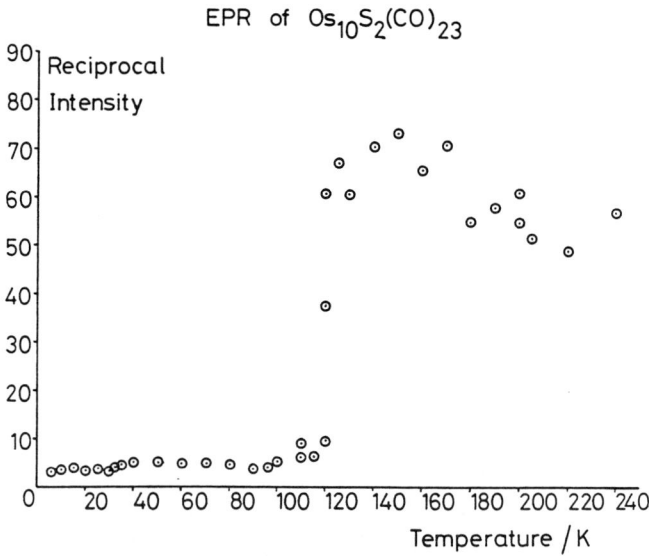

Fig. 5. Temperature-dependence of the intensity of the EPR spectrum of solid $Os_{10}S_2(CO)_{23}$. The data suggest a phase change at 120 K.

was a sharp increase in spectral intensity, giving evidence of a phase change (Figure 5). Very little attention has been paid to this aspect of the crystal chemistry of metal carbonyl clusters. On cooling below 100 K, the resonance moved to lower field, reaching $g = 3.0$ at 25 K [30]. This behaviour resembles that of some antiferromagnetic compounds [64]. Crystals of $Os_{10}S_2(CO)_{23}$ contain a high concentration of osmium atoms. The two sulphur atoms bonded to the cluster create 'holes' in the envelope of carbonyl ligands which would

TABLE 3

EPR data for osmium carbonyl clusters [49].

Cluster	$g_\perp$	$g_\|$
$Os_{10}C(CO)_{24}I_2$	2.12	
$Os_{10}C(CO)_{24}(AuPMe_2Ph)_2$	2.18	2.21
$Os_{10}C(CO)_{24}[Cu(NCMe)]_2$	2.11	2.40
$H_6Os_{10}(CO)_{24}$	2.30	
$H_2Os_{11}C(CO)_{27}$	2.09	2.31
$Os_{11}C(CO)_{27}[Cu(NCMe)]_2$	2.11	2.37
$[Os_{20}Hg(C)_2(CO)_{48}]^{2-}$	2.08	2.50
$H_2Os_{20}Hg(C)_2(CO)_{48}$	2.19	2.42
$[Os_{48}Hg_3(C)_4(CO)_{96}]^{2-}$	2.18	

otherwise completely surround each cluster molecule [16]. This provides a potential pathway for magnetic interactions between clusters, which are not possible in the other Os_{10} compounds.

The EPR of several other high-nuclearity osmium carbonyl clusters, containing up to 40 osmium atoms as well as atoms of metals such as copper, gold, and mercury, have also been reported by the Cambridge group [49]. The g-values of these compounds are given in Table 3. All are significantly shifted from free-spin, and separate perpendicular and parallel g-values were resolved from clusters with axial or other non-cubic symmetries.

It has also been found that the decanuclear clusters $[Os_{10}C(CO)_{24}]^{2-}$ and $[H_4Os_{10}(CO)_{24}]^{2-}$ have an extensive redox chemistry in solution, and that several redox states are accessible by chemical and electrochemical methods. This supports the idea that these cluster molecules have closely spaced electronic energy levels in a developing band structure. The EPR g-values of some of the odd-electron cluster radicals derived from these compounds, recorded at low temperatures in frozen solutions [65], are almost identical to the 2.280 of $H_2Os_{10}C(CO)_{24}$.

It is worth drawing attention to the cluster $Os_{10}C(CO)_{24}I_2$. We did not find any EPR signal from solid samples of this cluster, and our magnetic susceptibility measurements showed it to be diamagnetic at all temperatures [21]. This is understandable in terms of its structure. The effect of the two iodine atoms is to open out the cluster structure, so that the close-packed core of the cluster contains only eight osmium atoms. This is insufficient to give frontier orbital spacings close enough together to give paramagnetism. Similarly, the cluster $Os_8(CO)_{23}$ is EPR silent. However, the Cambridge group did later report an EPR signal in $Os_{10}C(CO)_{24}I_2$ [49], so the situation is unclear.

9.2.4. EPR Studies of Rhodium Carbonyl Clusters

Our EPR studies were subsequently extended to rhodium carbonyl clusters [44]. The even-electron $[Rh_{17}S_2(CO)_{32}]^{3-}$ as its polycrystalline $[N(PPh_3)_2]^+$ salt gave EPR below 150 K. Comparison of these spectra with those of $H_2Os_{10}C(CO)_{24}$ showed some significant differences. The principal g-value was 2.040 ± 0.002; the smaller g-shift from free-spin than in the osmium clusters such as $H_2Os_{10}C(CO)_{24}$ is consistent with the weaker spin-orbit coupling in rhodium. The EPR lineshape reflected the axial symmetry of the Rh_{17} cluster. The linewidth at 10 K was 83 ± 2 G, implying an electronic spin-spin relaxation time T_2 of the order of 10^{-9} sec, similar to that found in the osmium carbonyl clusters. However, in contrast to the case of $H_2Os_{10}C(CO)_{24}$, the EPR linewidth of $[Rh_{17}S_2(CO)_{32}]^{3-}$ was independent of temperature between 4 and 75 K. This suggested that the spectrum of the rhodium cluster was inhomogeneously broadened by unresolved hyperfine coupling to ^{103}Rh nuclei. The amount of s-orbital character of the spin density is limited by this lack of resolution of coupling to the $I = 1/2$ rhodium nuclei. The failure to resolve a multi-line hyperfine splitting pattern caused by the rhodium nuclei meant that one of the key aims of the EPR experiment could not be achieved, because it could not be determined whether the orbitals containing the spin density were fully delocalised over all 17 rhodium atoms or partly localised on just a few of them. Similar temperature-independence of the linewidth, showing inhomogeneous broadening, was encountered in the EPR spectrum ($g = 2.09$) of a radical, believed to be $[Rh_6(CO)_{16}]^+$, which was generated by the chemical oxidation of $Rh_6(CO)_{16}$ in sulphuric acid [44].

Inhomogeneous broadening of the EPR linewidth has also been observed in the $[H_4Os_{10}(CO)_{24}]^-$ radical [65], where it is caused by coupling to ^{1}H nuclei.

The EPR spectral intensity of $[Rh_{17}S_2(CO)_{32}]^{3-}$ followed the Curie-law up to 75 K. At higher temperatures, the linewidth began to increase, but the intensity reduced more than given by a simple reciprocal temperature dependence, following instead the Curie-Weiss law with a Weiss temperature of about 80 K. It is noteworthy that this intensity variation is almost identical to that of $H_2Os_{10}C(CO)_{24}$, despite the completely different temperature dependence of the spectral linewidths of the two clusters [44].

As an illustration of a more successful example of the use of EPR hyperfine splitting patterns to characterise the spin density in metal cluster radicals, we can quote the study of the odd-electron cluster $Co_3(CO)_9(\mu_3\text{-PPh})$ and the radical anion $[Co_2Fe(CO)_9(\mu_3\text{-PPh})]^-$ [66]. Their EPR spectra showed well-resolved multi-line patterns (22 and 15 lines respectively, with coupling constants about 30 G) resulting from hyperfine coupling to ^{59}Co nuclei ($I = 5/2$). However, no hyperfine coupling to ^{31}P nuclei could be detected within the experimental resolution of 2 G. It was concluded that the semi-occupied molecular orbitals in these clusters are fully delocalised over the metal atom triangles, but do not interact with the triply-bridging phosphorus group [66].

9.2.5. EPR STUDIES OF OTHER HIGH-NUCLEARITY CLUSTERS

The magnetic properties of the giant palladium cluster $Pd_{561}(phen)_{36}O_{200}$ are discussed elsewhere in this book, but it is worth noting here that the EPR of this material [67] showed spectral features quite similar to those of $[Rh_{17}S_2(CO)_{32}]^{3-}$; an asymmetric lineshape with g-values 2.010 and 2.035, and a temperature-independent linewidth. There are literature data on the EPR of palladium metal [68] and colloids [69] which can be compared with the results from the Pd_{561} molecular cluster.

We can also mention the results of attempts to study the EPR of the cluster $[H_2Pt_{38}(CO)_{44}]^{2-}$ as its polycrystalline $[N(PPh_3)_2]^+$ salt [70, 71]. A very broad signal (linewidth several hundred Gauss) with $g = 2.00$ was detected, but this was very weak and poorly resolved. In this compound, there was definite evidence of decomposition; this platinum cluster is known to be much more air-sensitive than the osmium carbonyl compounds. Nonetheless, it is worth noting that the EPR g-value was the same as that reported for 22 Å colloidal platinum particles in a gelatin matrix [72].

The gold cluster $Au_{55}(PPh_3)_{12}Cl_6$ is an odd-electron compound, and should be EPR-active. However, we have been unable to observe reproducible EPR spectra from solid samples of this material [22, 73, 74]. A signal observed at room temperature in air had a g-value of 2.00, linewidth 1.5 G, but this disappeared on evacuating the sample tube for low-temperature studies, and no EPR signal could then be detected at temperatures down to 4 K. Although it is possible that the spectrum was genuine, and that the sample desorbed ligands under high vacuum [75], it is much more likely that it arose from a volatile radical contaminant.

Recently, an EPR spectrum from $Au_{55}(PPh_3)_{12}Cl_6$ at 4.4 K, with $g = 1.9204$ and linewidth 670 G, has been reported [76]. The intensity followed the Curie–Weiss law, and the g-value and linewidth were independent of temperature. The temperature-independence of the linewidth suggests inhomogeneous broadening by unresolved hyperfine coupling to ^{31}P and ^{197}Au nuclei; this behaviour is similar to that described above for $[Rh_{17}S_2(CO)_{32}]^{3-}$ [44].

For comparison, we can consider what is known about conduction electron spin resonance in colloidal gold particles. As in the case of osmium, a large g-shift from the free-spin value of 2.0023 may be anticipated. The electronic relaxation time T_2 is expected to be very short, giving an EPR line too broad to be observed except at cryogenic temperatures. However, if gold behaves as an s-band metal (as evidenced by the plasma resonance absorption in the optical spectrum of gold colloids – see below), a smaller g-shift would result, and the line would be narrower. A theoretical g-value of 2.021 has been calculated for gold metal [77], but since this calculation gives an incorrect Fermi energy, its reliability is uncertain.

The experimental picture is also unclear. Gold colloids in the size range 20–40 Å (somewhat larger than $Au_{55}(PPh_3)_{12}Cl_6$) have been variously reported to give EPR spectra with $g = 2.26$, linewidth 200 G, between 180 K and room temperature [78], or $g = 2.0024$ with a linewidth of only 6–9 G at room temperature [79]. An experimental g-value of 2.11 has been obtained for Au foil below 20 K [80]. The whole subject clearly merits further experimental investigation before it can be decided which (if any) of the EPR signals from gold clusters and colloids are genuine.

Apart from relaxation time considerations, we can consider two other possible reasons why EPR might not be observed in solid $Au_{55}(PPh_3)_{12}Cl_6$ [74]. Firstly, the influence of local magnetic fields from unpaired electrons on neighbouring clusters in the solid might broaden the spectra beyond detection. Secondly, there may be a disproportionation of charge in the solid to give even-electron species. This could be related to the electron-hopping found from the AC and DC electrical conductivity measurements [52], but the disproportionation energy would probably be much too high for the effect to be significant at low temperatures.

9.3. Electronic (UV-visible-NIR) Spectra

The electronic absorption spectra (UV-visible-NIR) of metal cluster compounds contain valuable information on their electronic structure and bonding. They illustrate the way in which metallic properties begin to evolve in clusters of large enough size. Unfortunately, there has been little systematic study of the electronic spectra of metal cluster compounds. In many synthetic laboratories, UV-visible spectroscopic measurements do not form part of the routine characterisation methods employed on these compounds, which are dominated by X-ray crystallography and infra-red (vibrational) spectroscopy. Where electronic spectra of metal cluster compounds are reported in the literature, they are generally included only in the initial report of the preparation of a new compound. There has been little attempt to draw together all the available data for use in the characterisation of how electronic structure evolves with cluster size.

In this section we summarise some of our recent experimental measurements of cluster electronic spectra in solution and the solid state, and draw attention to some unsolved problems [54].

9.3.1. ONE-ELECTRON ABSORPTIONS

Low-nuclearity cluster compounds in solution show relatively simple electronic absorption spectra which are typical of molecules with well-spaced electronic energy levels. Absorptions within the ligands (CO, PPh_3, etc.) occur in the ultra-violet region, and can be compared with the absorptions of the free ligands, to study the effect of coordination to metal atoms. Metal cluster absorptions occur mainly in the UV and visible regions, extending in some cases into the near infra-red.

In principle, each absorption can be assigned (as $\sigma \rightarrow \sigma^*$, etc.) with the longest wavelength absorption corresponding to the HOMO-LUMO gap if this transition is symmetry-allowed. In practice, even with the use of sophisticated molecular orbital calculations such as X-alpha, this can currently be achieved only for spectra of relatively small clusters such as $Os_3(CO)_{12}$ (Figure 6a), [24, 23] and $Ir_4(CO)_{12}$ [81]. Figure 1 shows the frontier orbital structure of $Os_3(CO)_{12}$, derived on this basis [22].

By combining UV-visible spectroscopy with magnetic circular dichroism studies, the frontier orbitals of clusters such as $[Au_8(PPh_3)_8]^{2+}$ can be fully characterised [82].

As the clusters get larger, the frontier orbital separation becomes smaller, and the absorption bands move across the visible region towards the near-infrared. In $[Os_{10}C(CO)_{24}]^{2-}$ in solution, the longest wavelength absorption has λ_{max} about 770 nm (1.6 eV) (Figure 6b) [83]. As discussed above, other estimates of the HOMO-LUMO gap in Os_{10} clusters are also about 1 eV, raising the possibility that their ground state electronic structures may be high-spin and paramagnetic.

The quantitative intensities of the longest wavelength absorptions in $[Os_{10}C(CO)_{24}]^{2-}$ and clusters such as $Au_{11}(PPh_3)_7Cl_3$ are low. It is not clear why the relevant integrals governing these transitions have lower values than those for higher-energy transitions, and more theoretical work to characterise the frontier molecular orbitals is necessary.

It must be noted that, in cluster compounds with π-bonding ligands such as carbonyl groups, not all of the one-electron absorption bands in the UV-visible spectra can be assigned to transitions between metal-metal bonding and antibonding orbitals. Some bands may correspond to transitions between metal-metal bonding orbitals and carbonyl ligand $2\pi^*$ orbitals [84]. Even a cluster as small as $Fe_3(CO)_{12}$ can have an absorption band at a wavelength as long as 607 nm [14, 24]; this is probably associated with the fact that two of the carbonyl ligands in this cluster bridge one of the metal-metal bonds.

On cooling, one-electron absorption bands sharpen, because of the effect of the population of vibrational energy levels. λ_{max} may also change, according to the Frank–Condon principle, if the metal-metal distances are not the same in the ground and excited states. This has given information on the bonding character of the relevant orbitals of $Ru_3(CO)_{12}$, $Os_3(CO)_{12}$ and derivatives [24].

9.3.2. INTERBAND TRANSITIONS

In larger clusters, the resolution of one-electron bands is lost, and a broad, continuous electronic absorption evolves, spanning the visible region into the near-infrared. This absorption reflects the overall density of states of the electronic energy levels in the cluster, with the HOMO now being equivalent

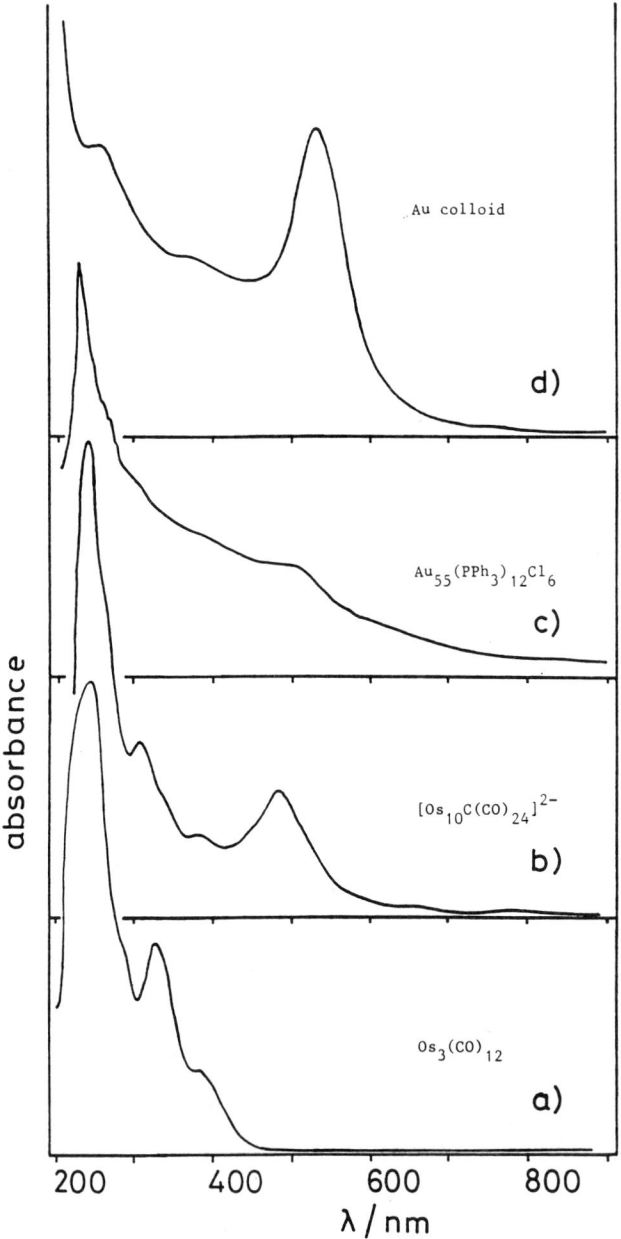

Fig. 6. UV-visible spectra of metal clusters and colloids [83]. (a) $Os_3(CO)_{12}$ in CH_2Cl_2 solution; (b) $[Os_{10}C(CO)_{24}]^{2-}[N(PPh_3)_2]_2^+$ in CH_2Cl_2 solution; (c) $Au_{55}(PPh_3)_{12}Cl_6$ in CH_2Cl_2 solution (the spectrum of $Au_{55}(PPh_2C_6H_4SO_3Na)_{12}Cl_6$ in H_2O solution is almost identical); (d) Colloidal Au, 150–200 Å diameter, in aqueous solution.

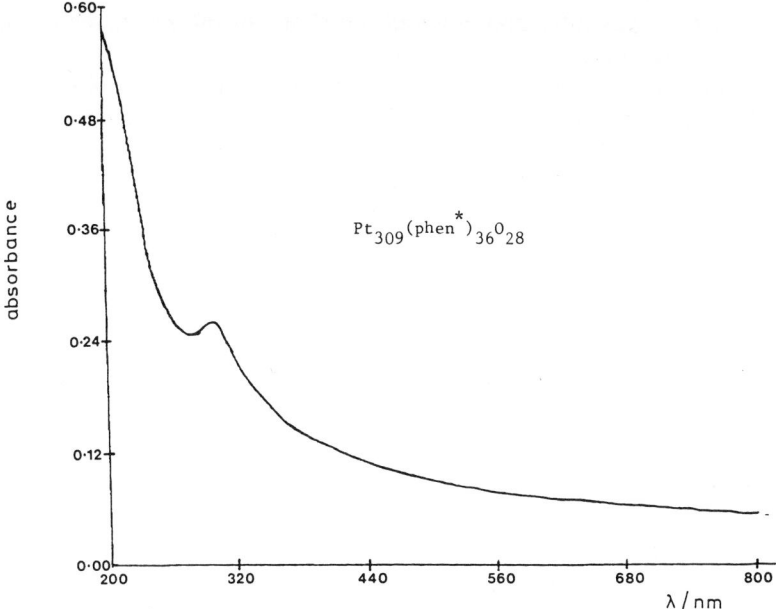

Fig. 7. UV-visible spectrum of $Pt_{309}(phen^*)_{36}O_{28}$ in aqueous solution [54].

to the Fermi energy. This trend is already apparent in $[Os_{10}C(CO)_{24}]^{2-}$ (Figure 6b) and is clearly seen in the spectra of clusters such as $Au_{55}(PPh_3)_{12}Cl_6$ (Figure 6c) [83] and $Pt_{309}(phen^*)_{36}O_{28}$ (Figure 7), which resemble the 'interband' absorptions of the corresponding metals. In colloidal gold (Figure 6d), this is a $5d \rightarrow 6s, 6p$ transition.

At the low-energy end of the spectrum, we can now use the onset wavelength λ_{onset} to estimate the 'band gap' energy difference between the HOMO and LUMO in the cluster. This is the usual practice in the study of small clusters of semiconductors such as CdS [85], but has not been applied to the electronic spectra of metal cluster compounds. The absorption in $[Os_{10}C(CO)_{24}]^{2-}$ extends to 830 nm (1.5 eV). Those in $Au_{55}(PPh_3)_{12}Cl_6$ and its water-soluble derivative $Au_{55}(PPh_2C_6H_4SO_3Na)_{12}Cl_6$ extend to at least 1350 nm (0.9 eV), with no narrow molecular-like one-electron absorptions to 4000 nm [86]. This highlights an unsolved problem in understanding the quantitative density of states at the Fermi level in clusters and colloids, because in colloidal gold the $5d \rightarrow 6s, 6p$ interband transition has comparable intensity only to 1050 nm.

At the high-energy end of the spectrum, the shape of the interband absorption in cluster compounds is often obscured by UV absorptions within the ligands. Difference spectra between $Au_{55}(PPh_3)_{12}Cl_6$ and compounds such as $(PPh_3)AuCl$ in solution suggest that the interband absorption in $Au_{55}(PPh_3)_{12}Cl_6$ may have a maximum around 250–300 nm. This type of measurement gives information about the shape of the lower region of the 5d-band in large gold clusters.

Changing the temperature has little effect on absorption bands of this type. For example, the spectrum of $Au_{55}(PPh_3)_{12}Cl_6$ is almost unchanged between 300 and 2 K [87]. There may be some change in λ_{onset} as the thermal population of levels around the Fermi energy alters.

9.3.3. PLASMA RESONANCE ABSORPTIONS

Delocalised, mobile conduction electrons within a metal particle have a characteristic collective oscillation frequency. This 'plasma resonance' is seen as an absorption band in the UV-visible spectrum of metal colloids. Metals with s-band electronic structures show well-defined plasma resonances in the visible region [88, 89]; for silver and gold, these occur respectively at wavelengths of 390 and 520 nm. These absorptions weaken as the particle size is reduced, but can be observed for silver and gold colloids of diameter as small as 10–20 Å [90, 91].

We can therefore ask whether the recently-characterised large organometallic clusters of silver and gold might show plasma resonance absorptions in their electronic spectra. This has been a topic of considerable recent controversy.

We suggested a weak absorption at 510 nm in the solution spectrum of $Au_{55}(PPh_3)_{12}Cl_6$ (Figure 6c) as a possible molecular plasma resonance [83]. The same assignment was subsequently proposed by Marcus [92] after measurements on independently prepared samples. This 55-atom cuboctahedral gold cluster has an overall diameter (from vertex to opposite vertex) of about 14 Å. The mean first-nearest-neighbour coordination number of the metal atoms, $\bar{N}_1$, which is a parameter correlating well with measures of metallic behaviour [93], is 7.85.

A similar suggestion had previously been made by Teo for $Ag_{20}Au_{18}(PPh_3)_{12}Cl_{14}$, which in solution shows a strong absorption at 495 nm, the same wavelength as the plasma resonance in bimetallic AgAu colloids of similar composition [94]. We doubted the assignment of this band as a plasma resonance [83], because it seemed unlikely that a plasma resonance absorption could be so strong in $Ag_{20}Au_{18}(PPh_3)_{12}Cl_{14}$ but so weak in $Au_{55}(PPh_3)_{12}Cl_6$. The $Ag_{20}Au_{18}$ cluster contains fewer metal atoms than the Au_{55} one, and is not a close-packed cluster, having a structure based on three 13-atom icosahedra aggregated into an oblate spheroid; its $\bar{N}_1$ value is 7.18. Moreover, the effect of particle shape on the plasma resonance frequency had not been taken into account in comparing the $Ag_{20}Au_{18}$ molecular cluster with the bimetallic colloids. In an oblate cluster such as this, any plasma resonance absorption should occur at a significantly longer wavelength than that in a spherical particle of the same composition, and another effect of the departure from spherical symmetry is to split the resonance into two non-degenerate absorption bands [83, 88, 89].

In turn, Kreibig demonstrated [86, 87] that the 510 nm absorption in $Au_{55}(PPh_3)_{12}Cl_6$ probably arises from aggregates of cluster molecules in solution, and that any plasma resonance in individual $Au_{55}(PPh_3)_{12}Cl_6$ molecules is too weak to be observed above the interband absorption. The water-soluble derivative $Au_{55}(PPh_2C_6H_4SO_3Na)_{12}Cl_6$ behaves in an almost identical manner. This limits the number and delocalisation of the $6s$ electrons in the Au_{55} clusters.

Kreibig has discussed three possible reasons for the absence of a clear plasma resonance absorption in $Au_{55}(PPh_3)_{12}Cl_6$ [95]. These are: that the $6s$ electrons are too localised for collective behaviour to occur; that the plasmon is broadened and dampened by interaction with the interband absorption, which acts as a decay channel; and that increased $6s$ electron density in the clusters, caused by the contraction of their Au-Au distances relative to that in bulk gold [92, 96], results in a broadening and shift of the plasmon absorption to higher frequency. The first and third of these reasons are, however, inconsistent with the extensive Mössbauer and photoelectron spectroscopy measurements on $Au_{55}(PPh_3)_{12}Cl_6$ [74].

As a further possibility, we can mention that the quantitative intensity C_{abs} of a plasma resonance absorption depends on the third power of the particle diameter a [88, 89]:

$$C_{abs} = \frac{8\pi^2 a^3}{\lambda} \text{Im}\left(\frac{\varepsilon - 1}{\varepsilon + 2}\right)$$

where λ is the wavelength in the medium, and ε is the complex relative permittivity of the metal relative to that of the surrounding medium. Even if the cluster electrons are fully delocalised and able to undergo a dipole resonance, the plasma resonance absorption in $Au_{55}(PPh_3)_{12}Cl_6$ might simply be too weak to detect above the background interband absorption, which is quite intense at 520 nm (Figure 6c). So there may be no need to invoke any special broadening or damping mechanisms [74].

In metals having a less free-electron behaviour than silver and gold, absorption maxima in the UV-vis-NIR spectrum cannot be assigned as pure plasma resonances, because they also have considerable interband character [88, 89]. In palladium colloids, the UV-visible band corresponding most closely to a plasma resonance absorption occurs in the UV at about 230 nm [88, 89]. The electronic absorption spectrum of $Pd_{561}(phen)_{36}O_{200}$ (cluster diameter $\approx$ 30 Å, with $\bar{N}_1 = 10.05$, assuming an ideal 561-atom cuboctahedral structure) is therefore uninformative, because this region of the spectrum is obscured by electronic transitions within the π-systems of the ligands [22].

A similar situation has now been encountered with the platinum cluster $Pt_{309}(phen^*)_{36}O_{28}$ (cluster diameter $\approx$ 25 Å, with ideal $\bar{N}_1 = 9.63$). This shows an absorption in solution with λ_{max} at 292 nm above the interband absorption (Figure 7). A plasma resonance is not usually detected from platinum colloids [97], but the theoretical position of the absorption band most

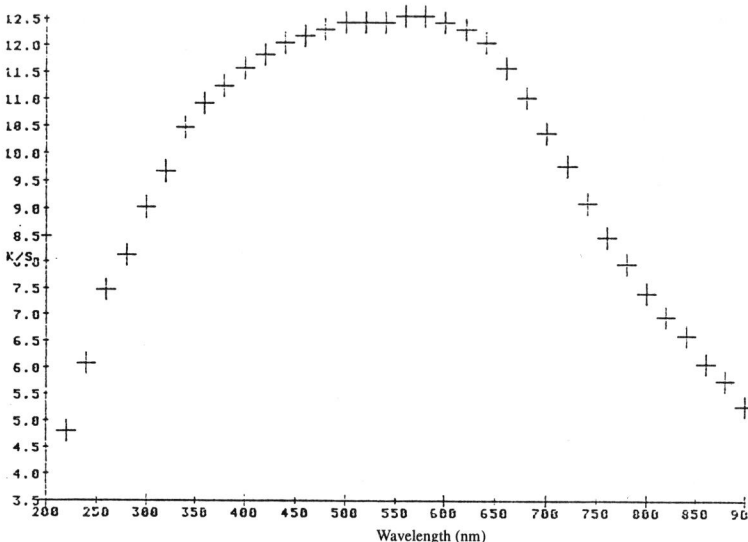

Fig. 8. UV-visible spectrum of solid $Au_{55}(PPh_3)_{12}Cl_6$ (diffuse reflectance, Kubelka-Munk transformed) [54].

closely corresponding to one is about 240 nm [88, 89]. However, the 292 nm band of the Pt_{309} cluster is much more likely to be a ligand $\pi \rightarrow \pi^*$ transition, which occurs in the free ligand (a water-soluble derivative of phenanthroline) at 277 nm.

Thus, at present, no definite example is known of a plasma resonance absorption in a molecular metal cluster compound. This contrasts with the close resemblance between the interband absorptions of large molecular clusters and colloidal particles of the corresponding metals, and illustrates the important fact that different physical properties of clusters can converge towards metallic behaviour at completely different cluster sizes.

The position and intensity of a plasma resonance absorption band is unlikely to depend on temperature. This gives the possibility of distinguishing experimentally between this type of absorption and one-electron absorptions in the same region of the spectrum.

9.3.4. CHARGE TRANSFER IN THE SOLID STATE

There are few solid-state electronic spectra of metal cluster compounds reported in the literature; the great majority of UV-visible spectra have been recorded on clusters in solution. We have recently been considering the possibility of electronic charge-transfer in metal cluster compounds in the solid state and have used diffuse reflectance UV-visible-NIR spectroscopy as an experimental probe [53, 54].

The electronic spectrum of $Au_{55}(PPh_3)_{12}Cl_6$ in the solid state, measured by diffuse reflectance, is quite different from that obtained from the same cluster in solution. A very broad, intense absorption is found (Figure 8), with λ_{max} about 500 nm, and extending across the visible and near-infrared to 2000 nm. The $\pi \rightarrow \pi^*$ absorption of the PPh$_3$ ligands, responsible for the peak at 229 nm in the solution spectrum (Figure 6c), is not resolved. As the solid cluster sample is progressively diluted with an electronically inert material such as paraffin wax, the absorption band moves to shorter wavelength, and the ligand absorption reappears. The spectrum of dilute samples resembles that of Au$_{55}$(PPh$_3$)$_{12}$Cl$_6$ in solution.

We have suggested [53, 54] that this absorption at 500 nm arises from electronic charge-transfer between cluster molecules in the solid state; i.e. a disproportionation of electronic charge caused by irradiation with visible light. The position of the absorption band is in reasonable agreement with a calculation of the energy at which such a charge-transfer should take place, made by estimating the ionisation energy and electron affinity of the clusters and the Coulomb energy of the charged clusters in the undiluted solid. On diluting the solid cluster sample, the average distance between the cluster molecules increases, reducing the Coulomb stabilisation of the charged clusters and causing the shift of the charge-transfer absorption to higher energy. The intensity of the charge-transfer absorption is also reduced, causing the spectrum to resemble that of the cluster in solution.

Au$_{55}$(PPh$_3$)$_{12}$Cl$_6$ is an odd-electron molecule if correctly formulated. A similar charge-transfer idea has previously been proposed for the odd-electron molecule V(CO)$_6$ [98]. In solution, this material is yellow, but in the solid state it is black, because of an additional broad absorption band centred at 580nm. The position of this band approximately fits the expected energy for charge-transfer based on ionisation energy, electron affinity and Coulomb energy, when allowance is made for the polarisation of the solid by charged species [98]. On dilution of the solid V(CO)$_6$ with inert material, a similar behaviour was observed to that we have found in Au$_{55}$(PPh$_3$)$_{12}$Cl$_6$ [98].

The idea of charge-transfer in Au$_{55}$(PPh$_3$)$_{12}$Cl$_6$ in the solid state can be related to the electrical conductivity of this material, which shows semiconducting behaviour with a non-Ohmic electron-hopping mechanism similar to that of granular metals [52]; these conductivity measurements are discussed in detail elsewhere in this book.

One problem to be addressed is that a preliminary measurement did not succeed in detecting photoconductivity in thin films of Au$_{55}$(PPh$_3$)$_{12}$Cl$_6$ above the level of the (substantial) dark current [99].

To summarise this section, it can be seen that a great deal of information on the electronic structures of metal cluster compounds can be derived from careful study of their UV-visible-NIR spectra in solution and in the solid state. The field is ripe for further rapid progress in characterising the molecular orbital structures of these metal clusters and relating them to the physical properties of the cluster materials.

9.4. Conclusion

The experimental measurements (magnetic susceptibility, electron paramagnetic resonance and UV-visible-NIR spectroscopy) described in this chapter have given a consistent picture of the way in which the electronic structures of organometallic cluster molecules begin to develop, as the cluster size is increased, away from the simple covalent molecular type of bonding towards a much more delocalised metallic character.

The transition from the molecular to the metallic type of bonding is not a sharp one. There is not a sudden nonmetal-metal transition as the cluster size increases. Neither is the transition necessarily a smooth one; high-nuclearity cluster molecules may have their own characteristic properties which are distinct from those found in either of the limiting bonding regimes. These properties can be related to those of small, ligand-free metallic particles.

We have introduced the term 'metametallic' to describe the distinctive behaviour of metal cluster compounds and metallic particles in the size range where quantum-size effects are important [29, 30, 31, 32].

With the recent availability of much larger clusters than had been synthesised at the time this work was begun, there is the opportunity for many more experiments to clarify this type of behaviour further. The development of much more powerful theoretical computational methods, such as local density functional theory [43], also offers the prospect of reaching in the near future a much more satisfactory understanding of the electronic structure and physical properties of high-nuclearity metal cluster compounds.

Acknowledgements

I am grateful for financial support for this research from the Royal Society, the U.K. Science and Engineering Research Council, and the Commission of the European Communities. Thanks are due to Profs. J. Lewis, B. F. G. Johnson and P. P. Edwards for their role in the inauguration of the work; Drs. M. D. Vargas, W. J. H. Nelson, and J. M. Mace for supplying samples of the osmium carbonyl clusters; Prof. G. Schmid for samples of $Au_{55}(PPh_3)_{12}Cl_6$ and related clusters; Dr. J. A. Creighton for sharing his expertise in colloid chemistry; and Prof. L. J. de Jongh and his colleagues at Leiden for their continuing interest in this work.

References

1. H. M. Powell and R. V. G. Evans, *J. Chem. Soc.* A **286** (1939).
2. L. F. Dahl, E. Ishishi and R. E. Rundle, *J. Chem. Phys.* **26** (1957) 1750.
3. P. Chini, *J. Organomet. Chem.* **200** (1980) 37.
4. A. L. Mackay, *Acta Cryst.* **15** (1962) 916.
5. M. R. Hoare and P. Pal, *Adv. Phys.* **20** (1971) 161.
6. S. Ino, *J. Phys. Soc. Japan* **21** (1966) 346.

7. J. G. Allpress and J. V. Sanders, *Surf. Sci.* **7** (1967) 1.
8. M. F. Gillet, *Surf. Sci.* **67** (1977) 139.
9. K. P. Hall and D. M. P. Mingos, *Prog. Inorg. Chem.* **32** (1984) 237.
10. E. L. Muetterties and R. M. Wexler, *Survey Prog. Chem.* **10** (1983) 61.
11. E. L. Muetterties, T. N. Rhodin, E. Band, C. F. Brucker and W. R. Pretzer, *Chem. Rev.* **79** (1979) 91.
12. K. Wade, *Inorg. Nucl. Chem. Lett.* **8** (1972) 559.
13. D. M. P. Mingos, *Nature (Phys. Sci.)* **236** (1972) 99.
14. P. Chini, *Pure Appl. Chem.* **23** (1970) 409.
15. M. Tachikawa and E. L. Muetterties, *Prog. Inorg. Chem.* **28** (1981) 203.
16. R. E. Benfield and B. F. G. Johnson, *J. Chem. Soc., Dalton Trans.* (1980) 1743.
17. J. Lewis and B. F. G. Johnson, *Pure and Applied Chem.* **54** (1982) 97.
18. J. H. Van Vleck, *Science* **201** (1978) 113.
19. P. W. Anderson, *J. Appl. Phys.* **50** (1979) 7281.
20. R. E. Benfield, P. P. Edwards and A. M. Stacy, *J. Chem. Soc. Chem. Commun.* (1982) 525.
21. D. C. Johnson, R. E. Benfield, P. P. Edwards, W. J. H. Nelson and M. D. Vargas, *Nature* **314** (1985) 231.
22. R. E. Benfield, *J. Organomet. Chem.* **372** (1989) 163.
23. B. Delley, M. C. Manning, D. E. Ellis, J. Berkowitz and W. C. Trogler, *Inorg. Chem.* **21** (1982) 2247.
24. D. R. Tyler, R. A. Levenson and H. B. Gray, *J. Amer. Chem. Soc.* **100** (1978) 7888.
25. R. S. Ruoff, D. Beach, J. Cuomo, T. McGuire, R. L. Whetten and F. Diederich, *J. Phys. Chem.* **95** (1991) 3457.
26. K. Kimura and S. Bandow, *Phys. Rev. Lett.* **58** (1987) 1359.
27. J. M. van Ruitenbeek and D. A. van Leeuwen, *Phys. Rev. Lett.* **67** (1991) 640.
28. J. M. van Ruitenbeek, *Zeitschrift für Physik* **D19** (1991) 247.
29. R. E. Benfield, *J. Phys. Chem.* **91** (1987) 2712.
30. R. E. Benfield, in P. Jena, B. K. Rao and S. N. Khanna (eds.), *Physics and Chemistry of Small Clusters*, Plenum, New York (1987), p. 401.
31. L. J. de Jongh, H. B. Brom, G. Longoni, P. R. Nugteren, B. J. Pronk, G. Schmid, H. H. A. Smit, M. P. J. van Staveren, and R. C. Thiel, in P. Jena, B. K. Rao and S. N. Khanna (eds.), *Physics and Chemistry of Small Clusters*, Plenum, New York (1987), p. 807.
32. L. J. de Jongh, H. B. Brom, G. Longoni, B. J. Pronk, G. Schmid and M. P. J. van Staveren, *J. Chem. Res.* (1987) 150.
33. R. Kubo, *J. Phys. Soc. Japan* **17** (1962) 975.
34. W. D. Knight, *J. Vac. Sci. Tech.* **10** (1973) 705.
35. R. F. Marzke, *Catal. Rev. Sci. Eng.* **19** (1979) 43.
36. R. Kubo, *J. de Physique* **38** (1977) C2-69.
37. J.-L. Millet and J.-P. Borel, *Surface Science* **106** (1981) 403.
38. R. F. Marzke, W. S. Glaunsinger and M. Bayard, *Solid State Commun.* **18** (1976) 1025.
39. B. J. Pronk, H. B. Brom, L. J. de Jongh, G. Longoni and A. Ceriotti, *Solid State Commun.* **59** (1986) 349.
40. B. K. Teo, F. J. DiSalvo, J. V. Waszczak, G. Longoni and A. Ceriotti, *Inorg. Chem.* **25** (1986) 2262.
41. J. W. Lauher, *J. Amer. Chem. Soc.* **100** (1978) 5305.
42. G. Ciani and A. Sironi, *J. Organomet. Chem.* **197** (1980) 233.
43. N. Rösch, L. Ackermann and G. Pacchioni, *J. Amer. Chem. Soc.* **114** (1992) 3549.
44. R. E. Benfield, *Zeitschrift für Physik* **D12** (1989) 453.
45. C. J. Ballhausen, *Introduction to Ligand Field Theory*, McGraw-Hill, New York (1962), p. 147.
46. L. E. Orgel, *Introduction to Transition Metal Chemistry*, 2nd edition, Methuen, London (1966), Chapter 3.

47. C. J. Ballhausen, *Molecular Electronic Structures of Transition Metal Complexes*, McGraw-Hill, London (1979), pp. 144–146.
48. F. A. Cotton and G. Wilkinson, *Advanced Inorganic Chemistry*, 3rd edition, Wiley, London (1972), p. 565.
49. S. R. Drake, P. P. Edwards, B. F. G. Johnson, J. Lewis, E. A. Marseglia, S. D. Obertelli and N. C. Pyper, *Chem. Phys. Lett.* **139** (1987) 336.
50. J. A. O. de Aguiar, A. Mees, J. Darriet, L. J . de Jongh, S. R. Drake, P. P. Edwards, B. F. G. Johnson and J. Lewis, *Solid State Commun.* **66** (1988) 913.
51. A. Kawabata, *J. de Physique* **38** (1977) C2-83.
52. M. P. J. van Staveren, H. B. Brom and L. J. de Jongh, *Physics Reports* **208** (1991) 1.
53. A. P. Maydwell, Ph.D. Thesis, University of Kent (1993); R. E. Benfield and A. P. Maydwell (to be published).
54. R. E. Benfield, A. P. Maydwell, J. M. van Ruitenbeek and D. A. van Leeuwen, *Zeitschrift für Physik D* **D26** (1993) S4.
55. B. F. G. Johnson, unpublished work; quoted in L. J. Farrugia, *Adv. Organomet. Chem.* **31** (1990) 301.
56. R. J. Elliott, *Phys. Rev.* **96** (1954) 266.
57. G. A. Ozin, *J. Amer. Chem. Soc.* **102** (1980) 3301.
58. S. R. Drake, P. P. Edwards, B. F. G. Johnson, J. Lewis, D. Obertelli and N. C. Pyper, *J. Chem. Soc. Chem. Commun.* (1987) 1190.
59. G. Feher and A. F. Kip, *Phys. Rev.* **98** (1955) 337.
60. F. J. Dyson, *Phys. Rev.* **98** (1955) 349.
61. S. C. Guy, R. N. Edmonds and P. P. Edwards, *J. Chem. Soc., Faraday Trans. II* **81** (1985) 937.
62. B. F. G. Johnson and R. E. Benfield, in B. F. G. Johnson (ed.), *Transition Metal Clusters*, Wiley, Chichester, U.K. (1980), p. 471.
63. S. R. Drake, personal communication.
64. P. R. Elliston, *J. Phys.* **C7** (1974) 425.
65. S. R. Drake, B. F. G. Johnson, J. Lewis and R. C. S. McQueen, *J. Chem. Soc., Dalton Trans.* (1987) 1051.
66. H. Beurich, T. Madach, F. Richter and H. Vahrenkamp, *Angew. Chem. Intl.* **18** (1979) 690.
67. J. A. O. de Aguiar, H. B. Brom, L. J. de Jongh, and G. Schmid, *Zeitschrift für Physik* **D12** (1989) 457.
68. P. Monod, *J. de Physique* **39** (1978) C6-1472.
69. F. Blatter and K. W. Blazey, *J. Phys. Chem. Solids* **52** (1991) 629.
70. B. J. Pronk, H. B. Brom, A. Ceriotti and G. Longoni, *Solid State Commun.* **64** (1987) 7.
71. R. E. Benfield, unpublished work.
72. D. A. Gordon, R. F. Marzke and W. S. Glaunsinger, *J. de Physique* **38** (1977) C2-87.
73. J. M. van Ruitenbeek, unpublished work.
74. R. C. Thiel, R. E. Benfield, R. Zanoni, H. H. A. Smit and M. W. Dirken, *Structure and Bonding* **81** (1993) 1; *Zeitschrift für Physik* **D26** (1993) 162.
75. L. R. Wallenberg, J. O. Bovin and G. Schmid, *Surf. Science* **156** (1985) 256.
76. G. Goll, H. v. Löhneysen, U. Kreibig and G. Schmid, *Zeitschrift für Physik* **D20** (1991) 329.
77. C. Schober, G. Kurz, H. Wonn, V. V. Nemoshkalenko and V. N. Antonov, *Physica Status Solidi* **B136** (1986) 233.
78. R. Dupree, C. T. Forwood and M. J. A. Smith, *Physica Status Solidi* **24** (1967) 525.
79. R. Monot, A. Chatelain and J.-P. Borel, *Phys. Lett.* **A34** (1971) 57.
80. P. Monod and A. Janossy, *Low-Temperature Physics* **26** (1977) 311.
81. G. F. Holland, D. E. Ellis, D. R. Tyler, H. B. Gray and W. C. Trogler, *J. Amer. Chem. Soc.* **109** (1987) 4276.
82. H-R. C. Jaw and W. R. Mason, *Inorg. Chem.* **30** (1991) 3552.

83. R. E. Benfield, J. A. Creighton, D. G. Eadon and G. Schmid, *Zeitschrift für Physik* **D12** (1989) 533.
84. S. R. Drake, B. F. G. Johnson, J. Lewis and R. G. Woolley, *Inorg. Chem.* **26** (1987) 3952.
85. X. K. Zhao, Y. Yuan and J. H. Fendler, *J. Chem. Soc. Chem. Commun.* (1990) 1248.
86. K. Fauth, U. Kreibig and G. Schmid, *Zeitschrift für Physik* **D20** (1991) 297.
87. K. Fauth, U. Kreibig and G. Schmid, *Zeitschrift für Physik* **D12** (1989) 515.
88. D. G. Eadon, Ph.D. Thesis, University of Kent (1988).
89. J. A. Creighton and D. G. Eadon, *J. Chem. Soc., Faraday Trans.* **87** (1991) 3881.
90. H. Abe, W. Schulze and B. Tesche, *Chem. Phys.* **47** (1980) 95.
91. U. Kreibig, *J. de Physique* **38** (1977) C2-97.
92. M. A. Marcus, M. P. Andrews, J. Zegenhagen, A. S. Bommannavar and P. Montano, *Phys. Rev.* **B42** (1990) 3312.
93. R. E. Benfield, *J. Chem. Soc., Faraday Trans.* **88** (1992) 1107; H-G. Fritsche and R. E. Benfield, *Zeitschrift für Physik* **D26** (1993) S15.
94. B. K. Teo, K. Keating and Y.-H. Kao, *J. Amer. Chem. Soc.* **109** (1987) 3494.
95. U. Kreibig, K. Fauth, C.-G. Granqvist and G. Schmid, *Z. Phys. Chem., Neue Folge* **169** (1990) 11.
96. M. C. Fairbanks, R. E. Benfield, R. J. Newport and G. Schmid, *Solid State Commun.* **73** (1990) 431.
97. D. N. Furlong, A. Launikonis, W. H. F. Sasse and J. V. Sanders, *J. Chem. Soc., Faraday Trans. I* **80** (1984) 571.
98. G. F. Holland, M. C. Manning, D. E. Ellis and W. C. Trogler, *J. Amer. Chem. Soc.* **105** (1983) 2308.
99. J. W. Couves, J. D. Wright and R. E. Benfield, unpublished work.

10. MAGNETIC PROPERTIES OF METAL CLUSTER COMPOUNDS

J. M. VAN RUITENBEEK, D. A. VAN LEEUWEN and L. J. DE JONGH
Kamerlingh Onnes Laboratorium, Leiden University,
P.O. Box 9506, NL–2300 RA Leiden, The Netherlands

10.1. Introduction

The properties of a metal atom are very different from those of a bulk metal. The question of how such properties evolve in going from one atom to an infinite lattice of atoms is a central theme in the research on metallic clusters. Apart from learning how the sometimes very disparate properties of the two limits connect together in the finite cluster regime, one often discovers properties that are unique to the clusters, being neither atomic, nor metallic properties. The dominant parameter in this discussion is the energy level separation, ΔE. For atoms, this energy scale is large compared to both the temperature, $k_B T$, and magnetic energies, $\mu_B B$. In metals, the energy level separation is so small that the distribution of levels can be regarded as a continuum. In this case the magnetic field imposes a quantization of the energy levels, and thus determines the relevant energy scale. Many of the interesting properties of clusters are found in the range where these energies, ΔE, $k_B T$, and $\mu_B B$, are of comparable magnitude. In this chapter, the magnetic properties of clusters are discussed, namely the magnetization and the magnetic susceptibility.

The magnetic properties of clusters have been studied intensively in the past. The seminal paper by Kubo [1] in 1962 discussed the unusual thermodynamic properties of clusters, in particular the specific heat and the magnetic susceptibility. The many experiments and theories that followed are concerned mainly with the averaged properties of the spin susceptibility of ensembles of clusters, with a wide distribution in sizes and shapes. A comprehensive review of these developments is given by Halperin [2]. More specific information is obtained, when the properties of one cluster of unique size and shape can be studied. For this purpose, two approaches are available. One approach is to produce clusters in metal vapor jets in ultra high vacuum, and subsequently mass select the ionized clusters. Recently, the magnetic properties of these size-selected clusters have been studied by de Heer *et al.* [3], by Bloomfield *et al.* [4] and others. The advantages of this approach are that it can be applied to all metals, the clusters are isolated from the environment and a wide range of sizes is accessible. On the other hand, the technique only allows in-situ experiments, i.e. in the beam itself, and it is difficult to obtain any information on the shape of the clusters. Also, the sensitivity of the Stern-Gerlach type of device, used to measure the magnetic moment, limits the application of this technique to clusters with sufficiently large magnetic moments. The other approach is the use of metal cluster compounds, which is the subject of this volume. These have the advantage that macroscopic quantities are available for the measurements, and sensitivity seldom forms a limitation. The size and shape are often well determined and a wide variety of compounds for many types of metal atoms have been produced. The drawback in this case is the fact that the clusters do not have a free surface. They are separated from each other by the ligand shell, which is chemically bonded to the surface of the cluster and strongly influences its properties. In addition, the cluster molecules are stacked with their ligand shells touching, so that in principle electrons may tunnel (or hop) from one cluster to the other. These two aspects necessarily form a part of the description of the properties of the cluster compounds.

In Section 2 a brief qualitative discussion is presented of the origin of magnetic properties in atoms and bulk metals. The transition between the two limits is then described in terms of the magnetic properties of metal clusters. The discussion of this subject differs from most previous treatments in that averaging over size distributions is not considered, in view of the uniformity of the material, and that the contribution of the orbital susceptibility is given as much attention as the spin susceptibility. In fact, size effects in the orbital susceptibility has received little attention, until recently, in the field of metal clusters. The effect of the ligand-bonding on the surface of the clusters is discussed at the end of this section. Section 3 discusses the experimental results in this field. It is not intended to be an exhaustive review, but rather to illustrate the interesting aspects of the problem with a number of experiments.

10.2. Magnetic Properties: Atoms, Metals and Clusters

Consider a system of electrons (in an atom, metal or cluster), described by the Hamiltonian $\hat{\mathcal{H}}_0$ without an external magnetic field. In an external, homogeneous magnetic field, B, this Hamiltonian is modified in two ways. The orbital motion of the electrons is influenced through the replacement of the momentum operators $\hat{\mathbf{p}}_i$ by $\hat{\mathbf{p}}_i + e\hat{\mathbf{A}}_i$. Here the index i runs over all the electrons in the system, the electron charge is $-e$, and $\hat{\mathbf{A}}_i = \hat{\mathbf{A}}(\mathbf{r}_i)$ is the vector potential at the site of electron i, with $\mathbf{B} = \nabla \times \mathbf{A}$. In addition, the energy of the spin states is shifted by $g_0 \mu_B \mathbf{B} \cdot \mathbf{S}$, where g_0 is the free electron gyromagnetic ratio and S the spin. Thus the Hamiltonian becomes:

$$\hat{\mathcal{H}} = \frac{1}{2m_e}\sum_i (\hat{\mathbf{p}}_i + e\hat{\mathbf{A}}_i)^2 + \sum_i g_0 \mu_B \mathbf{B} \cdot \hat{\mathbf{S}}_i + \hat{\mathcal{H}}_C + \hat{\mathcal{H}}_{so}. \tag{1}$$

The Coulomb interactions between the electrons and the nuclei and the mutual Coulomb interaction between the electrons are represented by $\hat{\mathcal{H}}_C$, and $\hat{\mathcal{H}}_{so}$ is the spin-orbit interaction, which is a relativistic effect. These last two terms are not modified by the field. The term in brackets can be expanded as

$$(\hat{\mathbf{p}}_i + e\hat{\mathbf{A}}_i)^2 = \hat{\mathbf{p}}_i^2 + 2e\hat{\mathbf{p}}_i \cdot \hat{\mathbf{A}}_i + e^2\hat{\mathbf{A}}_i^2, \tag{2}$$

provided that $\hat{\mathbf{p}}_i$ and $\hat{\mathbf{A}}_i$ commute. This is the case if we choose a gauge $\hat{\mathbf{A}}_i = \frac{1}{2}\mathbf{B} \times \hat{\mathbf{r}}_i$, which, in addition, is a convenient choice for systems with rotational symmetry. We will come back to the question of the appropriate choice of gauge below. For the second term in this expansion we have $2e\hat{\mathbf{p}}_i \cdot \hat{\mathbf{A}}_i = e\hbar \mathbf{B} \cdot \hat{\mathbf{L}}_i$, where $\hbar\hat{\mathbf{L}}_i$ is the angular momentum operator for electron i. The last term in Equation (2) may be rewritten as $e^2\hat{\mathbf{A}}_i^2 = (e^2/4)B^2\hat{r}_{i,\perp}^2$, with $\hat{r}_{i,\perp}$ the component of the position vector perpendicular to the magnetic field. Taking the field parallel to the z-direction for convenience, and inserting these replacements in Equation (1), we have for the Hamiltonian of our system in a magnetic field,

$$\hat{\mathcal{H}} = \hat{\mathcal{H}}_0 + \sum_i \mu_B(\hat{L}_{i,z} + g_0\hat{S}_{i,z})B + \frac{e^2}{8m_e}B^2\sum_i(\hat{x}_i^2 + \hat{y}_i^2), \tag{3}$$

where $\hat{\mathcal{H}}_0 = \frac{1}{2m_e}\sum_i \hat{\mathbf{p}}_i^2 + \hat{\mathcal{H}}_C + \hat{\mathcal{H}}_{so}$. The second term produces a paramagnetic response, the last term gives diamagnetism as is discussed below for several specific examples. The separation of the orbital contributions in a term $\mu_B\hat{L}_z B$ and a term $eB^2(\hat{x}^2+\hat{y}^2)/8m_e$ is conventional, but artificial. The two terms produce two intimately connected aspects of orbital magnetism, and are fundamentally inseparable in the sense that the one cannot be considered without taking the other into account. Below, we will come back to this problem.

The magnetic thermodynamic functions can be derived with the help of the appropriate free energy, say F, which can be calculated on the basis of the eigenvalues of Equation (3). The type of free energy depends on the specific problem that we are studying. The magnetization, M, and the magnetic susceptibility (per unit of volume), χ, can be calculated with the formulae

$$M = -\frac{1}{\Omega}\frac{\partial F}{\partial B}, \tag{4}$$

$$\chi = -\frac{\mu_0}{\Omega}\frac{\partial^2 F}{\partial B^2}, \tag{5}$$

where Ω is the volume of the system. The validity of using B instead of the field H in these equations is based on the fact that $\chi \ll 1$ in the problems considered here. In the following sections, these equations are first used to illustrate briefly how the magnetic properties of atoms and metals are derived. Only those results that are relevant for a comparison to clusters are mentioned. For a detailed treatment we refer to standard textbooks on these subjects [5, 6, 7]. These sections are followed by a more detailed discussion of the cluster properties.

10.2.1. ATOMS

In the atom the electrons move in the central Coulomb force field of the nucleus, so that the problem has spherical symmetry. In that case the angular momentum is conserved, so that it is a good quantum number and Equation (3) can be applied. In the case of a single isolated atom, the separation ΔE between the ground state and the first excited level is usually much larger than $k_B T$ and $\mu_B B$. Thus, we need only consider the ground state, and the field in Equation (3) can be treated as a small perturbation.

Let us first look at the situation when the eigenvalue of total angular momentum $\hat{\mathbf{J}} = \hat{\mathbf{L}} + \hat{\mathbf{S}}$ is non-zero, so that the ground state is degenerate. This degeneracy is lifted by the magnetic field due to the second term in Equation (3), according to

$$E_m(B) = E_m(0) + g\mu_B m B. \tag{6}$$

Here, m are the eigenvalues of the z-component of $\hat{\mathbf{J}}$, and g is the Landé gyromagnetic constant, which can be expressed [5] in terms of g_0, L, S, and J. Equation (6) describes the Zeeman splitting of magnetic levels and gives rise to Curie paramagnetism. This magnetic response is calculated from Equation (6) via the free energy,

$$F = F_0 - k_B T \ln \sum_{m=-J}^{J} \exp\left(-\frac{g\mu_B m B}{k_B T}\right). \tag{7}$$

With Equation (4) we obtain

$$M = \frac{g\mu_B J}{\Omega} B_J \left(\frac{g\mu_B J B}{k_B T}\right), \tag{8}$$

with $B_J(x)$ the Brillouin function [5], which saturates at 1 for large x. For low fields $g\mu_B J B \ll k_B T$ we have for the susceptibility

$$\chi = \frac{\mu_0}{\Omega} \frac{g^2 \mu_B^2 J(J+1)}{3 k_B T}, \tag{9}$$

which is known as Curie's law.

Second order contributions in B become important when $J = 0$. Applying standard perturbation theory, the energy of the ground state without field,

$E_0(0)$, is shifted as,

$$E_0(B) - E_0(0) = +\frac{e^2}{8m_e}B^2 \sum_i \langle 0|\hat{x}_i^2 + \hat{y}_i^2|0\rangle$$

$$- \mu_B^2 B^2 \sum_i \sum_{n\neq 0} \frac{|\langle 0|\hat{L}_{i,z} + g_0\hat{S}_{i,z}|n\rangle|^2}{E_n - E_0}. \quad (10)$$

The index n runs over all excited states of the atom. The two terms in Equation (10) both result in a field and temperature independent susceptibility, but of opposite sign. The first describes Larmor (or Langevin) diamagnetism, which depends on the extent of the wave functions and the number of electrons in the atom. It is diamagnetic because $\chi \sim -\partial^2 F/\partial B^2 = -\partial^2 E/\partial B^2 < 0$. The second term in Equation (10), on the other hand, produces a positive contribution to the susceptibility, which is known as Van Vleck paramagnetism. It can only be relevant if $|n\rangle$ is not an eigenfunction of $\hat{L}_z$ and $\hat{S}_z$. Different eigenstates of $\hat{L}_z$ for the atom may be mixed due to external electric fields, e.g., when the atom is incorporated in a host lattice, or in a molecule. $\hat{S}_z$ and $\hat{L}_z$ eigenstates are mixed in general as a result of spin-orbit coupling. Van Vleck paramagnetism is of particular importance when the atom has some low lying excited states, as is seen from the denominator in Equation (10).

The Van Vleck susceptibility is thus seen to arise from that part of the moment due to the admixing of excited states into the ground state by means of the Zeeman effect. It will be particularly apparent if the ground state itself has no moment, for instance if the degeneracy of the orbital ground state is removed by crystal field effects from the surroundings of the magnetic atom in a solid (the orbital moment is then 'quenched', since the expectation value of $\hat{L}_z$ is zero for any nondegenerate orbital state). The Van Vleck susceptibility is only independent of temperature if the energy difference between the excited state and the ground state is sufficiently large compared to $k_B T$. Otherwise it will also acquire a $1/T$ dependence.

It may also be noted from eqs. (3) and (10) that there is an ambiguity in the diamagnetic susceptibility, in that it appears to depend on the position of the origin. This arises from the fact that the separation into diamagnetic and Van Vleck contributions is ambiguous, because both depend on the gauge chosen. This was already noted by Van Vleck [8], who has shown that the sum of these two contributions is independent of the choice of gauge. In case of spherical symmetry, as for the single atom considered above, the origin can be taken at the centre of symmetry. However, for polyatomic molecules and particles without spherical symmetry both contributions to the susceptibility should be evaluated together, since only their sum has a physical meaning. We shall come back to this point below, when discussing the orbital susceptibility of independent electrons in confined geometries.

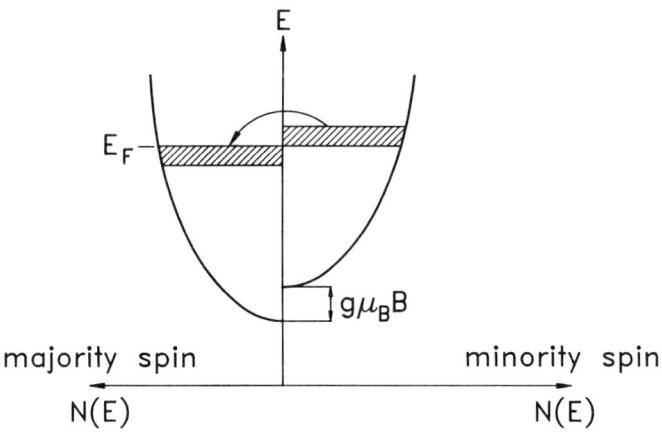

Fig. 1. For a free electron model of a metal the spin-up and spin-down bands are shifted in a magnetic field by an amount $g\mu_B B$. The transfer of electrons to the lower lying states, in order to align the Fermi energies, produces a net magnetisation, known as Pauli paramagnetism.

10.2.2. METALS

For a discussion of the magnetic properties of metals it is convenient to start from the simplest approximation, i.e. the free and independent electron model, and introduce the interactions required for the description of real metals as successive approximations. For free and non-interacting electrons the energies are given by $E(\mathbf{k}) = \hbar^2 k^2/2m_e$, which are occupied up to the Fermi energy, E_F. When a magnetic field is applied, this strongly modifies the eigenstates and the energies, because it forms the only potential that the electrons feel. Although the spin and orbital effects can be treated very elegantly at the same time, for future purposes we prefer to discuss the two effects separately. The spin degeneracy of the electron energies is lifted due to the term in $\hat{S}_z$ in Equation (3). All states with spin up are shifted down, and those with spin down are shifted up in energy. The ground state, therefore, has more spins up than down (see Figure 1), by an amount $\Delta n = \mu_B B N(E_F)$, where $N(E)$ is the density of levels per unit energy per unit volume. $N(E) = (m_e/\hbar^3 \pi^2)\sqrt{2m_e E}$ for the free electron gas. The magnetization follows from $M = \mu_B \Delta n$ giving a temperature independent susceptibility $\chi = \mu_0 \partial M/\partial B$,

$$\chi_{\text{Pauli}} = \mu_0 \mu_B^2 N(E_F), \tag{11}$$

which is the well known Pauli spin susceptibility.

The orbital part is slightly more complicated. In order to estimate it, we may consider the motion of a free electron in a magnetic field, which we take again parallel to the z-axis. For the gauge we choose now $\mathbf{A} = (-By, 0, 0)$.

Then the effect of the field is to transform the kinetic term, $\hat{p}^2/2m_e$, in the Hamiltonian into

$$\hat{\mathcal{H}}_{\text{kin}} = \frac{(\hat{p}_x + eB\hat{y})^2}{2m_e} + \frac{\hat{p}_y^2}{2m_e} + \frac{\hat{p}_z^2}{2m_e}. \tag{12}$$

The magnetic field is seen to impose a harmonic potential, as may be recognized from the first term in Equation (12). The energy levels, ignoring spin for the moment, are given by

$$E_n(k_z) = (n + \tfrac{1}{2})\hbar\omega_c + \frac{\hbar^2 k_z^2}{2m_e}, \quad (n = 0, 1, 2, \ldots). \tag{13}$$

The first term represents the eigenvalues of the two dimensional harmonic oscillator, with $\omega_c = eB/m_e$ the cyclotron frequency. The states indexed by n are known as Landau levels. The eigenstates along the z- axis are not altered by the field.

The density of levels is very different from the zero field situation, which is best illustrated for two dimensions, disregarding the z-direction momentarily. The total energy is extremely sensitive to whether the Fermi energy is at, or between a Landau level. The degeneracy of these levels is such that, for each value of k_z, all states that were found between $E_n(k_z)$ and $E_{n+1}(k_z)$ before we applied the field, are now accommodated in these levels as indicated in Figure 2. The Landau levels move up when the field is increased (since the splitting increases), and, therefore, the density of levels at the Fermi energy will strongly oscillate. This causes an oscillation in all thermodynamic quantities, and in particular in the susceptibility. The oscillatory susceptibility is known as the De Haas van Alphen effect [9]. However, such oscillations are readily smeared out by temperature, or lifetime broadening of the levels, leaving only an average contribution. This steady value is called Landau diamagnetism and its value can be estimated as follows.

As a consequence of the redistribution of states illustrated in Figure 2, the levels can be subdivided into those that shift up in energy and those that shift down, similar as for the spin susceptibility. Since also the energy splitting $\hbar\omega_c$ is equal to $2\mu_B B$ (at least for free electrons) we expect a result of the same form as the Pauli spin susceptibility. For the exact analysis we refer to textbooks [6] on the subject, but the result is:

$$\chi_{\text{Landau}} = -\frac{1}{3}\chi_{\text{Pauli}}. \tag{14}$$

The free and independent electron approximation is an excellent one for the alkali metals, where one has just one s-orbital valence electron. In most metals, the electrons strongly interact with the periodic potential of the ions. As a result the eigenstates are modified from plane waves to the so-called Bloch waves,

$$\psi_{n\mathbf{k}}(\mathbf{r}) = e^{i\mathbf{k}\cdot\mathbf{r}} u_{n\mathbf{k}}(\mathbf{r}). \tag{15}$$

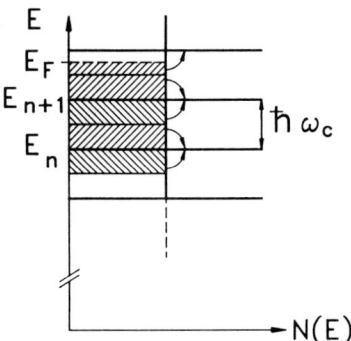

Fig. 2. The Landau diamagnetism for a free electron metal can be described by the redistribution of electrons due to the applied field. In two dimensions the density of levels is constant in zero field. For finite fields the electrons occupy Landau levels, which are separated by $\hbar\omega_c$.

The part of the wave function that has the periodicity of the lattice is described by $u_{n\mathbf{k}}$, the free electron like plane wave part by the exponential. In this way, the free electron results can be translated to the situation of independent electrons on a periodic lattice. Again there is a dispersion relation $E_n(\mathbf{k})$, now consisting of many branches, n, and in general very different from the simple parabolic free electron relation. This so called bandstructure, however, can be calculated and the density of states, $N(E_F)$, can be extracted. For this general density of states, the Pauli susceptibility Equation (11) is recovered. The density of states, may be much smaller than the free electron value, as in the case of semiconductors, or much larger, as in the case of energy bands derived from atomic d- or f-states.

The orbital susceptibility also survives the transition from free electrons to Bloch waves, although it is a little harder to derive rigorously [10]. The De Haas van Alphen effect is very sensitive to details in the bandstructure, and its measurement is one of the most important tools to obtain experimental information on the $E_n(\mathbf{k})$ relation. The steady Landau susceptibility is no longer given by relation Equation (14), it can even change sign. However, the principle of its calculation is the same as for free electrons, with the general $E_n(\mathbf{k})$ replacing the simple parabolic dispersion [11].

The next step in going from the free electron gas to real metals is to incorporate electron-electron interactions. These interactions are usually subdivided into correlation effects, due to the direct Coulomb interaction, and exchange effects, resulting from Coulomb repulsion combined with the requirement for the wave function to be antisymmetric.

The correlation effects (and also interactions with lattice vibrations) can be taken into account by defining quasi-particles, at least if they are not too strong. The quasi-particles are described by the Landau theory for Fermi liquids. They are the Bloch electrons plus the cloud of excitations around them, and close to the Fermi energy they are described by the same k vectors as the non-interacting Bloch electrons. The result is an enhancement of the density of levels $N(E_F)$, resulting with Equation (11) in an enhanced Pauli susceptibility. The enhancement in ordinary metals is less than about a factor of two. The effect of correlations on the Landau diamagnetism is believed to be negligibly small.

The exchange effects are responsible for the ferromagnetism in metals such as iron, nickel and cobalt. However, the effects are relevant to some extent in all metals. Since electrons are fermions we require that the total wave function is antisymmetric. In many cases, it is then favorable to make the spatial part of the wave function antisymmetric and thus the spin part should be symmetric (all spins parallel). The reason being that for an antisymmetric wave function the probability for two electrons to be at the same point in space $\psi(\mathbf{r}_1, \mathbf{r}_2) = -\psi(\mathbf{r}_2, \mathbf{r}_1) = 0$, which reduces the Coulomb repulsion energy. On the other hand, putting all electrons with parallel spins in different spatial states requires occupying states with much higher kinetic energies, than when we simply occupy all states with two spins. The trade-off between these energies, the exchange energy and the increased kinetic energy, results for most metals in no more than a tendency for spin alignment. This can be expressed in terms of a factor S multiplying the spin susceptibility of Equation (11), where S is the Stoner exchange enhancement factor. In most metals it is slightly above unity, notable exceptions being Pd and Pt, with $S = 9.4$ and 3.0, respectively. In just a few metals, Fe, Ni and Co, the effect is strong enough to induce a spontaneous magnetization at room temperature.

In all of this discussion we have used a fully itinerant electron model, as opposed to a description starting from nearly localized electrons. In fact, the atomic d and f electrons remain, to a large extent, localized around the atoms in the metal, and they are primarily responsible for the magnetism in metals. The question whether a description of the magnetism in metals should start from moments localized on the atoms or from moments associated with electrons that are free to move through the entire lattice is much discussed [12]. For instance, the well-known Hubbard Hamiltonian may be used to interpolate between these two limiting cases. However, many of the properties can be arrived at from both starting points and we choose to restrict ourselves in this discussion to the itinerant picture.

10.2.3. CLUSTERS

The aim, here, is to bridge the gap between the atomic and the bulk metal description, by modeling the properties of metal clusters. In particular we have to deal with the evolution of the atomic magnetic properties such as Curie paramagnetism and Larmor diamagnetism, into bulk metallic properties such as Pauli paramagnetism and Landau diamagnetism. One may anticipate some peculiar behavior in the intermediate (cluster) range of system sizes, e.g., by considering the last term in Equation (3). As we have seen, it leads to Larmor diamagnetism in atoms, which grows with the square of the extent of the wave

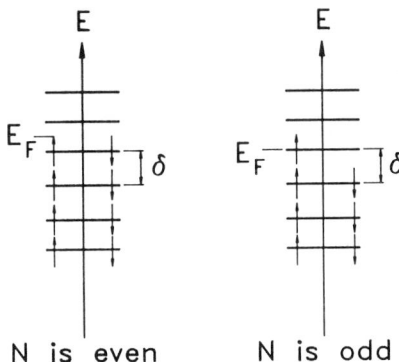

Fig. 3. Schematic representation of the spin occupation of a set of equally spaced levels in a cluster.

function. Assuming that the wave functions of electrons in metal clusters are delocalized over the entire cluster, then a giant diamagnetism would result for very large clusters. This example shows that the atomic properties connect to the metallic ones in a non-trivial way. In addition the transition is determined by the gradual development of electronic energy bands from the atomic states, in other words, by the decreasing influence of the surface. Electron-electron Coulomb interactions are not as effectively screened in clusters as they are in bulk metals. Exchange interactions leading to long range order in metals are expected to give rise to interesting cluster behavior, strongly influenced by fluctuations.

As in the summary of bulk metal magnetic properties, presented above, we start the discussion of spin and orbital susceptibilities from a model of free and independent electrons. Subsequently, the various complications are introduced: spin-orbit interactions and exchange interactions in clusters, and the effects of ligand coordination to the surface of the clusters.

10.2.3.1. The Spin Susceptibility in Metal Clusters

At this point our model of a metal cluster consists of a potential well filled with a number of independent electrons. We do not specify the shape of the well except that it is required to allow sufficient bound states and it must be singly connected (e.g., no torus). We assume that the stationary energy levels have been obtained [13] from the Hamiltonian, and we start occupying these levels with N electrons, in order of their energy, starting from the lowest. We occupy each level with two electrons, until we arrive at the last, and highest level, which is then occupied with one or with two electrons, depending on whether N is odd or even (Figure 3).

At low temperatures, $k_B T/\Delta E \ll 1$, with ΔE the separation between the highest occupied level and the lowest unoccupied level, the susceptibility is determined by the filling of the last level: For N odd, we have Curie paramagnetism, and for N even there is no contribution of the spin to the magnetism of the system.

At high temperatures, $k_BT/\Delta E \gg 1$, levels higher and lower than E_F are partially occupied as a result of thermal activation. Each of these partially occupied levels produces a Curie type paramagnetic susceptibility, and the number of such levels is of the order $k_BT/\Delta E$. Thus, we estimate the total high temperature susceptibility at,

$$\chi \sim \frac{C}{T} \cdot \frac{T}{\Delta E} \sim C \cdot N(E_F),$$

since $N(E_F) = 2/\Omega\Delta E$. This is just the bulk metal Pauli susceptibility. From these simple considerations we find that the above mentioned transition between atomic and metallic spin susceptibilities occurs for each cluster, as the temperature is raised. The transition temperature depends, of course, on the energy level separation, i.e., on the size of the cluster.

As an explicit example, let us consider a model system with an even number of electrons and with a uniform spacing ΔE between the energy levels. This model was considered by Denton et al. [14], and for this example we have the expression [15]

$$\chi_{\text{even}} = \frac{\mu_0\mu_B^2}{\Omega k_BT} \frac{8\sum_{n=0}^{\infty}(n+1)^2 \exp(-\Delta E(n+1)^2/k_BT)}{1 + 2\sum_{n=0}^{\infty} \exp(-\Delta E(n+1)^2/k_BT)}. \qquad (16)$$

The calculations and the resulting expressions are complicated by the fact that the number of electrons needs to be kept constant. In atomic systems this poses no problem, since only a few levels are relevant, and for bulk metals the number of electrons is so large that we can resort to the grand canonical ensemble for the calculations, introducing a chemical potential which keeps the average number constant. At high temperatures, $k_BT/\Delta E \gg 1$, Equation (16) reduces to

$$\chi_{\text{even}} = \frac{2\mu_0\mu_B^2}{\Omega\Delta E} \qquad (17)$$

which is just the Pauli susceptibility, since $2/\Omega\Delta E$ is the density of levels. Also at low temperatures, $k_BT/\Delta E \ll 1$, Equation (16) behaves according to expectation:

$$\chi_{\text{even}} = \frac{8\mu_0\mu_B^2}{\Omega k_BT} \exp(-\Delta E/k_BT). \qquad (18)$$

The full temperature dependence of Equation (16) is given in Figure 4, together with that for an odd number of electrons.

This very simple model may give a good first order description of the actual spin magnetism of a sample containing identical clusters. The energy

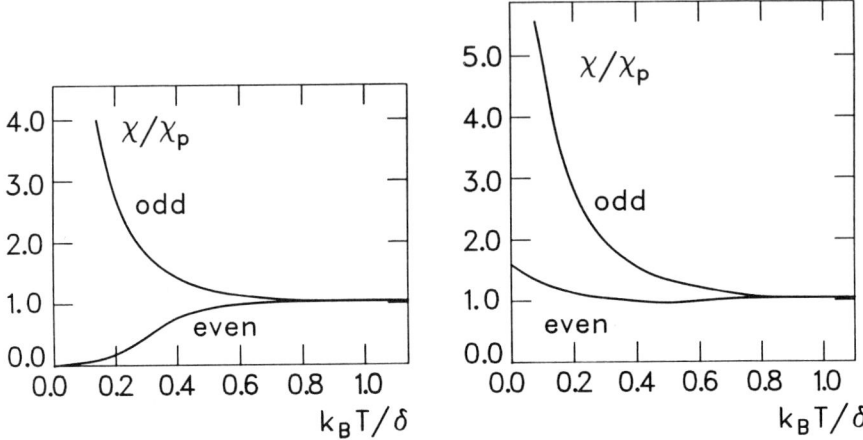

Fig. 4. Susceptibility for a model of a cluster (left) with equally spaced levels and (right) for a Poisson distribution of levels. There is a clear distinction between an even and an odd number of electrons. (Adapted from Denton et al. [14]).

levels are, of course, not equally spaced, but the behavior will be dominated by the separation between the occupied and unoccupied levels, which sets the value for ΔE. In more general discussions of this subject, distributions in size and shape of clusters are considered. There, the energy levels are described by statistical ensembles, the type of ensemble being determined by the symmetry of the problem [14, 16]. It is then argued that surface roughness should be sufficient to induce a strong level repulsion, leaving only Kramer's degeneracy, in the absence of an external magnetic field. The metal cluster compounds form an exception. Even if the clusters are not exactly identical, as may be anticipated in particular for the very large cluster compounds, the symmetry of the cluster is very high, allowing many degenerate or near-degenerate levels to be found. Under the assumption that deviations from perfect cluster symmetry are small, the appropriate energy level distribution is the Poisson distribution [16, 17],

$$P(\Delta E) = \frac{1}{\delta} \exp(-\Delta E/\delta), \tag{19}$$

where δ is the average level spacing. This would be a unique property of metal cluster compounds. The susceptibility for a Poisson level distribution was calculated by Denton et al. [14] and is also given in Figure 4. Note that the susceptibility for an even number of electrons does not vanish at low temperatures. The reason is that at each temperature there is a finite number of clusters having an energy level separation smaller than $k_B T$.

10.2.3.2. The Orbital Susceptibility in Metal Clusters

The problem of the orbital magnetic susceptibility of metal clusters has received relatively little attention. A review of developments related to this subject can be found in [18]. Some important aspects of the problem can be learned from very simple model systems. We shall distinguish systems with a high symmetry, allowing degenerate states, and systems with no degeneracy. In this section we ignore the spin of the electrons.

As an example of the first let us look at a spherical potential well, with hard walls, containing a number of otherwise free and independent electrons. The electron wave functions are the product of spherical harmonics and spherical Bessel functions, the eigenvalues being determined by the hard wall boundary condition. In a magnetic field the energies E_{nl} are shifted according to Equation (3):

$$E_{nlm}(B) = E_{nl}(0) + m\mu_B B + \alpha_{nlm} B^2, \tag{20}$$

in strong analogy with the atomic description. Here, α_{nlm} are the diagonal matrix elements due to the last term in Equation (3), and m is the magnetic quantum number. A correct treatment of the full temperature dependence of the susceptibility would require using a canonical ensemble, but the essentials can be found from the grand canonical ensemble, which is much simpler to use. Inserting the energies of Equation (20) in the partition function and calculating the susceptibility in the limit $B \to 0$, using Equation (5), gives:

$$\chi = \frac{\mu_0}{\Omega} \sum_{nlm} f_0 \left[\frac{m^2 \mu_B^2}{k_B T}(1 - f_0) - 2\alpha_{nlm} \right]. \tag{21}$$

The first term in Equation (21) is Curie like at low temperatures, the last term gives the temperature independent diamagnetism. The Van Vleck term is absent for this spherical problem. The factors f_0 and $(1 - f_0)$ describe the temperature dependent occupation of the (n, l) quantum numbers, where $f_0(E_{nl}, T)$ is the Fermi distribution at $B = 0$. The α_{nlm} can be expressed in analytical form [19, 20] and then it is straight forward to evaluate Equation (21) numerically. The calculation was performed for all levels up to $2 \times 10^4 \hbar^2/2m_e a^2$, with a the radius of the sphere, which involves n and l values up to 45 and 145, respectively. The chemical potential was adjusted at each temperature to keep N constant to within 10^{-5}.

Figure 5 shows the evolution of χ as a function of $k_B T/\Delta E$ for various numbers of electrons N in the spherical box. $N = 20$ and 93 are closed shell configurations, which are diamagnetic at low T. For the other examples we find a fast decay of χ with T, faster than $1/T$. The inset shows for $N = 50$, how the paramagnetic term evolves from low temperature Curie like paramagnetism to a constant value at high T. This high temperature behavior is analogous to the Pauli spin susceptibility for a degenerate electron gas. It is striking to see how this constant orbital paramagnetism is almost

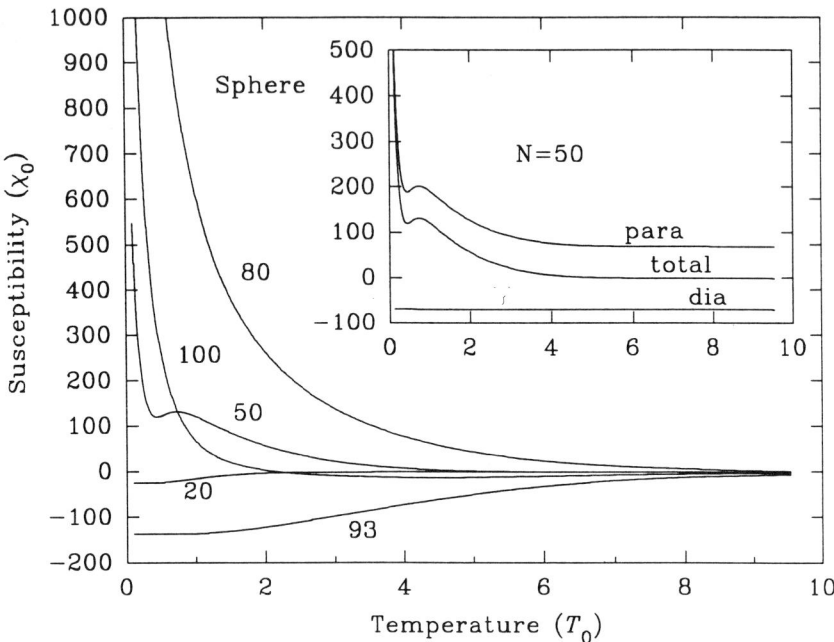

Fig. 5. The orbital susceptibility as a function of T for a spherical potential well, filled with N electrons; the number N is indicated for each curve. The inset shows the two para- and diamagnetic contributions separately for $N = 50$. The units for χ and T are $\mu_0 e^2/24\pi m_e a$ and $\hbar^2/2m_e a^2 k_B$, where a is the radius of the sphere.

exactly compensated by the temperature independent diamagnetism. This illustrates the fact, already mentioned in Section 2, that the paramagnetic and diamagnetic contributions are intimately related and inseparable.

The total orbital susceptibility at high temperatures is not exactly zero, but approaches the bulk Landau diamagnetic value, which is small on the scale of Figure 5. We find that the size of the cyclotron orbit is not relevant for Landau diamagnetism. The relevant parameter is the temperature, which needs to be larger than the scale of the size-induced fluctuations in the density of levels, $k_B T > \hbar v_F/L$, with L the size of the system, in order to observe bulk behavior [18].

A realistic model for small particles, will in general not be spherical. In order to illustrate what changes occur when the degeneracies are lifted as a result of the shape of the potential well, we next consider a rectangular box containing N electrons. With the field along the z- axis, we choose the lengths $L_z = L_x = \sqrt{1.1}L_y$, with the aim of making the model as simple as possible, but leaving no degeneracies which can be lifted by applying the field. The wave functions are simple standing plane waves,

$$\psi(x,y,z) = \sqrt{\frac{8}{\Omega}} \sin\frac{n_x \pi x}{L_x} \sin\frac{n_y \pi y}{L_y} \sin\frac{n_z \pi z}{L_z}. \tag{22}$$

Here $\Omega = L_x L_y L_z$ is the volume of the system. It is a straightforward matter to calculate the field dependent perturbation of the energy levels, analogous to Equation (10). The diamagnetic susceptibility for a wave function with quantum numbers (n_x, n_y, n_z) can be written as

$$\chi_{\text{DIA}}(n_x) = -\frac{1}{12}\mu_0 \frac{e^2}{m_e \Omega} L_x^2 \left(1 - \frac{6}{\pi^2 n_x^2}\right), \tag{23}$$

and the Van Vleck paramagnetic susceptibility is given by:

$$\chi_{\text{VV}}(n_x, n_y) = -\mu_0 \frac{64 e^2}{\pi^6 m_e \Omega} \gamma L_x^2 \sum_{k_x, k_y}^{\infty'} n_y^2$$

$$\times \frac{\left(\frac{1}{n_y+k_y} - \frac{1}{n_y-k_y}\right)^2 \left(\frac{1}{(n_x+k_x)^2} - \frac{1}{(n_x-k_x)^2}\right)^2}{n_x^2 - k_x^2 + \gamma(n_y^2 - k_y^2)}. \tag{24}$$

The primed sum denotes a summation over values of k_x and k_y for which $n_x + k_x$ is odd and $n_y + k_y$ is odd, and $\gamma = L_x^2/L_y^2$ which is here equal to 1.1. The total susceptibility follows by summing Equations (23) and (24) over all occupied levels. Figure 6 shows the result at $T = 0$ as a function of the number of electrons in the box. Both χ_{DIA} and χ_{VV} grow with the size of the system (i.e., with N if the electron density is constant). However, Figure 6 demonstrates that the two contributions nearly cancel, even at $T = 0$. The erratic structure on χ_{VV} is due to the variations in the energy level spacing at E_F. Figure 7 shows how the steady Landau diamagnetism survives again at high temperatures. The large paramagnetism for $N = 1303$ is a result of a small energy separation to the next level (Equation (10)), and for $k_B T$ larger than this energy the behavior for these near degenerate levels is identical to that of the degenerate levels in Figure 5.

The intimate connection between χ_{DIA} and χ_{VV}, leading to these near cancellations, is in fact very fundamental. Expressions (23) and (24) depend on the free choice that was made for the vector potential **A** (the gauge). By changing our choice for **A**, the values for χ_{VV} can be made arbitrarily large. Any physical observable should not depend on the choice of gauge, and indeed the accompanying changes in χ_{DIA} make the sum $\chi_{\text{DIA}} + \chi_{\text{VV}}$ gauge independent [8, 22, 21]. Therefore, it is unphysical and incorrect to take only one of the two into account. Only in special situations, the separation $\chi = \chi_{\text{DIA}} + \chi_{\text{VV}}$ can be useful. One example is that for spherical systems with filled l-shells (and for proper choice of the gauge), $\chi_{\text{VV}} = 0$ so that $\chi = \chi_{\text{DIA}}$. In general, however, we must consider the two contributions together.

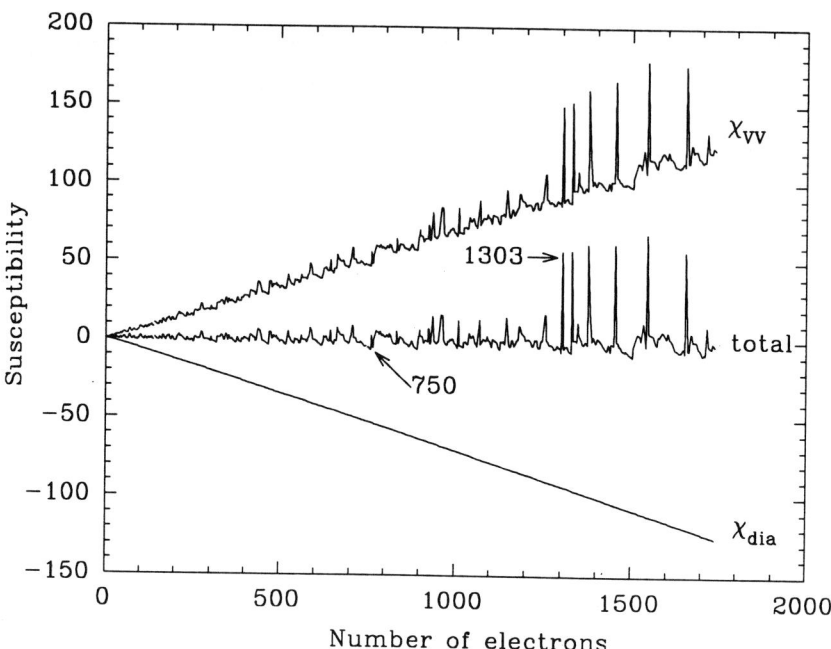

Fig. 6. The orbital susceptibility for a rectangular box at $T = 0$ as a function of the number of particles in the box. The dia- and paramagnetic contributions approximately cancel. Here, the ratio of sizes along the x- and y- axes is $(L_1/L_2)^2 = 1.1$. The units for χ are $\mu_0 e^2/m_e L_1$.

There seem to be no simple rules to predict whether the total susceptibility will be positive or negative. Only for some specific highly symmetric systems a few predictions can be made. When a set of degenerate levels is partially filled, Curie paramagnetism results; when it is completely filled the closed shell configuration is diamagnetic. An example for the latter is benzene (see also [23]). Six π-electrons of the benzene molecule are strongly delocalized over the carbon ring. The orbitals can be approximated by one dimensional plane waves $\psi(\theta) = (2\pi)^{-1/2} \exp(ik\theta)$, with $(k = 0, \pm 1, \pm 2, \ldots)$. Two electrons occupy the $k = 0$ state, the other four are in the degenerate $k = \pm 1$ states. The closed shell structure is indeed strongly diamagnetic, as are all known (4n+2)- annulenes. Rings with other numbers of carbon atoms, or charged rings, are paramagnetic; the paramagnetism is not Curie type but rather Van Vleck type, however, as a result of a Jahn–Teller distortion of the ring which lifts the orbital degeneracy.

When we consider a non-uniform distribution of clusters, we obtain a remarkable prediction from the models described above. For a distribution of cluster sizes and shapes, the high paramagnetic excursions in Figure 6 do not average out, as is shown in Figure 8. For sufficiently wide distributions the result is always paramagnetic and may be much larger than $|\chi_L|$ [21].

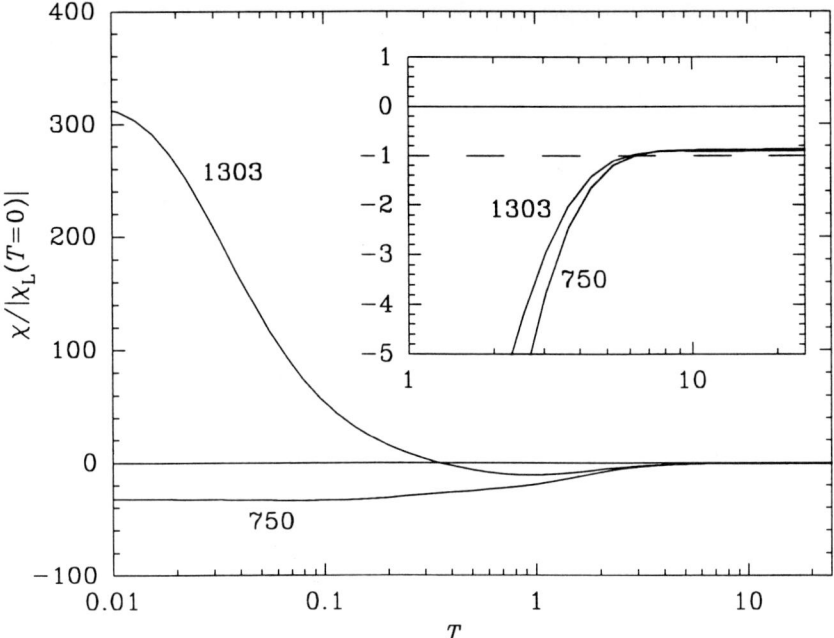

Fig. 7. Temperature dependence for $N = 750$ and $N = 1303$ (first high paramagnetic excursion in Figure 6) of the rectangular box. The inset shows how the bulk Landau value is approached.

When the temperature is raised such that $k_B T$ becomes comparable to the energy level spacing, such size effects as discussed here are strongly reduced. For a full suppression of all size effects, however, the temperature must be raised much further. The larger temperature scale is set by the size-induced fluctuations in the density of levels, which exist on a scale $\hbar v_F/L$, where v_F is the Fermi velocity [20]. Only for $k_B T \gg \hbar v_F/L$ it is possible to observe the bulk spin and orbital susceptibility.

Although the appearance of this new energy scale, $\hbar v_F/L$ may appear surprising, it finds a simple explanation in the tendency of the energy levels of free electrons in small particles to bunch together in 'shell' structures. This phenomenon has been the subject of intensive theoretical and experimental studies in the last few years [24, 25, 26, 27]. As a consequence of the bunching, two different energy scales need to be considered, namely the average separation of levels, and the average separation of shells. Only when the thermal energy exceeds the latter the behavior of the system becomes bulk metal like. The existence of these shells has been beautifully demonstrated in the experiments on the mass spectra of alkali metal clusters by Martin *et al.* [25].

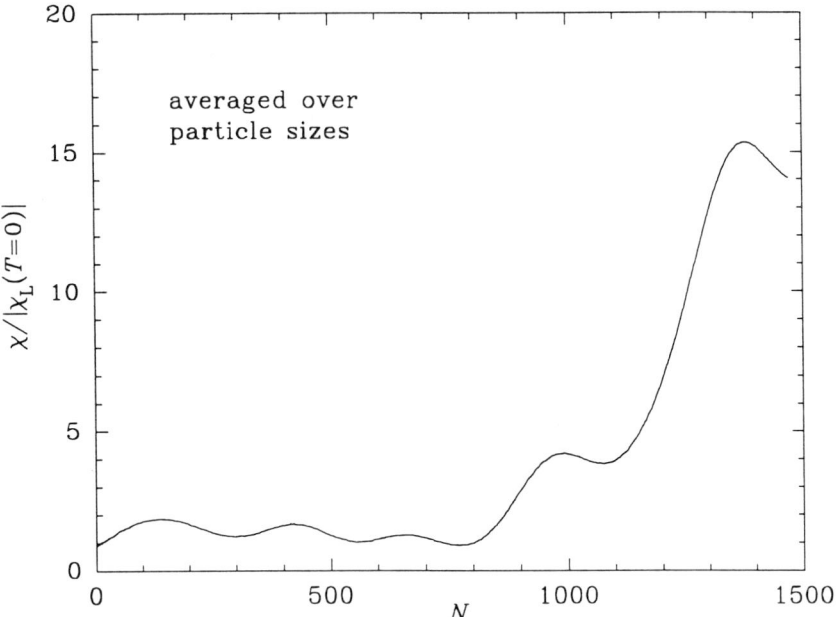

Fig. 8. Average susceptibility of a gaussian distribution of boxes with different N values, of width $\sigma = 90$ as a function of the center of the distribution N_0.

How relevant may these predictions based on models of free electrons in a box be for the experimental metal clusters? Obviously, these will be in many cases much more complicated than a free electron model. Apart from exchange and correlation effects and the spin-orbit interaction, which will be briefly discussed below, it is clear that the lattice potential due to the ions that constitute the cluster strongly scatter the electrons, and thereby profoundly change the energy level structure. This is particularly true for the transition metal atoms, for which the d-band is quite narrow. In general, a numerical molecular orbital calculation for each particular cluster species is required, for a realistic approach to the energy level structure. However, for some metals, notably those with only s- and p-valence electrons, the free electron model works remarkably well to describe the bulk properties. For clusters of those metals, the conclusions presented above should be relevant. In general, also for the more complicated metals, the message to remember is that a systematic increase as a function of cluster size of diamagnetism or paramagnetism of orbital origin, is not expected. Whether more realistic models will produce such systematics remains to be investigated, but at present there is neither theoretical, nor experimental evidence.

Finally, we would like to point out that the cancellation of the large paramagnetic and diamagnetic terms, that is found throughout, is required by the correspondence principle: For any classical system the Bohr-van Leeuwen theorem [28] maintains that the orbital moment is zero. At $k_B T \gg \Delta E$ the quantization is washed out and the classical result should be obtained. The reason for the small Landau diamagnetism, χ_L, to persist is that χ_L does not depend directly on the size of $\hbar$, but the quantum nature enters via the existence of a Fermi energy. Indeed, for $k_B T > E_F$, χ_L vanishes as $\hbar^2$.

10.2.3.3. Exchange and Correlation Effects
The effects of exchange and correlation that are important for the description of bulk metals, have been applied to clusters in general [14, 32] or to specific realizations of clusters by explicit spin-polarized energy level calculations [29, 30, 31, 33]. Clusters of the transition metals Fe, Ni and Co are calculated to be ferromagnetic, with moments per atom comparable to the bulk metal, even for clusters of only a few atoms. Ligands bonded to the surface have the effect of suppressing the magnetism, as is discussed below.

An interesting cluster effect was predicted as a result of the symmetry of the clusters. Clusters of 13 atoms of transition elements such as Fe, Pd, Rh and Rh, have a highly symmetric shape, both when packed as icosahedra or as cuboctahedra (fcc). The high symmetry results in a density of levels which is enhanced compared to the bulk density. The exchange interaction is very sensitive to the density of states and the higher density results in magnetic moments which are much larger than for bulk metals. Such giant magnetic moments were obtained in local density calculations [33].

The giant moments as considered in [33] are an amplification of the exchange interactions of the d-electrons of the transition metals, which exist also in bulk metals and this is an interaction which is strongly localized on the atoms. In highly symmetric free electron clusters, however, large moments may result as a consequence of the analog of the Hund rules in atoms. Consider, as an illustrative example, a spherical potential well, with otherwise free electrons. When a particular l-shell is only partially filled, electron-electron interactions will favor filling different m orbital quantum numbers with all spins *parallel*. Such mechanisms have been proposed to explain magnetic moments in even-electron Os clusters [34].

A high degeneracy or high density of levels can in practice be avoided by the clusters, since a lower energy state can be found through a slight deformation of the clusters, which lowers the symmetry (Jahn-Teller distortion).

10.2.3.4. The Spin-Orbit Interaction
In small metallic particles of the heavier metals the spin-orbit interaction cannot be neglected. An introductory discussion is given by Halperin [2]. As a consequence, the magnetic moment of an odd-electron cluster is reduced due to the mixture of states: spin up and down cease to be eigenstates of the problem. Also the even-electron character is suppressed due to this admixture of states and for very strong spin-orbit coupling the distinction between even

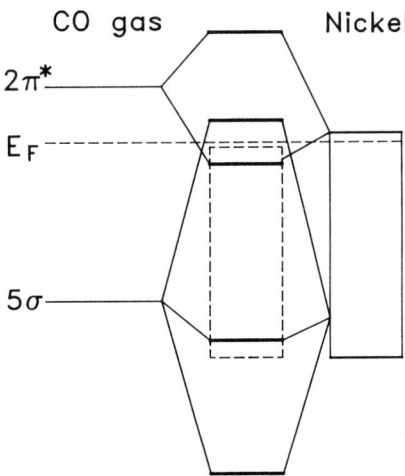

Fig. 9. Schematic energy level diagram for the interaction between Ni atoms and carbon monoxide (adapted from Raatz and Salahub [29]).

and odd electron clusters is removed and the value of the Pauli paramagnetic susceptibility is found at all temperatures.

10.2.3.5. Ligand Coordination: Suppression of Magnetism
The influence of the bonding of the ligands to the surface has particularly dramatic effects in the case of magnetic transition metal clusters. It is known that chemisorption of molecules, such as O_2 and CO, suppress the magnetization at the surface of transition metal particles. This phenomenon was investigated theoretically in detail for CO bonding on Fe, Ni and Co clusters by Holland, Ellis and Trogler [30], by Salahub and Raatz [29], and by Pacchioni, Rösch and coworkers[31]. They use spin polarized calculations to investigate the electronic structure of free metal cluster surfaces and the effect of CO bonding to the surface atoms. Figure 9 shows the schematic energy structure that Salahub and Raatz propose to explain the suppression of magnetism. The 5σ orbitals of the CO molecule form bonding and antibonding levels with the Ni d-states. The antibonding levels are pushed above the chemical potential and emptied into the Ni d-band. The result is that the Ni d-states, which were almost filled, are now completely filled. As a consequence of the complete filling of all d-states the magnetism is destroyed. The suppression of magnetic moments is a very local effect: for a Ni_{13} cluster the magnetism on the central Ni atom survives CO chemisorption to the 12 surface atoms.

Recently, all electron calculations have been perormed for a series of Ni carbonyl clusters, closely approximating the structure of the actual experimental species [31]. Calculations for clusters containing up to 44 Ni atoms have been performed. The magnetic properties of these clusters will be compared to the calculated results below.

10.3. Experiments on Metal Cluster Compounds

The analysis of the magnetic susceptibility often starts with a separation of terms of the form

$$\chi = \chi_0 + \chi_\infty + \frac{C}{T - \theta}. \qquad (25)$$

Here χ_0 is a constant diamagnetic contribution due to the organic ligands and the core electrons of the metal atoms. Reliable values to calculate χ_0 can be found in the literature. In most cases it forms a minor contribution, so that details of the type of bonding do not severely influence the results. The temperature independent contribution, χ_∞, is a result of orbital diamagnetism, Van Vleck paramagnetism and Pauli spin paramagnetism of the valence electrons of the metal atoms. Note that in calculating χ_0 it is important to take only the core electrons of the metal atoms into account. Would we also include the valence electron diamagnetism of the neutral atom, then the resulting χ_∞ does not describe the magnetism of the valence electrons in the cluster in the absolute sense, but that relative to the diamagnetism of the free atoms.

The temperature dependent part of the susceptibility is usually fitted by the last term in Equation (25). The Curie constant $C = \mu_B^2 \mu_{\text{eff}}^2 / 3k_B$ measures the uncompensated magnetic moment per cluster. The effective moment μ_{eff} can be expressed in terms of the gyromagnetic constant and the total angular momentum quantum number, J: $\mu_{\text{eff}} = g\sqrt{J(J+1)}$. The Curie–Weiss temperature θ is a result of possible (anti-) ferromagnetic interactions between the clusters in a solid (or within the cluster).

The field dependence of the magnetic moment of an isolated cluster ($\theta = 0$), with a total angular momentum quantum number J is given by

$$M(B) = (\chi_0 + \chi_\infty) B/\mu_0 + gJ\mu_B B_J(x), \quad (x = gJ\mu_B B/k_B T). \quad (26)$$

Here, $B_J(x)$ is the Brillouin function, and the first term in Equation (26) gives a linear background to this function. Expressions (25) and (26) describe the simplest possible situation. They are mainly used as a starting point in order to identify the interesting deviations.

10.3.1. THE SPIN SUSCEPTIBILITY

10.3.1.1. The Even-Odd Dichotomy
The metal cluster compounds are particularly well suited to investigate the distinction between even and odd clusters, since they allow the study of clusters of uniform size. For metal cluster compounds of a few metal atoms the expected behavior is indeed observed. For Os clusters [34] up to Os_7 the temperature independent diamagnetic susceptibility is consistent with

the even number of electrons in the clusters, and excitation energies much larger than $k_B T$ at room temperature. In a series of Nb_6 and Ta_6 halide clusters [36] which can be produced in the charge states 2+, 3+ and 4+, it is found that for the even electron clusters the susceptibility is temperature independent, and that the Curie paramagnetism observed for the 3+ odd electron clusters is very well described by one independent electron moment per cluster. For higher nuclearity metal cluster compounds, however, the experimental information is much more complicated [34, 35, 37, 38, 39]. The clusters of the magnetic elements shall be discussed separately, below. Apart from those, the most remarkable property observed for the higher nuclearity metal cluster compounds is a Curie type paramagnetism corresponding to a fraction of a spin per cluster, independent of the electron count. An example of this behavior is given in the chapter by Benfield for the Os_{10} cluster.

The origin of these fractional moments is still unclear, in particular in view of the fact that the number of electrons is even, in the majority of the examples given. The obvious interpretation would be to ascribe it to some paramagnetic contamination. Indeed, many cluster compounds readily react with small amounts of oxygen or water in the atmosphere. An intrinsic mechanism for the fractional moments has been proposed by Drake et al. [34]. Essentially, they propose that the ground state of the clusters is paramagnetic, as a result of the exchange and correlation effects discussed in Section 10.2.3.3, and they further have to assume that the resulting moment is strongly reduced by the spin-orbit interaction (Section 10.2.3.4).

Other possibilities for interpretation include slight variations in the structure or composition of the clusters, as for $Ni_{38}Pt_6$ (see below). Many clusters contain hydrogens as interstitials or surface atoms. The number of hydrogens may be different from one cluster to the next, implying that the number of electrons may also be different. The most reliable approach to a homogeneous sample is to work on single crystals, when available. Single crystals also reduce the risk of contamination by reaction with an impure atmosphere.

10.3.1.2. Pauli Paramagnetism

As metal cluster compounds of sufficiently large size begin to come available it is possible to observe Pauli spin paramagnetism. A systematic size dependence of the Pauli susceptibility was observed in a series of large Pd clusters and colloids [40]. The study involves a $Pd_{561}Phen_{36}O_{200}$ 'Schmid'-cluster. The metal core of this compound consists of a central atom surrounded by 5 layers of close packed nearest neighbors. Further, a similarly produced mixture of 7 and 8 shell clusters was used, in addition to a 150 Å colloid stabilized using the same type of ligands. For the smallest system in this series, Pd_{561}, the average energy level separation $\Delta E/k_B$ can be estimated to be of order 20 K. Above this temperature Pauli paramagnetism develops. Observation of quantum size effects below this temperature is probably inhibited by the strong spin-orbit coupling. The susceptibility as a function of

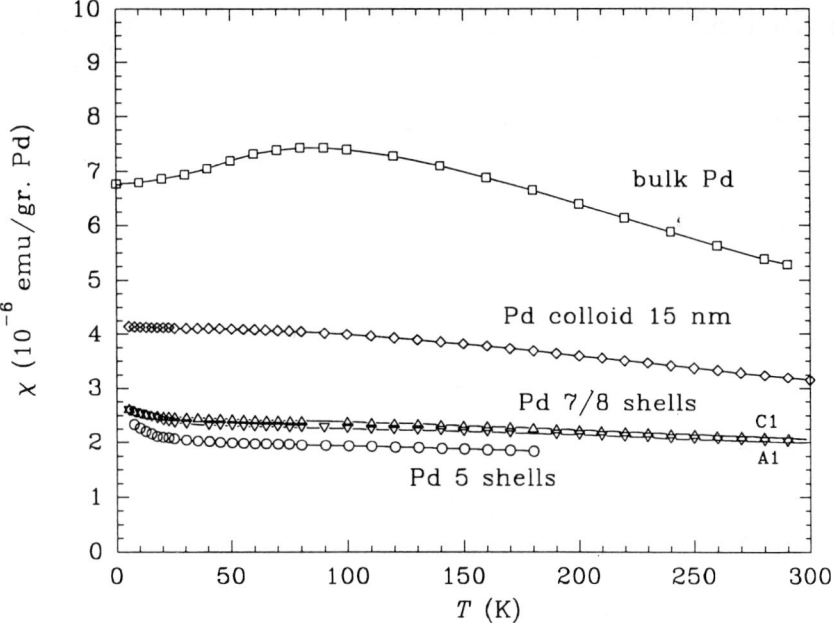

Fig. 10. The temperature dependent susceptibility (in 0.1 T) of various Pd-clusters compared to bulk Pd. The values are normalized to the (estimated) weight of the Pd cores, and corrected for various diamagnetic contributions. The bulk measurements are adapted from Manuel and St Quinton [41].

temperature measured for these three systems, using a SQUID magnetometer, is reproduced in Figure 10, and compared to the behavior of bulk Pd.

As was mentioned in Section 10.2.2, the Pauli susceptibility of Pd is exchange enhanced by a Stoner factor $S = 9.4$. The temperature dependence is a result of a marked energy dependence in the density of states. The susceptibility of the clusters shows an exchange enhancement, but one which is reduced with respect to the bulk. A temperature dependence is also observed, again reduced with respect to the bulk. The size dependence was described with a model which assumes a reduction of the density of states at the surface as a result of the ligand bonding, in analogy to similar effects on nickel surfaces described in Section 10.2.3.5. Accordingly, the susceptibility without enhancement effects, was taken to depend on the radial coordinate as

$$\chi_{\text{Pauli}}(r) = \chi_{\text{Pauli}}(\text{bulk})(1 - Ae^{(r-R)/\lambda}). \tag{27}$$

Here, A is the reduction factor at the surface, which heals back over a characteristic length λ, away from the surface of the cluster, and R is the cluster radius. Then, the Stoner enhancement factor is calculated from $S = 1/(1 - I\bar{\chi}_{\text{Pauli}})$, where $\bar{\chi}_{\text{Pauli}}$ is the average of $\chi_{\text{Pauli}}(r)$ over the cluster, and I is the interaction constant determined from the bulk susceptibility. A

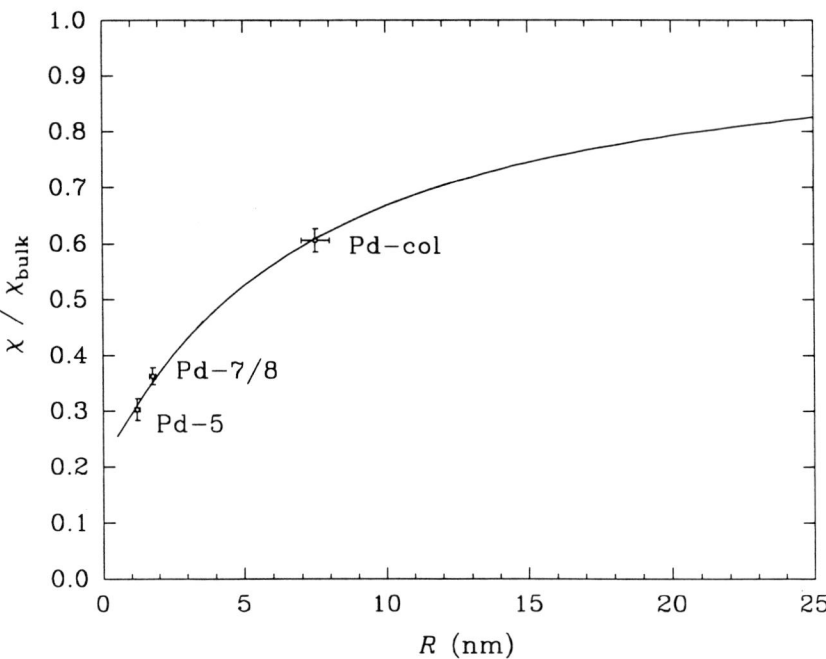

Fig. 11. The $T = 0$ extrapolated values of the susceptibility in Figure 10. The curve is a fit of the model describing the size dependence.

fit of this model to the low temperature data (Figure 11) gives $A = 0.32$ and $\lambda = 0.68$ nm, which appear to be reasonable numbers. Note that the resulting size-induced reduction of χ extends to very large diameters. In view of the limited number of points, it is reassuring that the temperature dependence $d\chi/dT$ scales as S^2, in agreement with the model. Still, measurements on a larger range of cluster sizes are required to test the validity of the model.

Indeed, for sufficiently small clusters the Stoner model, based on a smooth density of states, is expected to break down. Experiments on a series of Pt clusters, smaller in size than the Pd clusters of the previous work, showed a much stronger suppression of the susceptibility [20], see Figure 12. The clusters all show a weak Curie paramagnetic contribution, corresponding to much less than an average of one spin per cluster. These signals are attributed to impurities and subtracted in order to find the temperature independent part. For clusters of 55 Pt atoms or less, the susceptibility is diamagnetic and equal to that of the electrons in the atom core only. Presumably, the energy level spacing is large enough in order to quench the susceptibility of the valence electrons. For a Pt_{309} 4-shell Schmid cluster the susceptibility becomes paramagnetic, but a factor ~ 5 less than for bulk Pt. A Pt colloid of 3 nm diameter shows about half the bulk value. The resulting variation of the susceptibility as a function of the diameter is too rapid to be described by the

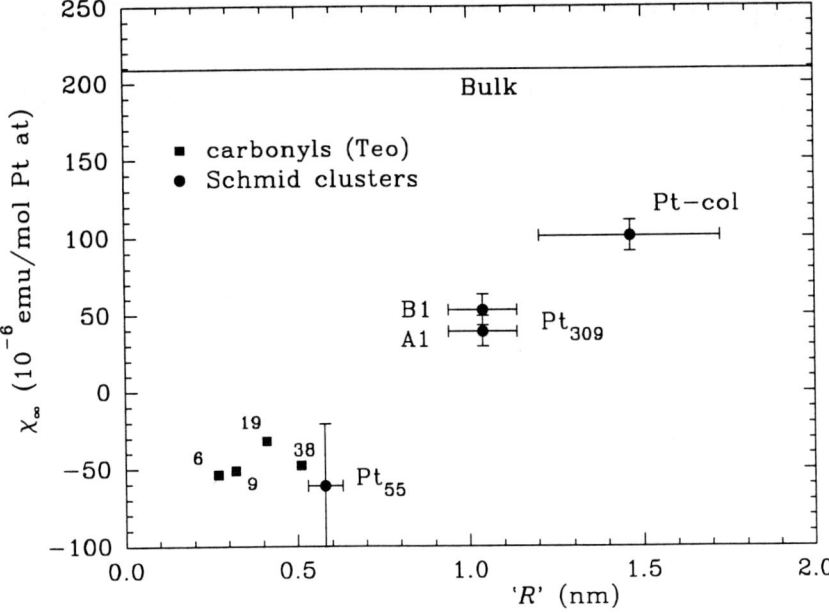

Fig. 12. The temperature independent susceptibility as a function of cluster size for a series of Pt clusters. For the Pt carbonyl clusters the numbers indicate the cluster nuclearity [38].

model employed for Pd clusters, as seen in Figure 13. The full transition from the atomic behavior, observed here for Pt, to the size-reduced bulk behavior, found for Pd, remains to be explored with clusters of one type of metal only.

10.3.1.3. Clusters Containing Ni or Co

For ferromagnetic metals the exchange interactions tend to align the spins, so that large paramagnetic moments may be formed in a cluster. As is discussed in Section 10.2.3.5 the ligands act to suppress the moments of the atoms at the surface. Many of the smaller Ni carbonyl clusters (Ni_9, Ni_{10}, Ni_{12}) are observed to have no paramagnetic moment, if a small (much less than 1 spin per cluster) impurity signal is discarded [20]. As all Ni atoms in these clusters are at the surface, and bonded to the CO ligands, this is to be expected, and is also found in the calculations [31]. Previous experiments [37] reporting moments of the order of 1 spin per cluster for Ni_9 and Ni_{12} probably suffered from a degradation of the samples by reaction with the atmosphere at the surface, although care was take to handle and store the samples in inert gasses. The former work was done on single crystals, reducing the risk of such contaminations.

For the larger Ni carbonyl clusters, made by the Italian groups of Longoni and Ceriotti, Curie magnetic moments of the order of several Bohr magnetons have been measured [37]. The largest moment is observed for

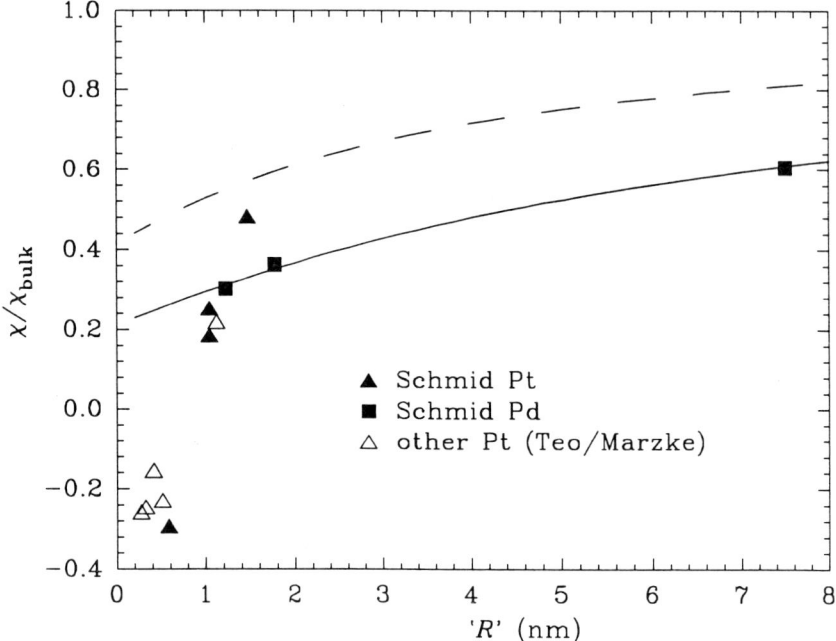

Fig. 13. The temperature independent susceptibility data for both Pd and Pt measurements. The full curve is the fit through the Pd cluster data, and the broken curve uses the same parameters for A and λ as for the Pd fit, but a Stoner factor $S = 2.95$, appropriate for Pt. The open symbols represent data for Pt taken from the literature [38]. They are Pt carbonyl clusters with 6, 9, 19 and 38 Pt atoms in the core, and a Pt colloid in gelatin of 22.4 Å average diameter.

$[Ni_{34}(CO)_{38}C_4H](NEt_4)_5$, corresponding to about 4 unpaired spins per cluster. The $Ni_{38}Pt_6$ clusters, which are larger in size than the Ni_{34} cluster, have smaller moments. This is in agreement with the structure of the cluster, being built from a Pt cuboctahedron at the centre, surrounded by the Ni atoms, which are all at the surface. From the model calculations on a Ni_{44} cluster of the same structure (the Pt atoms being replaced by Ni) Rösch *et al.* show that only the moments on the central Ni atoms survive CO bonding [31]. As in the actual compound the central positions are occupied by Pt, the magnetic moment for the real structure is expected to be much smaller, or even zero.

In [37] it was observed that the magnetization of the larger Ni clusters as a function of magnetic field is quite unusual: the moment does not saturate even in fields as high as 30 Tesla. A detailed study on single crystals of one of these clusters, $[Ni_{38}Pt_6(CO)_{48}H](AsPh_3)_2(NBu_4)_3$, was recently performed, using a 6 Tesla SQUID magnetometer [20]. The low temperature moment in a field of 6 T is as small as 0.5 μ_B per cluster, but continues to increase with field, as observed by Pronk *et al.* The anomalous field and temperature dependence, as reproduced in Figure 14, was interpreted as follows.

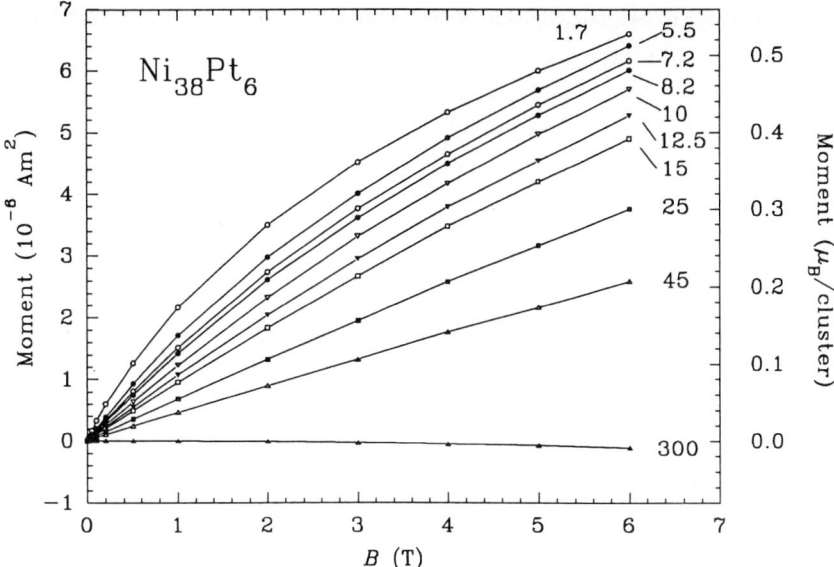

Fig. 14. Magnetic moment vs. applied field for the $Ni_{38}Pt_6$ crystal. The temperature for each curve is given in Kelvins.

From X-ray crystallography there is strong evidence for Pt on Ni positions. This structural disorder suggests that also some clusters may have Ni atoms in central positions, surrounded by metal atoms only. Assuming a reasonable value for an antiferromagnetic exchange interaction between two neighboring clusters, carrying moments due to central Ni positions, a model of a diluted random antiferromagnet was applied to the data. All characteristic features could be accounted for, assuming 20% of the clusters carry a moment of 5 μ_B and an exchange interaction strength of 10 K.

Apart from the carbonyl clusters an interesting series of Ni and Co selinide or telluride clusters are being investigated by and Steigerwald et al. [39]. Despite the presence of the ferromagnetic metal atoms the bulk structures are non-magnetic semimetals. The charge state of a cluster such as $[Ni_9Te_6L_8](BF_4)_n$ can be adjusted [39] by the number of counter ions from $n = 0$ to $n = 2$. At low temperatures the moment on the nickel telluride cluster changes from small, to ~ 1 μ_B, back to small when the charge is changed from 0+ to 1+ to 2+. This is just what is expected for non-magnetic clusters. However, for cobalt telluride, the analogous series brings the moment from approximately 0, to 1, to 2 μ_B. The authors claim that the temperature dependence of the magnetic moments is highly anomalous, which would distinguish them from the other Ni cluster compounds studied so far.

10.3.2. THE ORBITAL MAGNETIC SUSCEPTIBILITY

Contributions of the electron orbital motion to the magnetic susceptibility for clusters have been discussed in a very limited number of papers. A prominent cluster in this context is C_{60}, which is not ordinarily regarded as a metallic cluster, although graphite is a metal. In analogy to the free-electron like susceptibility of benzene rings, it was speculated [42] that C_{60} would be strongly diamagnetic. However, detailed calculations [43] and subsequent experiments [44] agreed that the susceptibility was small and paramagnetic. In view of the discussion in Section 10.2.3.2, this is not surprising, as only for a filled shell geometry unusually large diamagnetism is expected. The 60 π-electrons of the cluster occupy orbits, which, in the approximation of free electrons on a spherical shell, are described [23] by spherical harmonics $Y_{lm}(\theta, \phi)$. The $l = 4$ shell is fully occupied, but the $l = 5$ shell has 22 states and only 10 electrons. The Zeeman splitting of these levels would give rise to strong Curie paramagnetism, however, the icosahedral structure of the cluster partially lifts the degeneracy, leaving a closed sub-shell structure with a separation of about 1 eV to the next levels, and the result is a small Van Vleck paramagnetic signal.

The magnetic properties for a Os-series of clusters were originally presented [34] as evidence for growing Van Vleck paramagnetism as a function of cluster size. However, the paramagnetism presented in ref.[34] is a result of the correction that was calculated for all the electrons of the Os metal atoms, i.e., including the valence electrons [21]. As explained in the introduction of Section 10.3, it is appropriate to correct only for the diamagnetism of the core electrons, if we are interested to learn of the behavior of the valence electrons in the cluster. With the latter correction applied the susceptibility as a function of cluster size is found to fluctuate around zero [21], in agreement with the arguments given in Section 10.2.3.2.

Kozik *et al.* have employed a very sensitive NMR technique to measure the change in susceptibility in a series of heteropoly blue complexes, when the number of electrons on the ion is changed [45]. Heteropoly blue complexes are ions of the type $[SiW_{12}O_{40}]^{6-}$, $[P_2W_{18}O_{62}]^{8-}$, etc., where the valence of the ions in the reduced state is given. They are readily oxidized to have two electrons less. For one additional electron the count is odd, and the complex is Curie-paramagnetic, as expected. In the $2e$ reduced state the complexes are found to be more diamagnetic than the oxidized parent ions. This behavior is observed for 7 different species. The diamagnetism was ascribed to the screening currents of the two added electrons, which are delocalized over the complex. The size of the signal is of the order of magnitude expected for the classical screening currents and the size of the complex. A full explanation of these intriguing results has not been given. As explained above, diamagnetic and paramagnetic contributions are equally expected for additional electrons. Possibly, the fully reduced state forms a closed shell configuration with a large energy separation to the next excited state, in which case diamagnetism is indeed expected. That this would occur for every one of the 7 species investigated is not a coincidence, as also the optical properties (blue) are similar.

References

1. R. Kubo, *J. Phys. Soc. Japan* **17** (1962) 975.
2. W. P. Halperin, *Rev. Mod. Phys.* **58** (1986) 533.
3. W. A. de Heer, P. Milani and A. Châtelain, *Phys. Rev. Lett.* **65** (1990) 488.
4. D. C. Douglass, J. P. Bucher and L. A. Bloomfield, *Phys. Rev. Lett.* **68** (1992) 1774.
5. See, e.g., L. D. Landau and E. M. Lifshitz, *Course of Theoretical Physics*, Vol. 3: *Quantum Mechanics*, 3rd ed., Pergamon Press (1977), pp. 461–468; N. W. Ashcroft and N. D. Mermin, *Solid State Physics*, Holt, Rinehart and Winston (1976), pp. 643–659.
6. See, e.g., L. D. Landau and E. M. Lifshitz, *Course of Theoretical Physics*, Vol. 5: *Statistical Physics, Part 1*, 3rd ed., Pergamon Press (1977), pp. 173–174.
7. C. Kittel, *Introduction to Solid State Physics*, John Wiley and Sons (1986).
8. J.H. Van Vleck, *The Theory of Electric and Magnetic Susceptibilities*, Oxford University Press (1932).
9. D. Shoenberg, *Magnetic Oscillations in Metals*, Cambridge University Press (1984).
10. J. Callaway, *Quantum Theory of the Solid State*, Academic Press (1974).
11. L. M. Roth, *Phys. Rev.* **145** (1966) 434.
12. See e.g., F. Gautier, in M. Cyrot (ed.), *Magnetism of Metals and Alloys*, North-Holland (1982).
13. It is interesting to note that, in classical mechanics, for a particle in a potential well of arbitrary shape, it is not always possible to find a stationary orbit. This remarkable difference between quantum and classical mechanics forms the subject of the field of quantum chaos, see e.g. M. V. Berry, *Proc. R. Soc. Lond. A* **413** (1987) 183.
14. R. Denton, B. Mühlschlegel, D. J. Scalapino, *Phys. Rev. B* **7** (1973) 3589.
15. J. A. A. J. Perenboom, Thesis, Nijmegen (1979).
16. J. A. A. J. Perenboom, P. Wyder and F. Meier, *Phys. Rep.* **78** (1981) 173.
17. S. Tanaka and S. Sugano, *Phys. Rev. B* **34** (1986) 740.
18. J. M. van Ruitenbeek and D. A. van Leeuwen, *Mod. Phys. Lett. B* **7** (1993), and references there in.
19. R. B. Dingle, *Proc. Roy. Soc.* **212A** (1952) 47.
20. D. A. van Leeuwen, Thesis, Leiden (1993).
21. J. M. van Ruitenbeek and D. A. van Leeuwen, *Phys. Rev. Lett.* **67** (1991) 640; J. M. van Ruitenbeek, *Z. Phys. D* **19** (1991) 247.
22. J. S. Griffith, *The Theory of Transition Metal Ions*, Cambridge University Press (1961).
23. J. M. van Ruitenbeek and D. A. van Leeuwen, in P. Jena, S. N. Khanna and B. K. Rao (eds.), *Physics and Chemistry of Finite Systems: From Clusters to Crystals*, Vol. 1, (NATO ASI Series, Vol. C374), Kluwer Academic Publishers (1992), p. 807.
24. V. Subrahmanyam and M. Barma, *J. Phys. A: Math. Gen.* **22** (1989) L489.
25. T. P. Martin, T. Bergman, H. Göhlich and T. Lange, *Z. Phys. D* **19** (1991) 25.
26. H. Nishioka, K. Hansen and B. R. Mottelson, *Phys. Rev. B* **42** (1990) 9377.
27. F. von Oppen and E. K. Riedel, *Phys. Rev. B* **48** (1993) 9170.
28. R. Peierls, *Surprises in Theoretical Physics*, Princeton University Press (1979).
29. D. R. Salahub and F. Raatz, *Intern. J. Quantum Chem.* **18s** (1984) 173; F. Raatz and D. R. Salahub, *Surface Science* **176** (1986) 219.
30. G. F. Holland, D. E. Ellis and W. C. Trogler, *J. Chem. Phys.* **83** (1985) 3507.
31. G. Pacchioni and N. Rösch, *Inorg. Chem.* **29** (1990) 2901; N. Rösch, L. Ackermann, G. Pacchioni and B. I. Dunlap, *J. Chem. Phys.* **95** (1991) 7004; L. Ackermann, N. Rösch, B. I. Dunlap and G. Pacchioni, *Int. J. Quant. Chem.* **26** (1992) 605.

32. P. Joyes, *Les Agrégats Inorganiques Élémentaires*, Les Éditions de Physique (1990).
33. B. I. Dunlap, *Phys. Rev. A* **41** (1990) 5691; B. V. Reddy, S. N. Khanna and B. I. Dunlap, *Phys. Rev. Lett.* **70** (1993) 3323.
34. S. R. Drake, P. P. Edwards, B. F. G. Johnson, J. Lewis, E. A. Marseglia, S. D. Obertelli and N. C. Pyper, *Chem. Phys. Lett.* **139** (1987) 336; D. C. Johnson, R. E. Benfield, P. P. Edwards, W. J. H. Nelson and D. Vargas, *Nature* **314** (1985) 231; see also this volume, Chapter 4.
35. J. A. O. de Aguiar, A. Mees, J. Darriet, L. J. de Jongh, S. R. Drake, P. P. Edwards, B. F. G. Johnson and J. Lewis, *Solid State Commun.* **66** (1988) 913.
36. J. G. Converse and R. E. McCarley, *Inorg. Chem.* **9** (1970) 1361.
37. B. J. Pronk, H. B. Brom, L. J. de Jongh, G. Longoni and A. Ceriotti, *Solid State Commun.* **59** (1986) 349.
38. B. K. Teo, F. J. DiSalvo, J. V. Waszczak, G. Longoni and A. Ceriotti, *Inorg. Chem.* **25** (1986) 2262; R. F. Marzke, W. S. Glaunsinger and M. Bayard, *Solid State Commun.* **18** (1976) 1025.
39. T. T. M. Palstra, M. L. Steigerwald, A. P. Ramirez, Y.-U. Kwon, S. M. Stuczynski, L. F. Schneemeyer, J. V. Waszczak and J. Zaanen, *Phys. Rev. Lett.* **71** (1993) 1768.
40. D. A. van Leeuwen, J. M. van Ruitenbeek, G. Schmid and L. J. de Jongh, *Phys. Lett. A* **170** (1992) 325.
41. A. J. Manuel and J. M. P. St. Quinton, *Proc. Roy. Soc.* **A273** (1963) 412.
42. H. W. Kroto, J. R. Heath, S. C. O'Brien, R. F. Curl and R. E. Smalley, *Nature* **318** (1985) 162.
43. V. Elser and R. C. Haddon, *Nature* **325** (1987) 792.
44. R. C. Haddon *et al.*, *Nature* **350** (1991) 46.
45. M. Kozik, N. Casan-Pastor, C. F. Hammer and L. C. W. Baker, *J. Am. Chem. Soc.* **110** (1988) 7697.

INDEX OF CHEMICAL COMPOUNDS

The organisation of the chemical index is as follows. First, a separate "carbonyl cluster index" is presented, in which all the compounds mentioned in the review chapter 2 have been collected. In a second index the chemical formulae to be found in the remaining chapters have been compiled.

I. Carbonyl Cluster Index (chapter 2) (as provided by Ceriotti et al.)

The index is organized by descending groups of the Periodic Table from left to right. The clusters are listed with respect to increasing nuclearity. Heterometallic clusters are listed under the heading of homometallic clusters (T means Table).

Molybdenum
$Mo_2\{O_2CCCo_3(CO)_9\}_4\{HO_2CCCo_3(CO)_9\}_2$ T.2
$Mo_4Hg_4\{Mo(CO)_3Cp\}_4$ T.2, 84

Rhenium
$[Re_7C(CO)_{21}]^{3-}$ 51
$[Re_7AgC(CO)_{21}]^{2-}$ 56
$[\{Re_7AgC(CO)_{21}\}_2Br]^{5-}$ T.2, 56, 82

Iron
$[Fe_4(CO)_{13}]^{2-}$ 57
$[HFe_6Pd_6(CO)_{24}]^{3-}$ T.2, T.6, 57, 75, 90
$[Fe_8Ag_{13}(CO)_{32}]^{4-}$ T.2, T.6, 57, 90

Ruthenium
$[HRu_3(CO)_{11}]^-$ 57
$Ru_5C(CO)_{15}$ 56
$[Ru_6C(CO)_{16}]^{2-}$ 55, 56
$[Ru_{10}C(CO)_{24}]^{2-}$ 51
$[Ru_{10}C_2(CO)_{24}]^{2-}$ 51, 88
$[\{Ru_6C(CO)_{16}\}_2Tl]^-$ T.1, 55, 83
$[Ru_6Pd_6(CO)_{24}]^{2-}$ T.2, T.6, 57, 75, 90
$[Ru_9Pt_6(CO)_{28}]^{4-}$ T.2, T.6, 57, 76
$\{Ru_5C(CO)_{14}Cl\}_2(HgCl)_2$ T.2, 56, 83
$[\{Ru_6C(CO)_{16}\}_2Hg]^{2-}$ T.2, 56, 83
$[\{Ru_9C(CO)_{21}\}_2Hg_3]^{2-}$ T.2, 57, 83

Osmium
$[HOs_3(CO)_{11}]^-$ 55

Os$_3$(CO)$_{10}$(NCCH$_3$)$_2$ 51
[Os$_{10}$C(CO)$_{24}$]$^{2-}$ 51, 55, 66, 84
[Os$_{11}$C(CO)$_{27}$]$^{2-}$ 59
[Os$_{17}$(CO)$_{36}$]$^{2-}$ T.1, T.6, 51, 75
[Os$_{20}$(CO)$_{40}$]$^{2-}$ T.1, T.6, 51, 72
Os$_6$Pt$_4$(CO)$_{22}$(cod)   59
Os$_6$Pt$_7$(CO)$_{21}$(cod)$_2$ T.2, 59, 88
{Os$_3$Hg(CO)$_{11}$}$_3$ T.2, 55, 83
[Os$_{10}$C(CO)$_{24}$(HgR)]$^-$ 56
Os$_{10}$C(CO)$_{24}${Au(AuPCy$_3$)$_3$}   T.2, 59, 84
[Os$_{11}$C(CO)$_{27}${Cu(NCCH$_3$)}]$^-$ T.2, 58, 88
Os$_{11}$C(CO)$_{27}${Cu(NCCH$_3$)}$_2$ 59
[Os$_{11}$C(CO)$_{27}$(AuPPh$_3$)]$^-$ 58
Os$_{11}$C(CO)$_{27}$(AuPPh$_3$)$_2$ 58
[{Os$_9$C(CO)$_{21}$Hg$_2$]$^{2-}$ T.2, 52, 83
[{Os$_9$C(CO)$_{21}$}$_2$Hg$_2$]$^{2-}$ T.2, 63, 83
[{Os$_9$C(CO)$_{21}$}$_2$Hg$_3$]$^{2-}$ T.2, 52, 57, 63, 83
[{Os$_{10}$C(CO)$_{24}$}$_2$Hg]$^{2-}$ T.2, 55, 84

Cobalt
Co$_3$(CO)$_9$CCl   58
[Co$_6$C(CO)$_{15}$]$^{2-}$ 52
[Co$_6$N(CO)$_{15}$]$^-$ 52
[Co$_{11}$C$_2$(CO)$_{22}$]$^{3-}$ 86
[Co$_{13}$C$_2$(CO)$_{24}$]$^{4-}$ T.1, T.6, 52, 62, 66, 86
[Co$_{13}$C$_2$(CO)$_{24}$]$^{3-}$ T.1, 62, 86
[Co$_{14}$N$_3$(CO)$_{26}$]$^{3-}$ T.1, T.6, 52, 85
[Co$_{14}$P$_2$(CO)$_{27}$]$^{4-}$ T.1
{Co$_8$As$_2$(CO)$_{16}$(AsPh)$_2$}$_2$ T.1
Co$_4$O{O$_2$CCCo$_3$(CO)$_9$}$_6$ T.1
[Co$_2$Ni$_{10}$C(CO)$_{20}$]$^{2-}$ T.2, T.6, 58, 89
[Co$_3$Ni$_9$C(CO)$_{20}$]$^{3-}$ T.2, T.6, 58, 62, 63, 89
[Co$_3$Ni$_9$C(CO)$_{20}$]$^{2-}$ T.2, 62, 63, 89
{Co$_3$Hg$_3$(CO)$_9$}$_2$Hg$_3$ T.2, 84
Zn$_4$O{O$_2$CCCo$_3$(CO)$_9$}$_6$ T.2

Rhodium
Rh$_4$(CO)$_{12}$ 53
Rh$_6$(CO)$_{16}$ 43
[Rh$_6$(CO)$_{15}$]$^{2-}$ 55
[HRh$_6$(CO)$_{15}$]$^-$ 55, 63
[Rh$_6$C(CO)$_{13}$]$^{2-}$ 61
[Rh$_6$C(CO)$_{15}$]$^{2-}$ 55, 56, 60, 61, 82, 86
[HRh$_6$C(CO)$_{15}$]$^-$ 55
[Rh$_6$N(CO)$_{15}$]$^-$ 52, 54, 56
[Rh$_7$(CO)$_{16}$]$^{3-}$ 55
Rh$_8$C(CO)$_{19}$ 56, 88
[{Rh$_6$(CO)$_{15}$}$_2$]$^{2-}$ T.1, 53, 55, 63, 82
H$_2$Rh$_{12}$(CO)$_{25}$ T.1, T.6, 63, 68
Rh$_{12}$C$_2$(CO)$_{25}$ T.1, T.6, 56, 86
[Rh$_{12}$C$_2$(CO)$_{24}$]$^{2-}$ T.1, T.6, 55, 59, 63, 66, 86

INDEX OF CHEMICAL COMPOUNDS 309

$[Rh_{12}C_2(CO)_{23}]^{4-}$ T.1, 63, 66, 86
$[Rh_{12}C_2(CO)_{23}]^{3-}$ T.1, 63, 86
$[HRh_{12}N_2(CO)_{23}]^{3-}$ T.1, 54, 87
$[Rh_{12}Sb(CO)_{27}]^{3-}$ T.1, T.6, 53, 80
$[Rh_{13}(CO)_{24}]^{5-}$ 64
$[HRh_{13}(CO)_{24}]^{4-}$ T.1, T.6, 59, 64, 68, 70, 78
$[H_2Rh_{13}(CO)_{24}]^{3-}$ T.1, 64, 68, 70, 78
$[H_3Rh_{13}(CO)_{24}]^{2-}$ T.1, 53, 64, 68, 70, 78
$[Rh_{14}(CO)_{25}]^{4-}$ T.1, T.6, 53, 59, 60, 61, 63, 64, 77
$[HRh_{14}(CO)_{25}]^{3-}$ T.1, 59, 64, 77, 78
$[Rh_{14}(CO)_{26}]^{2-}$ T.1, 53, 55, 77, 78
$[\{Rh_6(CO)_{14}(CN)_2\}_2\{Rh(CO)_2\}_2]^{2-}$ T.1, 82
$[\{Rh_6C(CO)_{15}\}_2\{Rh_2(CO)_3\}]^{2-}$ T.1, 60, 82
$[Rh_{14}N_2(CO)_{25}]^{2-}$ T.1, T.6, 56, 72
$[Rh_{15}(CO)_{27}]^{3-}$ T.1, T.6, 53, 59, 60, 61, 77
$[Rh_{15}(CO)_{30}]^{3-}$ T.1, T.6, 53, 76
$[Rh_{15}C_2(CO)_{28}]^{-}$ T.1, T.6, 56, 79
$[Rh_{17}(CO)_{30}]^{3-}$ T.1, T.6, 53, 68
$[Rh_{17}S_2(CO)_{32}]^{3-}$ T.1, T.6, 53, 84
$[Rh_{22}(CO)_{37}]^{4-}$ T.1, T.6, 53, 74
$[H_xRh_{22}(CO)_{35}]^{5-}$ T.1, T.6, 53, 78
$[H_{x+1}Rh_{22}(CO)_{35}]^{4-}$ T.1, 53, 78
$[Rh_{23}N_4(CO)_{38}]^{3-}$ T.1, 52, 92
$Rh_{10}C_2(CO)_{18}(AuPPh_3)_4$ T.2, T.6, 59, 88
$Rh_{10}C_2(CO)_{20}(AuPPh_3)_4$ T.2, T.6, 59, 88
$[Rh_4Pt(CO)_{12}]^{2-}$ 60
$[Rh_5Pt(CO)_{15}]^{-}$ 54, 60
$[Rh_9Pt_2(CO)_{22}]^{3-}$ 60
$[Rh_{11}Pt_2(CO)_{24}]^{3-}$ T.2, 53, 54, 56, 60, 70
$[Rh_{12}Pt(CO)_{24}]^{4-}$ T.2, T.6, 54, 70
$[Rh_{12}Pt_2(CO)_{26}]^{2-}$ T.2, T.6, 60, 78
$[\{Rh_6C(CO)_{15}\}_2Ag]^{3-}$ T.2, 55, 82
$[Rh_{12}C_2(CO)_{23}(AuPPh_3)]^{-}$ T.2, 59
$[Rh_{13}Pt(CO)_{25}]^{3-}$ T.2, 56, 78
$[Rh_{18}Pt_4(CO)_{35}]^{4-}$ T.2, T.6, 60, 78

Iridium
$Ir_4(CO)_{12}$ 54
$[Ir_6(CO)_{15}]^{2-}$ 54, 55, 56
$[Ir_{12}(CO)_{24}]^{2-}$ T.1, T.6, 54, 74
$[Ir_{12}(CO)_{26}]^{2-}$ T.1, T.6, 55, 68
$[Ir_{14}(CO)_{27}]^{2-}$ T.1, T.6, 56, 74

Nickel
$[Ni_6(CO)_{12}]^{2-}$ 55, 57, 58, 60, 64
$[Ni_7C(CO)_{12}]^{2-}$ 62
$[Ni_8C(CO)_{16}]^{2-}$ 61
$[Ni_9(CO)_{18}]^{2-}$ 55
$[Ni_9C(CO)_{17}]^{2-}$ 57, 61, 62

$[Ni_{10}C_2(CO)_{16}]^{2-}$ 57, 61, 87
$[Ni_{10}Ge(CO)_{20}]^{2-}$ 61
$[Ni_{11}Bi_2(CO)_{18}]^{2-}$ 66
$[Ni_{11}Bi_2(CO)_{18}]^{3-}$ 66
$[Ni_{11}Bi_2(CO)_{18}]^{4-}$ 66
$[Ni_{12}(CO)_{21}]^{4-}$ T.1, T.6, 54, 64, 70
$[HNi_{12}(CO)_{21}]^{3-}$ T.1, 55, 64, 70
$[H_2Ni_{12}(CO)_{21}]^{2-}$ T.1, 55, 64, 70
$[Ni_{12}C_2(CO)_{16}]^{4-}$ T.1, T.6, 61, 62, 87
$[Ni_{12}Ge(CO)_{22}]^{2-}$ T.1, T.6, 58, 61, 80
$[Ni_{12}Sn(CO)_{22}]^{2-}$ T.1, T.6, 58, 80
$[Ni_{11}Sb_2(CO)_{18}\{Ni(CO)_3\}_2]^{4-}$ 80
$[Ni_{11}Sb_2(CO)_{18}\{Ni(CO)_3\}_2]^{3-}$ T.1, 58, 66, 80
$[Ni_{11}Sb_2(CO)_{18}\{Ni(CO)_3\}_2]^{2-}$ T.1, 58, 66, 80
$Ni_{15}Se_{10}(CO)_3(Cp^*)_8$ T.1
$Ni_{15}Se_{10}(CO)Cl_2(Cp^*)_8$ T.1
$[Ni_{16}(C_2)_2(CO)_{23}]^{4-}$ T.1, T.6, 61, 88
$[Ni_{34}C_4(CO)_{38}]^{6-}$ 58
$[HNi_{34}C_4(CO)_{38}]^{5-}$ T.1, T.6, 58, 91
$[Ni_{35}C_4(CO)_{39}]^{6-}$ T.1, T.6, 58, 91
$[HNi_{38}C_6(CO)_{42}]^{5-}$ T.1, T.6, 58, 91, 92
$[Ni_3Pt_3(CO)_{12}]^{2-}$ 60
$[Ni_6Pt_6(CO)_{21}]^{4-}$ T.2, 60, 70
$[Ni_9Pt_3(CO)_{21}]^{4-}$ T.2, T.6, 57, 65, 70
$[HNi_9Pt_3(CO)_{21}]^{3-}$ T.2, 57, 65, 70
$[H_2Ni_9Pt_3(CO)_{21}]^{2-}$ 65
$[Au_6Ni_{12}(CO)_{24}]^{2-}$ T.2, T.6, 57, 89, 90
$[Ni_{36}Pt_4(CO)_x]^{6-}$ T.2, T.6, 57, 72
$[Ni_{38}Pt_6(CO)_{48}]^{6-}$ 65
$[HNi_{38}Pt_6(CO)_{48}]^{5-}$ T.2, T.6, 57, 65, 72
$[H_2Ni_{38}Pt_6(CO)_{48}]^{4-}$ T.2, 57, 65, 72
$[H_3Ni_{38}Pt_6(CO)_{48}]^{3-}$ 65

Palladium
$Pd_{10}(CO)_{12}(PEt_3)_6$ 62
$Pd_{16}(CO)_{13}(PEt_3)_9$ T.1, T.6, 62, 81
$Pd_{23}(CO)_{22}(PEt_3)_{10}$ T.1, T.6, 62, 72
$Pd_{23}(CO)_{20}(PEt_3)_8$ T.1, T.6, 62, 77
$Pd_{38}(CO)_{28}(PEt_3)_{12}$ T.1, T.6, 62, 92

Platinum
$Pt_5(CO)_6(PEt_3)_4$ 52
$[Pt_6(CO)_{12}]^{2-}$ 53, 60
$[Pt_9(CO)_{18}]^{2-}$ 51, 52, 84
$[Pt_{12}(CO)_{24}]^{2-}$ T.1, T.6, 51, 52, 53, 61, 63, 84
$[Pt_{15}(CO)_{30}]^{2-}$ T.1, T.6, 52, 61, 84
$H_xPt_{15}(CO)_8(PBu_3^t)_6$ T.1, 70
$Pt_{17}(CO)_{12}(PEt_3)_8$ T.1, T.6, 52, 80
$[Pt_{18}(CO)_{36}]^{2-}$ T.1, T.6, 52, 53, 61, 63, 84
$[Pt_{19}(CO)_{22}]^{4-}$ T.1, T.6, 51, 56, 66, 80

INDEX OF CHEMICAL COMPOUNDS 311

$[Pt_{24}(CO)_{30}]^{2-}$ T.1, T.6, 51, 66, 70, 71
$[Pt_{26}(CO)_{32}]^{2-}$ T.1, T.6, 51, 56, 66, 70
$[Pt_{38}(CO)_x]^{2-}$ T.1, T.6, 56, 66, 70, 71, 72
$[Pt_{52}(CO)_x]^{2-}$ T.1, T.6, 56, 70, 71

II. General Index for remaining chapters

$Ag_{20}Au_{18}(PPh_3)_{12}Cl_{14}$ 268
Au-clusters 169
Au colloid 110ff, 189, 266
Au/Pt colloid 113ff
Au/Pd colloid 113ff
$HAuCl_4$ 109
$Au_4(PPh_2Ch_2PPh_2)_3I_2$ 190
$Au_5(PPh_2CH_2PPh_2)_3(PPh_2CHPPh_2)^{2+}$ 190
Au_6^{2+}-clusters 147
$[Au_8L_8](PF_6)_2$ 190
$[Au_9(PAr_3)_8](PF_6)_3$ 190
$[Au_9L_8](NO_3)_3$ 183
$Au_{11}L_7X_3$ 25, 169–171, 173, 183, 190, 265
$[Au_{13}(PPhMe_2)_{10}Cl_2]^{3+}$ 6
$Au_{13}(dppm)_6(NO_3)_4$ 190
$Au_{55}(PPh_3)_{12}Cl_6$ (abbreviated: Au_{55}) 6, 20, 25, 121, 122, 124, 125, 132, 171, 183, 221, 245,
    263, 264, 266, 267, 268, 269, 270, 271
    photoemission spectra 171ff
    conductivity 25ff
    synthesis 120
    Mössbauer spectroscopy 192ff
    specific heat 194ff, 221ff
    UV-visible spectrum 266
$Au_{55}(PPh_2C_6H_4NaSO_3 \times 2H_2O)_{12}Cl_6$ 190, 198ff, 267, 269
$Au_{55}(P[C_6H_4CH_3]_3)_{12}Cl_6$ 190, 198ff
$Au_{55}[(C_6H_5)_2P(m - C_6H_4SO_3Na)]_{12}Cl_6$ 122, 124

B (boron-structure) 5

C_{60} 6, 18, 24, 37, 251, 304
$[Co_2Fe(CO)_9(\mu_3 - PPh]^-$ 262
$Co_3(CO)_9(\mu_3 - PPh)$ 262
$Co_{55}(PMe_3)_{12}Cl_{20}$ 22
Cu_2Se-clusters 8
$[Cu_{70}Se_{35}(PEt_3)_{22}]$ 7, 9
$[Cu_{146}Se_{73}(PPh_3)_{30}]$ 7

$Fe_2(CO)_9$ 249
$Fe_3(CO)_{12}$ 265
Fe_{13}-cluster 148, 150

Ir-clusters 168, 169
IrO_2 173, 174
$Ir_4(CO)_{12}$ 173, 174, 265

312   INDEX OF CHEMICAL COMPOUNDS

K_n clusters 144
K_2NiF_4-structure 4, 5
K_3C_{60} 18

Li_n clusters 141
Li_nMg clusters 141
Li_nAl clusters 141

Metal colloids, ligand stabilized 109ff
Mo_2W-clusters 169
MoW_2-clusters 169

Na_n clusters 144, 145
Na_8Zn cluster 145
Nitrogen ligand
    $P-H_2N-C_6H_4-SO_3Na$ 111–113
Ni_9Cu_5-cluster 150
$[Ni_9Te_6L_8](BF_4)_n$ 303
$Ni_{14-n}Cu_nCO$ cluster 149
$[Ni_{32}C_6(CO)_{32}]^{n-}$ 149
$[Ni_{34}(CO)_{38}C_4H](NEt_4)_5$ 8, 302, 22
$Ni_{38}Pt_6$ 298, 302
$[Ni_{38}Pt_6(CO)_{48}H_2](NEt_4)_4$ 22
$[Ni_{38}Pt_6(CO)_{48}H_2](PPN)_4$ 25
$[Ni_{38}Pt_6(CO)_{48}H](AsPh_3)_2(NBu_4)_3$ 22, 302ff
$[Ni_{44}(CO)_{48}]^{n-}$ 18, 149, 151, 152

Osmium carbonyl clusters, 169
    susceptibility and EPR 249ff
$Os_3(CO)_{12}$ 250ff, 176
$H_2Os_3(CO)_{10}$ 251ff
$H_4Os_4(CO)_{12}$ 250ff
$Os_7(CO)_{21}$ 251ff
$Os_{10}S_2(CO)_{23}$ 260
$Os_{10}C(CO)_{24}I_2$ 251ff, 261
$H_2Os_{10}C(CO)_{24}$ 250ff, 251, 253–259, 262
$[Os_{10}C(CO)_{24}]^{2-}$ 250, 259 261, 265, 266, 267
$[Os_{10}C(CO)_{24}]^{2-}[N(PPh_3)_2]_2^+$ 250, 259
$[Os_{10}C(CO)_{24}AuPPh_3]^-[PPh_3Me]^+$ 259, 260
$Os_{10}C(CO)_{24}AuPMe_2Ph)_2$ 261
$Os_{10}C(CO)_{24}[Cu(NCMe)]_2$ 261
$[Os_{10}C(CO)_{24}I]^-[N(PPh_3)_2)_2^+]^+$ 259
$[H_4Os_{10}(CO)_{24}]^-$ 262
$H_6Os_{10}(CO)_{24}$ 261
$H_2Os_{11}C(CO)_{27}$ 261
$Os_{11}C(CO)_{27}[Cu(NCMe)]_2$ 261
$[Os_{20}Hg(C)_2(CO)_{48}]^{2-}$ 261
$H_2Os_{20}Hg(C)_2(CO)_{48}$ 261
$[Os_{48}Hg_3(C)_4(CO)_{96}]^{2-}$ 261

$PbMO_6S_8$ 5, 137, 138
Pd colloid 112ff, 269, 299
Pd clusters 167, 220

INDEX OF CHEMICAL COMPOUNDS 313

H_2PdCl_4 112
$Pd_{561}Phen_{36}O_{200}$ (abbreviated: Pd_{561}) 7, 22, 23, 25, 26, 117, 221, 223, 245, 263, 269, 298
    conductivity 25ff
    specific heat 223ff
    susceptibility 298ff
    synthesis 117
    UV-visible spectrum 269
$Pd_{570\pm30}L_{60\pm3}(OAC)_{180\pm10}O_{190\pm10}$ 7
$Pd_{1415}Phen_{54}O_{1000}$ 7, 22, 299
$Pd_{2057}Phen_{78}O_{1600}$ 7, 22, 117, 299
Phosphine ligands:
    $P(m\text{-}C_6H_4SO_3Na)_3$ 110–112
    $(C_6H_5)_2P(m\text{-}C_6H_4SO_3Na)$ 110, 111, 122
    $P(C_6H_5)_3$ 122
Pt-clusters 168, 169, 220, 227, 235
Pt colloid 111ff, 300
H_2PtCl_4 112
Pt_3Au-clusters 169
Pt_3Sn-clusters 169
$Pt_4(CO)_5(PMe_2Ph)_5$ 142
$Pt_4(O_2CCH_3)_8$ 143
$[Pt_{24}(CO)_{30}]^n$ 37
$[Pt_{26}(CO)_{32}]^n$ 37
$[Pt_{38}(CO)_{44}]^n$ 37
$[H_2Pt_{38}(CO)_{44}]^{2-}[N(PPh_3)_2]_3^+$ 263
$Pt_{309}Phen^*_{36}O_{30}$ (abbreviated: Pt_{309}) 7, 22–24, 119, 120, 216, 221, 228, 237, 267, 269, 300,
    ^{195}Pt-NMR 237ff
    specific heat 222f
    Mössbauer spectroscopy 201ff
    UV-visible spectrum 267
    susceptibility 300ff
Pt_nH_{2n}-clusters 143

Rh-clusters 168, 169
$Rh_4(CO)_{12}$ 178
$[Rh_6(CO)_{16}]^+$ 178, 262
$[Rh_6(CO)_{14}(\eta^3\text{-}C_3H_5)]^-$ 6
$[Rh_{13}(CO)_{24}H_{5-n}]^{n-}$ 6
$[Rh_{17}S_2(CO)_{32}]^{3-}$ 262, 263
$[Rh_{17}S_2(CO)_{32}]^{3-}[N(PPh_3)_2]_3^+$ 262, 263
$Rh_{55}[P(tert\text{-}Bu)_3]_{12}Cl_{20}$ 129
Ru-clusters 169
Ru_2Rh-clusters 169
$Ru_3(CO)_{12}$ 265
$Ru_{55}(P(t\text{-}Bu)_3)_{12}Cl_{20}$ 242

Si_n clusters 145

INDEX OF SUBJECTS

^{197}Au Mössbauer spectroscopy 183
Ab initio theory 16, 144
Absorption cross-section 137
Alkali metal clusters 145
Anderson transition 33
Anderson-Hubbard model 31
Annulenes 292
Antibonding orbitals 136
Atomization energies 142
$Au^{2+}{}_6$ cluster, molecular orbital calculations 147
Auger electrons 165

Band structure calculations 14–16, 118, 139, 284
Band-narrowing (at surfaces or in clusters) 166–168, 196, 197
Benzene 292
Bimetallic colloids 112
Binding energy 160ff
Bohr-van Leeuwen theorem 294
Bonding orbitals 136
Boron 5, 6
Broadening:
  homogeneous 257
  inhomogeneous 262
  in photoemission 176
  in ESR and NMR 230ff

^{13}C-NMR 237
C_{60} (fullerenes) 6, 18, 24, 37, 42, 304
Catalysis 43, 127ff, 148, 174, 175, 189, 127, 249
Ceramic metals 28
Charge transfer (inter-cluster) 14, 18, 24, 28, 28ff, 29, 270, 271
Charging effects (in photoemission spectroscopy) 176
Chemical shift:
  in photoemission spectroscopy 162ff
  in Mössbauer spectroscopy 185
Chemisorption (on metal surfaces) 3, 18
Chevrel phases 4, 5, 42, 137, 138
Cluster valence electrons 93ff

Cluster valence molecular orbitals 94
Cluster solids 4, 7, 13, 15
Cluster-support interactions 175, 177–179
Cluster-surface analogy 4, 249, 172–174
Colloids (see: metal colloids)
Conductivity (electrical) 25ff, 256, 258, 264, 271
Configuration-interaction method 16
Continuum approximation:
  for phonons 212, 222
  for electrons 22, 219, 230
Core-level photoemission spectra (Au clusters) 171, 172
Corel-level shifts 162, 165, 166, 173, 197
Correlation effects 15, 16, 141, 284, 295
Correlations (on-site, Hubbard-U) 31ff
Coulomb correlations, see: correlations
Covalent bonding 142
Crystal-field (ligand-field) 19
Curie-type paramagnetism 253, 255, 256, 259, 262, 280, 285, 287
Curie-Weiss law 259, 262
Cyclotron orbit 290
Cyclovoltametry 65ff

De Haas van Alphen effect 283, 284
Debye temperature 25, 187, 193, 195, 204, 212, 224
Debye formula 212
Debye-Waller factor 188, 221
Defect energy 140
Degradation process 124
Density matrix 29
Density Of States (DOS):
  for electrons 16, 17, 19, 20, 137, 168, 185, 206, 284
  for iron in RTB model 140
  for phonons 24, 217
Density Functional theory 15, 16, 18, 144
Diamagnetic susceptibilities 250ff, 279, 289, 291
Dielectric properties 25ff
Diffusive hopping 28
Dimensionality reduction 1

Dirac-Slater approach 147
Disproportionation reaction 33, 270
Distribution functions 211, 217, 218, 288
DSC measurements 124
Dysonian lineshape 257, 258

Effective atomic number rule 94
Eight-shell cluster 7, 117, 298, 299
Einstein relation 34
Einstein temperature 193
Einstein oscillators 215
Elastic continuum approximation 212
Electric quadrupolar splitting 186, 222
Electric field gradient (EFG) 186
Electrical conductivity, see: conductivity
Electrochemical activation, see: cyclovoltametry
Electron paramagnetic resonance (EPR), 230, 231, 235, 250ff
　EPR studies of decanuclear osmium clusters 256
　EPR lineshape 257, 259, 262
　EPR studies of other high-nuclearity clusters 263
　EPR studies of rhodium carbonyl clusters 262
Electron spin resonance (ESR) 230ff, 235
Electronic:
　spin-spin relaxation time T_2 257, 262
　spin-lattice relaxation time T_1 257
　energy-level structure 9, 10ff, 135ff, 159ff, 239ff, 252ff
　(UV-visible-NIR) spectra 264
　specific heat 217
Embrittlement of steels 152
Energy Dispersive X-Ray analysis (EDX) 113
Euler theorem 97
Exchange energy 285
Exchange-correlation, see: correlation effects
Extended pair approximation 29
Extended X-ray Absorption Fine Structure (EXAFS) 124, 196, 197
Extended Hückel Theory (EHT) 96, 136

F-factor 187
Fermi edge (of valence band) 169, 178
Fermi-energy (E_F) 11, 17, 20
Ferromagnetic clusters 295, 301ff
Final-state effects 163ff
Five-shell cluster 7, 117, 223, 298
Fluxionality 122
Four-shell cluster 7, 120
Free electron models 10, 218, 278ff

Fröhlich model 10
Full-shell clusters 108

G-value 256, 259, 261, 262, 263, 264
Gauge 279, 281, 291
Gaussian-type orbital (GTO) 142
Gold colloids 110, 263, 264, 266, 268
Goldanskii-Karyagin Effect 188 Grain boundaries (fracture at) 152, 153
GTO Variational 145
GVB theory 142

Hartree-Fock (HF) approach 141
Healing distance 233
Heat of decomposition 124
Heteroatomic clusters 142
Heterogeneous catalysis (see also: catalysis) 127
Heteropoly blue complexes 304
High resolution transmission electron microscopy (HRTEM) 111
High-spin/low-spin transition 19, 254
HOMO (Higest Occupied Molecular Orbital) 11, 17, 18, 19, 93
HOMO-LUMO gap (frontier orbital separation) 18, 24, 36, 45, 97, 251, 254, 255, 265
Homogeneous catalysis (see also: catalysis) 127
Hund rules 19, 295
Hydroformylation reactions 130
Hydrogenation 131
Hyperfine coupling 259, 262

Impurity atom (see also: defect energy) 140, 152
Initial-state effects (in photoemission) 163ff
Inter-cluster vibrations 24, 25, 194, 195, 212, 222
Interatomic potentials 149
Interband transitions 265, 269, 270
Intra-cluster vibrations 24, 25, 194, 195, 212, 221, 223
Ionization cross sections 160, 161
Ionization potentials 141, 142
Ionization energy 160
Isomer Shift (IS) 185

Jahn-Teller distortion 295
Jellium model 15, 16, 145

Knight-shift (K) 23, 24, 185, 231ff
Korringa law 24, 232ff
Kronig-Penney model 10, 13ff, 18, 24

INDEX OF SUBJECTS

Kubo model 10, 12

Lamb-Mössbauer factor 188
Landau levels 283
Landau diamagnetism 256, 283–285, 290, 291
Langevin diamagnetism 281
Larmor diamagnetism 252, 281, 285
Lattice specific heat 212ff
LCAO 139, 145
Level distribution (Poisson, Orthogonal, Unitary, Symplectic) 12, 13, 23, 288
Ligand shell 3, 108, 111, 296
Ligand stabilized clusters 116ff
Ligand-metal interactions 3, 18–20, 22, 137, 148, 198ff, 238ff, 297ff
Li_n, $Li_n Al$ and $Li_n Mg$ cluster 141
Linewidth 239, 256, 259, 263, 264
Local Density Functional Theory 143, 144, 151, 302
Localization 31, 32
Lowest Occupied Molecular Orbital (LUMO) 18

Magic numbers (see also: full-shell clusters) 6, 108
Magnetic properties (susceptibility, magnetization) 148ff, 277ff, 249ff
Melting temperature 198
Mesoscopic physics 2, 10, 14, 13, 15, 227
Metal colloids 109ff
Metal carbonyl clusters:
    synthesis, reactivity 42ff, 49ff,
    structure, electron counting 67ff, 93ff
Metal-ligand interactions (see: ligand-metal interactions)
Metal-nonmetal transition:
    disorder-induced 32
    size-induced 3, 22
Metal-nonmetal composites 28
Metallic behavior (as seen in XPS) 164, 166–169
Metallic thermodynamic behavior 22, 23
Metametallic 45, 253, 257, 272
Miller-Abrahams network 29
Mixed-valence charge transfer 32
Model Hamiltonians 136ff
Molecular dynamics approach 215
Molecular-orbital (calculations, theory) 10, 16, 17, 18, 20, 96, 136ff
Molybdenum chalcogenide clusters (see also: Chevrel phases) 4, 137
Monodisperse 2, 3
Mössbauer Effect Spectroscopy (MES) 183ff

Nanophase materials 8
Nanostructures (physical, chemical) 8
Navier equation 212
Nuclear Magnetic Resonance (NMR) 227ff
Non-ohmic effects (in conductivity) 25
Nuclear resonant fluorescence 183
Nuclear-spin lattice relaxation time (T_1) 231ff

One-electron approximation 15, 16
Optical spectra 249ff
Orbital magnetic susceptibility 304, 278, 279, 284, 289

^{31}P-NMR 122, 237, 243
^{197}Pt-NMR 235ff
Palladium colloid 298
Partial Densities of States (PDOS) 137
Pascal's constants 250, 251
Pauli spin susceptibility 253, 255, 282, 285, 287, 298
Phonon-assisted hopping (thermally activated hopping) 29, 32
Photodecomposition 176
Photoemission spectroscopy 159ff
Photoionization cross-sections 146
Plasma resonance 268, 269, 270
Platinum colloids 111, 300
Platinum particles 142
Poisson distribution 224, 239, 240, 288
Polaronic effects 28, 29, 33, 34
Polyhedral skeletal electron pair theory 94ff
Population analysis 136
Potential well:
    spherical 11, 233, 289
    box 10, 15, 290

Quadrupole moment 186
Quadrupole Splitting (QS) 186
Quantum dots (wells) 2, 9, 10
Quantum-size effect 10, 11, 20, 25, 206, 211, 216, 219, 229, 253, 257
Quenching of magnetic moment 19, 20, 149ff, 301ff

Random barriers 28, 29
Recoil free fraction, see: f-factor
Recursive Tight Binding (RTB) 139
Relaxation energy (polarization, screening) 163, 164, 165
Robin-Day classification 42

Scaling of conductivity 30
Scanning Tunneling Microscopy (STM) 119
Schottky-like anomaly 222

Screening effects 33, 34
  in photoemission 163ff
Secondary Ion Mass Spectroscopy (SIMS) 124
Self-Consistent-Field ($X\alpha$) 15, 16
Semiempirical approaches 136
Seven-shell cluster 7, 117, 298, 299
Shell-structure (of energy-levels) 12, 16, 293
Skeletal electron pair counts 249, 253
Specific heat (electronic, phonon) 22, 24, 195, 211ff
Spectroscopic properties 135
Spherical Bessel functions 214
Spherical potential well, see: potential well
Spheroidal oscillations 214
Spin-orbit interaction 164, 171, 172, 211, 279, 295
Spin-polarized LD calculations 151
Spin-spin relaxation 233
Splitting 187, 256
Spring model 222
Stochastic hopping 29
Stoner enhancement 225, 233, 285, 299

Submicron physics 1
Super-clusters 124, 126
Superconductor (superconducting, superconductivity) 5, 18, 37
Surface chemisorption 148, 159, 179
Surface effects 19, 166, 197, 211, 218, 229,
Surface modes 212, 217, 219, 220
Synchrotron radiation 161, 184

Temperature-independent paramagnetism 252
Thermal decomposition 124
Toroidal oscillations 214
Transition metal compounds 4ff, 19
Two-level (TLS) systems 244
Two-shell cluster 7, 120

Valence band photoemission spectra 163, 167, 168, 173, 174
Van Vleck paramagnetic susceptibility 251, 252, 281, 289, 291
Variable-range hopping 28

X-ray Absorption Spectroscopy (XAS) 169

Physics and Chemistry of Materials with Low-Dimension Structures

Previously published under the Series Title:
PHYSICS AND CHEMISTRY OF MATERIALS WITH LAYERED STRUCTURES

1. R.M.A. Lieth (ed.): *Preparation and Crystal Growth of Materials with Layered Structures.* 1977 ISBN 90-277-0638-7
2. F. Lévy (ed.): *Crystallography and Crystal Chemistry of Materials with Layered Structures.* 1976 ISBN 90-277-0586-0
3. T. J. Wieting and M. Schlüter (eds.): *Electrons and Phonons in Layered Crystal Structures.* 1979 ISBN 90-277-0897-5
4. P.A. Lee (ed.): *Optical and Electrical Properties.* 1976 ISBN 90-277-0676-X
5. F. Hulliger: *Structural Chemistry of Layer-Type Phases.* Ed. by F. Lévy. 1976
   ISBN 90-277-0714-6
6. F. Lévy (ed.): *Intercalated Layered Materials.* 1979 ISBN 90-277-0967-X

Published under:
PHYSICS AND CHEMISTRY OF MATERIALS WITH LOW-DIMENSIONAL STRUCTURES
SERIES A: LAYERED STRUCTURES

7. V. Grasso (ed.): *Electronic Structure and Electronic Transitions in Layered Materials.* 1986 ISBN 90-277-2102-5
8. K. Motizuki (ed.): *Structural Phase Transitions in Layered Transition Metal Compounds.* 1986 ISBN 90-277-2171-8

PHYSICS AND CHEMISTRY OF MATERIALS WITH LOW-DIMENSIONAL STRUCTURES
SERIES B: QUASI-ONE-DIMENSIONAL STRUCTURES

B1. P. Monceau (ed.): *Electronic Properties of Inorganic Quasi-One-Dimensional Compounds.* Part I: Theoretical. 1985 ISBN 90-277-1789-3
B2. P. Monceau (ed.): *Electronic Properties of Inorganic Quasi-One-Dimensional Compounds.* Part II: Experimental. 1985 ISBN 90-277-1800-8
B3. H. Kamimura (ed.): *Theoretical Aspects of Band Structures and Electronic Properties of Pseudo-One-Dimensional Solids.* 1985 ISBN 90-277-1927-6
B4. J. Rouxel (ed.): *Crystal Chemistry and Properties of Materials with Quasi-One-Dimensional Structures.* A Chemical and Physical Synthetic Approach. 1986
   ISBN 90-277-2057-6

Discontinued.

PHYSICS AND CHEMISTRY OF MATERIALS WITH LOW-DIMENSIONAL STRUCTURES
SERIES C: MOLECULAR STRUCTURES

C1. I. Zschokke (ed.): *Optical Spectroscopy of Glasses.* 1986 ISBN 90-277-2231-5
C2. J. Fünfschilling (ed.): *Relaxation Processes in Molecular Excited States.* 1989
   ISBN 07923-0001-7

Discontinued.

Physics and Chemistry of Materials
with Low-Dimension Structures

9. L.J. de Jongh (ed.): *Magnetic Properties of Layered Transition Metal Compounds.* 1990
   ISBN 0-7923-0238-9
10. E. Doni, R. Girlanda, G. Pastori Parravicini and A. Quattropani (eds.): *Progress in Electron Properties of Solids.* Festschrift in Honour of Franco Bassani. 1989
    ISBN 0-7923-0337-7
11. C. Schlenker (ed.): *Low-Dimensional Electronic Properties of Molybdenum Bronzes and Oxides.* 1989 ISBN 0-7923-0085-8
12. R. H. Friend (ed.): *Conducting Polymers.* 1991 (forthcoming)
13. H. Aoki, M. Tsukada, M. Schlüter and F. Lévy (eds.): *New Horizons in Low-Dimensional Electron Systems.* A Festschrift in Honour of Professor H. Kamimura. 1992
    ISBN 0-7923-1302-X
14. A. Aruchamy (ed.): *Photoelectrochemistry and Photovoltaics of Layered Semiconductors.* 1992 ISBN 0-7923-1556-1
15. T. Butz (ed.): *Nuclear Spectroscopy on Charge Density Wave Systems.* 1992
    ISBN 0-7923-1779-3
16. G. Benedek (ed.): *Surface Properties of Layered Structures.* 1992 ISBN 0-7923-1961-3
17. W. Müller-Warmuth and R. Schöllhorn (eds.): *Progress in Intercalation Research.* 1994 ISBN 0-7923-2357-2
18. L.J. de Jongh (ed.): *Physics and Chemistry of Metal Cluster Compounds.* Model Systems for Small Metal Particles. 1994 ISBN 0-7923-2715-2

KLUWER ACADEMIC PUBLISHERS – DORDRECHT / BOSTON / LONDON